A MANUAL OF
Field Hydrogeology

Laura L. Sanders
Northeastern Illinois University

Prentice Hall
Upper Saddle River, New Jersey 07458

Library of Congress Cataloging-in-Publication Data

Sanders, Laura L.
  A manual of field hydrogeology / Laura L. Sanders.
    p.   cm.
  Includes bibliographical references and index.
  ISBN 0-13-227927-4
  1. Hydrogeology—Field work.   I. Title
  GB1002.3.S26   1998
  551.49—dc21   97-45597
                                                    CIP

Executive Editor: Robert A. McConnin
Executive Managing Editor: Kathleen Schiaparelli
Assistant Managing Editors: Shari Toron / Lisa Kinne
Manufacturing Manager: Trudy Pisciotti
Art Director: Jayne Conte
Cover Designer: Karen Salzbach
Production Supervision / Composition: Aksen Associates
Cover Art: Aksen Associates / Pasini Graphics

© 1998 by Prentice-Hall, Inc.
Simon & Schuster/A Viacom Company
Upper Saddle River, NJ 07458

Printed in the United States of America
10 9 8 7 6 5 4 3 2 1

ISBN 0-13-227927-4

Prentice-Hall International (UK) Limited, *London*
Prentice-Hall of Australia Pty. Limited, *Sydney*
Prentice-Hall Canada Inc., *Toronto*
Prentice-Hall Hispanoamericana, S.A., *Mexico*
Prentice-Hall of India Private Limited, *New Delhi*
Prentice-Hall of Japan, Inc., *Tokyo*
Simon & Schuster Asia Pte. Ltd., *Singapore*
Editoria Prentice-Hall do Brasil, Ltda., *Rio de Janeiro*

For my families

Contents

8   Well Design and Installation   233

9   Geochemical Measurements   263

10   Hydrogeologic Mapping   305

11   Field Projects   335

Preface

The study of ground water is like the study of any other geologic field: It requires one to go outside and get dirty, observing firsthand the processes at work in and on the earth. As with most geologic specialties, most of the action takes place underground, out of sight. Successfully piecing together the puzzle of what is happening depends on the geologist's ability to think in four dimensions, to use imagination, and to extract the maximum amount of information possible from the skimpiest of clues.

In hydrogeology, as with other geologic specialties, we get some information by relying on "outcrops." Outcrops of water are springs, streams, lakes, and oceans. We also use test borings, geophysical studies, chemistry, physics, and loads of common sense to determine what is going on underground. But unlike most other geologic fields, in hydrogeology, the object of our study moves about, even as we study it. This presents a different set of challenges. It is made all the more compelling because of the environmental and social costs of ignorance of ground water movement. And as a young field, hydrogeology is only now defining its standard practice. Although standard texts exist that define the practice of field geology (notably Compton's *Geology in the Field*), few describe the practice of field hydrogeology. This book attempts to do so.

This text is intended for two audiences. It may be used as a textbook for courses in field hydrogeology at the junior/senior undergraduate or beginning graduate level. It also is intended to provide background for entry-level professional hydrogeologists who wish to review basic concepts before their first day on the job.

What This Book Is
This book describes basic field techniques in hydrogeology. It begins by explaining the steps and considerations in planning a field study. It devotes a chapter to field techniques in surface water, and then moves on to ground water. Following these are chapters on methods and considerations for describing soils and rocks, drilling boreholes, designing and installing wells, and sampling and analyzing well water. The book details construction of hydrogeologic maps, cross-sections, and flow nets. And, finally, it describes some typical field situations the entry-level hydrogeologic professional is likely to encounter.

What This Book Is Not
This book is not a basic textbook in hydrogeology. It assumes that the reader has completed an introductory course in hydrogeology. It does not define basic terms

or explain basic concepts. Nor is it a text in geology. The reader should be familiar with sediments, rocks, minerals, and geomorphological landforms.

This book is not an explication of regulations pertaining to ground water. A complex hierarchy of statutes and regulations on the federal, state, and local levels governs investigations into ground water supply, contamination, and remediation. For the ground water consultant or engineer, it is imperative that these be studied and adhered to. However, an overview of these regulations in this book would have to be so general as to be of limited use. It probably also would be out-of-date before the book's ink is dry, due to the rapidly–changing nature of the field. Hydrogeologists affected by these regulations should study them in another source.

This book is not a text on safety at field sites. Although it points out some commonsense safety rules and practices, many hazards and dangers exist that are specific to a field site, job, situation, and moment in time. Health and safety plans should be formulated by a qualified health and safety officer for any field job.

Knowledgeable readers will notice that some important topics have been left out of this text. To keep the book to a compact and field-portable size, the number of topics investigated was necessarily limited. Perhaps most notable among methods not included here are those in field geophysics. Readers are advised to consult texts in that field (for example, see Burger, 1992).

A Note to Students and Teachers

This book may be used as a text for a course in field hydrogeology. Topics explored in such a course may be tailored to suit the region and interests of the class, using this text as a guide. The class should be scheduled to include sessions at least six hours long, to provide ample time for travel to and from field sites and for performing field tests. Hydrogeological field camps are an ideal setting in which to learn hydrogeology. However, field hydrogeology can be taught and learned even in fairly conventional academic situations.

Why Field Hydrogeology?

My first field experience in hydrogeology was a trip led by Professor Sam Harrison, then at Allegheny College in northwestern Pennsylvania. He had been asked by a client to investigate and characterize the hydrogeologic conditions at a site, and he took our hydrogeology class there one rainy spring afternoon. I was excited by the prospect of seeing a real live hydrogeology project underway.

I am not sure what sort of drama and excitement I expected to find at the site. Perhaps I expected a network of wells, large enough for a 19-year-old to clamber into and watch water seep through the sides. Perhaps I imagined a huge trench, the wall of which would be a cross-section of the earth, layers of sediment and rock clearly delineated, and an upside-down triangle marking the water table.

Whatever I expected, I was sorely disappointed with the real thing. When we got to the site, it looked like a field. An empty field. No excitement; no drama. I learned a valuable lesson in that field: The real excitement of hydrogeology, the real drama of ground water, takes place in a hidden world beneath our feet. In a subterranean network of tiny pores, cracks, and solution cavities, cool, sparkling fluids seep and percolate, dreadful toxins glide silently along, dissolved cations aimlessly zigzag past one sand grain after another, then happen on a crystal of clay and are adsorbed for a millennium.

To play a part in the drama, or even to discern the plot, we must piece together the clues given to us in well-water levels, chemical concentrations, and borehole logs. This book provides some keys to unlocking the mysteries.

Acknowledgments

I am indebted to editor Robert A. McConnin for being persistent and providing guidance as this project developed. It was my good fortune to work with Bob in my first, and alas, one of his last, projects with Prentice Hall. Many thanks to Shelley Bannister, who read every word of every draft and was an invaluable critic, editor, and supporter. H. Lee Evans and Parvinder Sethi provided inspiration and helpful early reviews of the manuscript. Reviewers Tim Allen, Keene State College; S. Lawrence Dingman, University of New Hampshire; Charles W. Fetter, Univer-sity of Wisconsin/Oshkosh; Beverly L. Herzog, Illinois State Geological Survey; Keenan Lee, Colorado School of Mines; Marvin "Nick" Saines, OHM Remed-iation Services; John L. Sonderegger, Montana Tech; and Ian Watson, Florida Atlantic University helped clarify the text and pointed out errors and omissions. All errors remaining in the text are mine. It was a pleasure to work with Howard and Jo Aksen during the production phase. Thanks to Frances Skrabacz for typing portions of the manuscript. Finally, thanks to the students in my Field Hydrogeology classes, some of whom gave this material its first field test, and all of whom contributed to it in one way or another.

Laura L. Sanders

Introduction and Tools of the Trade

What Is Field Hydrogeology?

Hydrogeology is the study of the occurrence, character, and movement of water and its constituents in earth's subsurface. A great deal of the science of hydrogeology can be learned without any field work whatsoever. But to solve practical problems, theoretical knowledge is not enough: Field work is necessary. Field hydrogeology employs its own distinctive set of methods and practices. This fact, coupled with the unique challenges that frequently accompany field work, makes for an entirely new set of problems to be solved by the scientist.

This book identifies these problems and suggests potential solutions. It describes the methods used to investigate hydrogeologic systems and materials. It gives practical advice on planning and conducting a field hydrogeologic project. And it suggests sources for further study.

Chapter 1 introduces the tools that hydrogeologists use in the field. Chapter 2 identifies information sources vital to hydrogeologic reconnaissance; suggests steps in designing a health and safety plan and an action plan; and describes what to do before going to a site, upon reaching it, and before leaving it. As surface and ground water systems are inextricably linked, descriptions of hydrologic evaluation of springs, seeps, streams, lakes, ponds, and wetlands appear in Chapter 3. Chapter 4 gives methods for measuring infiltration, soil moisture, and hydraulic conductivity of the unsaturated zone. It also describes ground water systems and their delineation with piezometers and wells. Chapter 5 gives details of planning, performing, and analyzing results of aquifer and aquitard tests. Chapter 6 explains standard procedures for describing samples of soil, sediment, and rock; Chapter 7 explains sampling and drilling methods. Chapter 8 outlines the various aspects of well design, installation, development, and abandonment, with a special section on pumps. Chapter 9 provides methods for water sampling and describes field instrumentation for chemical and physical analyses of water. Chapter 10 tells how to convey hydrogeologic information on maps, cross-sections, and flow nets. Chapter 11 looks at common field projects and highlights important considerations for each.

The Practicing Field Hydrogeologist

As a practicing field hydrogeologist, you sooner or later will work on a variety of assignments. These projects will challenge, educate, and fascinate you. They may provide you with opportunities to travel and to work with talented colleagues and scientists from other disciplines. You will learn that hydrogeological field work is never routine, and that each new field site has something intriguing to offer those who study it. Some projects also might test the limits of your technical knowledge, physical comfort, patience, and ability to get along with others. It is very important that you establish yourself as a professional from the outset. A few simple rules will help you learn as you work, establish good working practice, and build a solid reputation.

- Maintain high professional and ethical standards.
- Agree to perform only those jobs for which you are qualified or able to complete.
- Keep in mind that your first responsibility is to your profession, and your second is to your client.
- Treat others in the field with respect.
- Maintain a positive attitude in the field.
- Remember that the field work in any project is the basis of all subsequent analyses and recommendations. The quality of the field work determines the quality of final decisions, predictions, or results of an investigation.

How to Use This Book

To make the most effective use of this book, first scan the Contents and familiarize yourself with the organization of topics. You could sit down and read the text from cover to cover, but it probably will be more efficient to turn first to chapters of primary interest. The extensive index will help you locate the topics most important to you. The Contents lists names and page numbers of specific tests so that you can find them quickly and easily. The References gives sources of more detailed information.

Beginning practitioners of hydrogeology will find it helpful to read Chapters 1 and 2 in their entirety before going into the field. Those who have had some field experience may wish to go directly to the chapters of interest.

Making Calculations

Field hydrogeologists work extensively with numbers, measurements, calculations, and a surprisingly large array of various units used to describe the same quantity. The practicing field hydrogeologist must be able to assess and describe a physical process in quantitative terms. Following these recommendations will simplify the operations:

- When appropriate, picture the situation in terms of simplified geometric forms, and recall the equations to describe those forms. Are you trying to determine a distance, an area, a volume, a velocity, or a discharge? Remember the simple formulas for geometric shapes. See Appendix 1 for some of these.
- Determine units of the desired quantity. For example, here are some commonly used units:

([L] refers to a length dimension; [T] refers to a time dimension.)

length	[L]	mm, cm, m, km, in, ft, mi, yd
area	$[L]^2$	yd^2, mm^2, cm^2, m^2, km^2, in^2, ft^2, mi^2, acre
volume	$[L]^3$	cm^3, m^3, in^3, ft^3, gal, acre-ft
velocity	$[L]/[T]$	cm/sec, ft/sec, mi/hr, ft/yr, m/sec, m/day
discharge	$[L]^3/[T]$	gal/min (gpm), ft^3/sec (cfs), gal/day (gpd), m^3/day, ac-ft/yr
time	$[T]$	sec, min, hr, day, yr

- Memorize a few conversion factors. Some of the most useful are these:

  1 in = 2.54 cm
  1 m = 3.281 ft
  1 ft = 12 in
  1 mile = 5,280 ft
  1 ft^3 = 7.48 gal
  1 acre-foot = 43,560 ft^3

- Sketch a diagram to illustrate the problem.
- Use standard equations and formulas when applicable, but be aware of their assumptions and limitations. For example, the Theis equation may be used to determine aquifer transmissivity, but in a strict sense, it applies only to perfectly confined aquifers, not unconfined or leaky aquifers.
- Use dimensional analysis to set up the problem and make conversions (see Appendix 2 for hints on using this method).
- Write down what is known and what is needed.
- Write units at every step of the problem.
- Describe in words what each step of the problem entails.
- Take a pocket calculator into the field. The calculator should be able to perform the following functions in addition to basic math functions: square root; exponential (scientific) notation; logarithms and inverse logarithms (both natural and base 10); sine, inverse sine, cosine, and inverse cosine of an angle. The calculator should have at least one memory and preferably more. A programmable calculator is not absolutely necessary, but can be a real time-saver.
- Recall 3 trigonometric definitions (Fig. 1.1):

$$\sin \alpha = \text{opposite/hypotenuse}$$
$$\cos \alpha = \text{adjacent/hypotenuse}$$
$$\tan \alpha = \text{opposite/adjacent}$$

- Recall the definition of a logarithm. An example: If you are given the number $10^{3.5}$, the logarithm (or "log") of that number is 3.5. To express a number in terms of logs, use a calculator to take the log of the number. It will display a result; the number you started with is 10 raised to the power of that result.

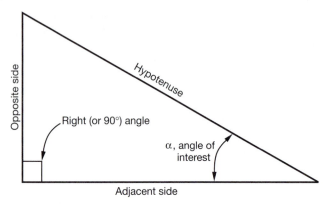

Fig. 1.1 Trigonometric relationships.

- Once you have an answer, evaluate it. Is it in the correct units of measurement? If not, there may be an error in the formula or calculations. Ask this critically important question: *Does the answer seem reasonable?* Practice estimating lengths, areas, volumes, velocities, and discharges. This will help determine if an answer is realistic or not. Check charts of common ranges of values. For example, if you have calculated a hydraulic conductivity of 10^1 cm/sec for a clay unit, either something is wrong in the calculations or you are dealing with a very unusual clay unit, because clays usually have a hydraulic conductivity that is much, much lower than 10^1 cm/sec. (A typical range for clays is 10^{-10} to 10^{-6} cm/sec.) Either way, the problem bears reevaluation.
- Check to be sure that the units are familiar and the quantities easily comprehended by others. For example, in most cases, rather than the answer "2160 minutes", it might make more sense to give the answer as "1.5 days."
- Round the answer to an appropriate number of significant digits. Do not round until the end of the calculation. Consider that natural processes are variable in rate and that measurements are by their nature limited in precision. To determine how to round the answer in an addition or subtraction problem, first find the least precise value being added or subtracted. The least precise value might appear in one of two ways: first, it might be the value with the least number of digits to the right of the decimal point. For example, in the following list of numbers,

    91.5   9.2252   81.359   0.104

    the number 91.5 is the least precise value. Second, if there are no digits to the right of the decimal point, the least precise value is the value with its rightmost place-holding digit (nonzero) to the farthest left of the decimal point. So in the following list of numbers,

    1045   32000   2530   3600

    the least precise value is 32000.

When adding or subtracting measurements, round the answer to the same number of decimal places as the least precise value added or subtracted. For example, when adding 43,000 + 1,200, round the answer to the thousands place, because 43,000 is the least precise value:

$$\begin{array}{r} 43,000 \\ + 1,200 \\ \hline 44,200 \end{array} \quad \textit{Round to} \quad 44,000$$

As another example, when subtracting 5.39 − 0.1, round the answer to the tenths place, because 0.1 is the least precise value:

$$\begin{array}{r} 5.39 \\ - 0.1 \\ \hline 5.29 \end{array} \quad \textit{Round to} \quad 5.3$$

- When multiplying or dividing, count the number of significant digits in the numbers multiplied or divided. Remember, a digit is significant if it is

  - a nonzero number to the right or left of the decimal point
  - a zero embedded anywhere between two other significant digits, or
  - a zero to the right of the decimal point if and only if the zero is to the right of a nonzero digit.

For example, all of the following numbers have four significant digits:

4.123   54,760   0.01009   1.000

Find the lowest number of significant digits in a multiplication or division problem, and round the answer to that number of significant digits. For example, when multiplying 150 × 1.18, round to two significant digits:

$$\begin{array}{r} 150 \\ \times 1.18 \\ \hline 177.00 \end{array} \quad \textit{Round to } 180$$

Using Geologic Tools

Although some of the tools of hydrogeology are unique to the field, some are standard tools in any geologic study. This section describes the use of the geologist's compass, geologist's hammer, measuring tape, and field notebooks.

Geologist's Compass

The geologist's compass, often called a Brunton compass, is a specialized tool with a variety of uses (Fig. 1.2). Some uses are particularly suited to hydrogeologic field work, and these are described in what follows. But a hydrogeologist should be able to use a geologist's compass for any standard geologic

Fig. 1.2 A geologist's compass.

measurement, because geologic mapping and determination of the attitude of rock units or fractures may be critical to a hydrogeologic study. This might be particularly important when the aquifer is composed of rock that is folded or faulted, or when the flow is controlled by the orientation of fractures in a rock unit. For information on other uses of the geologist's compass, see Compton's *Geology in the Field* (1985) or Freeman's *Procedures in Field Geology* (1991).

Find Directions
The base of the compass has an arrow which indicates magnetic North. The compass may be set to compensate for magnetic declination, which varies with location and time.

Survey a Line
Fold the compass so it is about one-quarter open and so the mirror can be used to view the compass base when the instrument is held horizontally. Fold out the sighting arm and fold up the tip of the sighting arm (Fig. 1.3). Now hold the compass at eye level and simultaneously do these things:

1. view the object at the other end of the line to be surveyed, through the window of the compass
2. line up the tip of the sighting arm with the hairline in the window
3. level the compass using the bull's-eye level, as reflected in the mirror
4. read the number, printed on the perimeter of the compass, to which the end of the arrow points. That gives the orientation of the line.

Alternatively, hold the compass open about three-quarters of the way, horizontally, at waist level (Fig. 1.4). Viewing the object of interest in the mirror, simultaneously do the following:

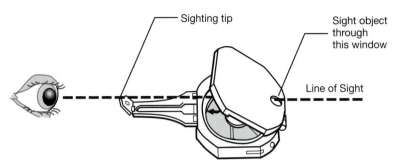

Sighting tip

Sight object
through
this window

Line of Sight

Fig. 1.3 To survey the direction of a line, hold the compass horizontal, at eye level, with the mirror open part way. Fold the arm out and the sighting tip up. Be sure the compass is level and the needle swings freely; check these by looking in the mirror.

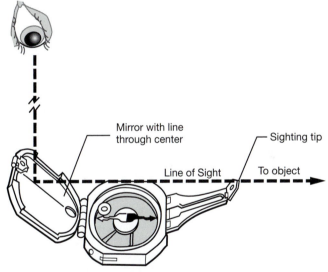

Mirror with line
through center

Sighting tip

Line of Sight

To object

Fig. 1.4 Another method for surveying a line is to hold the compass at waist level, open about three-fourths of the way. View the object in the mirror, lining up the object, the sighting tip, and the line on the mirror. Be sure the compass is level and the needle swings freely.

1. line up the object with the hairline in the mirror
2. line up the reflection of the tip of the sighting arm with the hairline in the mirror
3. level the compass using the bull's-eye level
4. read the number, printed on the perimeter of the compass, to which the end of the arrow points. That gives the orientation of the line.

Slope

Orient the compass vertically so that the clinometer is below the needle's pivot. Fold the compass so it is about one-quarter of the way open. Unfold the arm and fold in the tip of the sighting arm (Fig. 1.5).

Holding the compass at eye level, simultaneously do the following:

1. view the object being sighted through the small window
2. line up the object being sighted, the tip of the sighting arm, and the hairline on the mirror
3. using the arm on the outside of the compass' case, move the clinometer until the bubble level is leveled, as viewed in the mirror.

Then, read the value of the slope measurement from the clinometer. This may be read in degrees or percent slope.

Alternatively, open the compass completely, align the top of the compass' case so that it is parallel to the bed or surface of interest, and adjust the clinometer until the bubble is leveled. Then, read the value of the slope measurement from the clinometer.

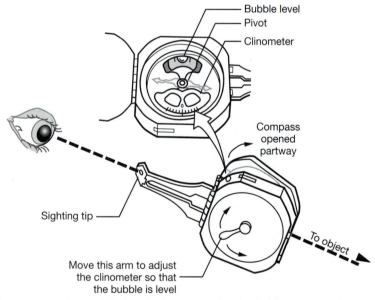

Fig. 1.5 To measure slope, orient the compass vertically so that the clinometer is below the needle's pivot. Open the compass about one-fourth of the way. Fold the arm out and the sighting tip in. View the object through the window, and, while viewing the clinometer in the mirror, adjust the clinometer until the bubble is centered in the level.

Hand Level

To use the compass as a hand level, set the clinometer to 0° and 0% slope. Orient the compass as described in the preceding "Slope." Line up the tip of the sighting arm and the hairline in the window. The object directly in line with these two things is level with the eye.

Slope also can be determined by using the Brunton compass as a hand level and sighting toward a stadia rod (a tall pole with easily seen markings of feet or meters) or other measuring device. The Brunton compass is limited in its precision, however, because it can only give readings to within one-half a degree. This makes it unsuited to measurements of shallow slopes. More information on measuring slope in the field is given in Chapter 3, in the section titled "Gradient."

Geologist's Hammer

Mapping potential aquifers sometimes requires using a hammer to break weathered rocks at outcrops to find a fresh exposure for proper identification. Rock hammers (or geologists' hammers) have a face that is specially hardened for use on rocks (Fig. 1.6).

One end of a rock hammer is shaped like either a pick or a chisel. Rock hammers come in various sizes and weights. For splitting particularly hard rocks, a rock chisel and "crack hammer" (2- to 4-pound blunt hammer) may be most useful. Some geologists working with shales find that a mason's hammer, with a broad, flat end, is most useful for splitting apart these layered rocks.

Whatever type of hammer is used, eye protection should be worn, and other people should be kept away during the hammering. Flying rock chips can penetrate the skin.

Fig. 1.6 Rock hammers: geologist's hammer (chisel and pick end), rock chisel, and crack hammer.

Measuring Tape

Measuring tapes come in various lengths and are made of various types of materials. A carpenter's steel tape, 25 or 50 ft long, is useful for many purposes. Longer distances in the field commonly are measured with tapes having lengths of 100 to 300 ft, or 25 to 100 m. These tapes may be fiberglass or steel. Fiberglass does not rust, but tends to stretch somewhat, especially on hot days. Steel expands and contracts as temperature changes, and may rust or kink. For measuring depth to water in a well, a steel tape with raised numbers and markings may be particularly useful.

Field notebook

Field notebooks have water-resistant paper that makes it possible to take notes even when the page is wet. They come in various sizes; one standard size is small enough to fit in a jeans pocket (Fig. 1.7). Often the pages have grids to help in drawing maps or diagrams to scale. Field notebooks may be purchased from vendors that supply materials to surveyors, foresters, and other outdoor workers. Suggestions for keeping field notations are given in Chapter 2.

Using Hand Tools

Field hydrogeologists frequently need to use common hand tools in the field. Although this may be second nature to some, anyone who has not grown up using tools may find them difficult and clumsy at first. Practical suggestions for identifying and using the most common tools follow. For further information, consult a home repair manual.

Pliers

Pliers are used to provide a strong grip (Fig. 1.8). To remember which way to turn a threaded nut, bolt, or screw, remember "righty–tighty, lefty–loosy." (This rule applies if the threads are normal, "right-handed" threads.) If the jaws of the pliers are offset, follow the turning rule under "Wrenches," which follows.

Fig. 1.7 Field notebooks come in various sizes and shapes.

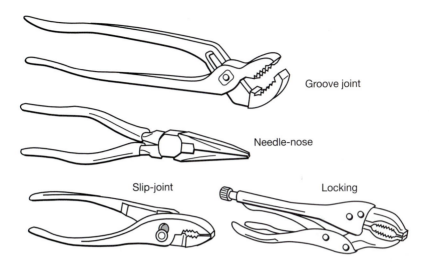

Fig. 1.8 Pliers: groove joint (Channelocks), needle nose, slip joint (water pump) and locking (vise grips).

Wrenches

Use a wrench to tighten or loosen a nut or bolt (Fig. 1.9). Use the appropriate size wrench. Note that some nuts and bolts are metric and will not fit an English unit (or "standard," in the United States) wrench without some slippage. Slippage of a wrench (or pliers) can round off the angled corners of a nut or bolt in no time flat, so use the right size.

Pipe wrenches are used on rounded objects, such as pipes. Drillers use them constantly for tightening and loosening lengths of drill stem or casing that are threaded together. Frequently, wrenches are used in tandem, with one wrench holding one piece still, and one wrench actively tightening or loosening the other piece. When pipes are laid flat on the ground, one wrench can be put around one pipe, so that the ground holds it, while the other wrench actively tightens or loosens (Fig. 1.10). For extra leverage, it may be tempting to use a "cheater bar." This is a piece of pipe slipped over the wrench handle to provide a longer handle, and, therefore, more leverage. Using a cheater bar may result in a bent or broken wrench handle. Instead of using a cheater bar, it generally is better simply to use a larger wrench.

Most socket wrenches are ratcheting, meaning that the wrench need not be repeatedly removed from the nut and put back on again while turning it. Direction of a ratcheting wrench can be reversed by flipping or turning a switch on the handle. Some ratchet handles have a push button to release the socket.

Using a wrench the right way is important for achieving maximum torque and efficiency. It also will keep drillers from snickering behind your back about your

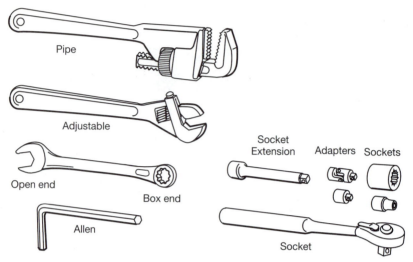

Fig. 1.9 Wrenches: pipe wrench, adjustable or crescent wrench, combination open end and box end wrench, Allen wrench, and socket wrench.

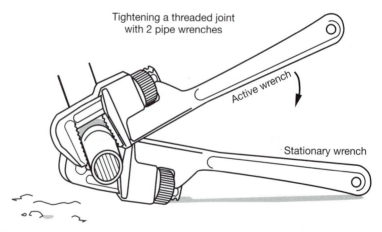

Fig. 1.10 Use pipe wrenches in tandem, one wrench on each pipe. Hold one wrench stationary while turning the other. The stationary wrench may be wedged against the ground.

inexperience. Here is how to use a wrench correctly: All wrenches (and many pliers) have jaws set at an angle so that they grab the bolt, nut, or pipe from the side. Look at that angle, and put the wrench on one way to loosen (Fig. 1.11), and the opposite way to tighten (Fig. 1.12).

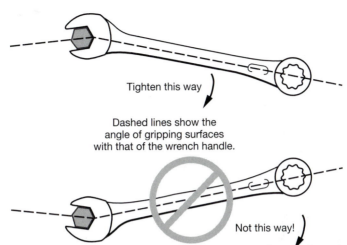

Fig. 1.11   Using a wrench to tighten. In the top diagram, the force is directed toward the nut. But in the bottom diagram, the wrench is likely to slip off the nut because of the angle between the gripping surfaces and the handle.

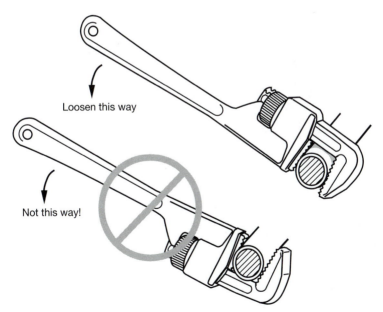

Fig. 1.12   Using a wrench to loosen. It is especially important to orient pipe wrenches correctly, as they will slip or simply fall off the pipe if oriented incorrectly.

Screwdrivers

Standard, or slotted screws often are used in carpentry, and Phillips head screws often are used in machinery. However, either type may be encountered in any given situation, so it is best to keep both types of screwdrivers on hand (Fig. 1.13). Look at the head of the screw, and use the right type and size of screwdriver for the job.

Saws

Use the appropriate saw for the material being cut (Fig. 1.14). When sawing, use the full length of the sawblade, not just a tiny fraction of it. This will make cutting go faster and more smoothly, and both the sawyer and the sawblades will wear out less quickly.

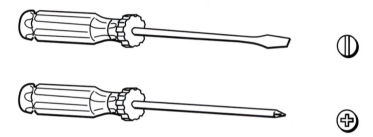

Fig. 1.13   Screwdrivers: standard (flat-blade) and Phillips head.

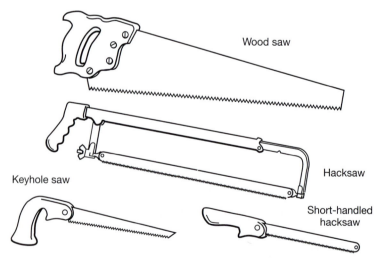

Fig. 1.14   Saws: wood saw, hacksaw, keyhole (compass) saw, and short-handled hacksaw.

Hammers

Use the proper hammer for the task (Fig. 1.15). Wear eye protection when using a hammer. For maximum force, hold the hammer near the end of its handle, not near the head.

Fasteners

Bolts and nuts, screws, and nails are various types of fasteners (Fig. 1.16). Remember "righty-tighty, lefty-loosy." In other words, to tighten a screw, nut, bolt, or piece of threaded pipe, turn clockwise. To loosen, turn counterclockwise. Occasionally, you may encounter "left-handed threads", which follow the opposite rule; these are used mainly on plumbing fixtures.

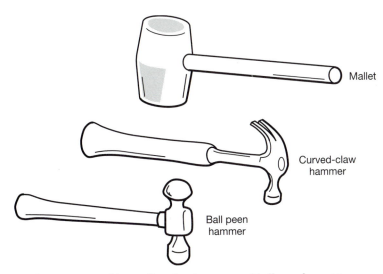

Fig. 1.15   Hammers: rubber mallet, claw hammer, and ball peen hammer.

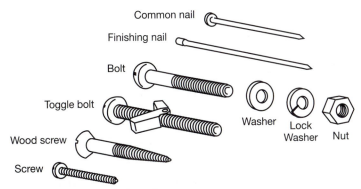

Fig. 1.16   Fasteners: nails, bolts and nuts, screws.

Field Operations

Introduction

Every field project is unique. However, every field project has some elements in common with every other. In this chapter, these common components of field operations are discussed: those elements that must be considered before going into the field, upon arrival at the site, and before leaving the site. Also described is the keeping of field notes.

Before Going Into the Field

Successful field projects begin with thorough planning and advance preparation. Before going into the field, rigorous site reconnaissance should be conducted to review and evaluate site records and other hydrogeologic data. An action plan and a health and safety plan should be drawn up. An equipment list should be compiled, and field equipment should be gathered and checked. In this section, each of these activities is described.

Site Reconnaissance

Most of this book describes methods for making observations and direct measurements in the field. However, it is wise to review the work of others before making one's own measurements. Knowing what has been found by others at a site makes it simpler to plan field work because an expected range of results may be anticipated. This makes it easier to choose the appropriate field equipment, stay within the budget and schedule, and generally be more successful in the field.

Ideally, the hydrogeologist should know a great deal about the field site before setting foot on it. Knowing something in advance about the topography and drainage, soils, geology, hydrogeology, surface-water hydrology, and land use can simplify the field work a great deal. In some projects, for example, site assessments for real estate transactions, reconnoitering the site may turn up information that "red flags" the need for further investigation in the field.

This section discusses some of the information that may be gleaned from a careful review of available sources. Some basic sources that will be most helpful are topographic maps, aerial photographs, information from geological surveys or other governmental agencies, and previous work done at the site. Because topographic maps and aerial photographs may provide information about several different aspects of a site, they are discussed in more detail at the beginning of this section. Other information sources, such as soil surveys and fire insurance maps, are discussed in the section about the aspect of site reconnaissance for which they are likely to be most helpful.

Topographic Maps

At a minimum, a topographic map shows elevation of the land surface. However, the topographic maps published by the United States Geological Survey (USGS) show a wealth of other information in addition to topography. USGS topographic maps show features such as streams and other bodies of water, dams and locks, roadways and bridges, urbanized areas, and, in some cases, individual buildings. They also may indicate power lines, tanks, towers, mines and quarries, and many other features.

The USGS is an excellent source of topographic maps. USGS maps may be obtained directly from the USGS map center, or from state geological surveys. Information may be obtained by calling 1-800-USA-MAPS, or by writing to the following address:

USGS Information Services
Box 25286
Denver, CO 80225

The USGS maintains information pages on the World Wide Web, as do many state surveys. Some maps are available in digital form and may be downloaded from USGS sites.

To find other sources of topographic maps at various scales, it may be helpful to contact county, municipal, or township administrators or engineers. Site plans or construction studies may have been completed during the planning of buildings, roadways, bridges, drainage or sewer projects, railroads, or other developments. Previous hydrogeologic or engineering studies may have been performed at a site. These often are a very good source of detailed and site-specific information.

Before relying heavily on any topographic map, field-check the map data. This is important because features may have changed after the map was made. Also, evaluate the quality of the map. Suggestions for doing this are given in Chapter 10.

Aerial Photographs

Aerial photographs can provide a great deal of information on land use, site activities, and even drainage and geological conditions. Aerial photographs may reveal all the information described under the preceding "Topographic Maps." But some features that appear on topographic maps may be even more obvious on aerial photographs. For example, drainage patterns, including drain tiles, may be quite visible in the photos; trails, paths, and vehicle tracks may be visible, too.

Aerial photos also show human-made features such as buildings, pipelines, air fields, cemeteries, golf courses, stained soil, distressed vegetation, mines and quarries, orchards, wells, storage tanks and lagoons, factories, water-treatment operations, utilities, windlasses, stockyards, large transformers, agricultural operations, landfills, waste piles, and storage drums, among others. Natural features also should be noted: outcrops, landforms, drainage patterns, coastlines, mass movement, erosional features, and evidence of sediment transport.

If a series of historical aerial photos is available, a sequence of land-use activities and human-made changes may be established, as can a history of geologic or other natural changes in an area.

Aerial photographs may be available as single-sheet photos or as stereo pairs. Stereo pairs are two photos of roughly the same area, taken moments apart from each other. When viewed with a stereo viewer, objects on these photos appear in three-dimensional view. A stereo viewer may be obtained from a geological or forestry supply house. Aerial photographs are available in black and white or color-infrared for the entire United States, most developed countries, and in many developing countries. Color photos are available for some areas.

Aerial photographs may be obtained from the U.S. Geological Survey, the U.S. Forest Service, or the Agricultural Stabilization and Conservation Service, which is a part of the U.S. Department of Agriculture. Information may be obtained from these agencies via the World Wide Web. Aerial photographs also may be available from commercial aerial survey companies (check the "Yellow Pages" telephone directory under "Photographers—Aerial") or municipal or regional planning commissions. They may be expensive, especially historical photos, and obtaining them may take weeks or even months. Nevertheless, they provide a great deal of information, and they should be highly regarded as a potential data source.

The following sections describe aspects of a site that should be investigated using documents before any field activities take place at the site.

Topography and Drainage

Topographic maps can be used to determine general site topography and relief, but have many additional uses. For example, topographic maps can be used to make a preliminary assessment of drainage. Larger drainage features such as streams are shown on these maps. Additional smaller drainage trends also may be determined indirectly from topographic maps, because surface water flows from higher elevation to lower. The delineation of drainage basin boundaries is described in Chapter 3.

Determining surface drainage directions from a topographic map can be complicated by several factors. In some areas, drainage is not well-developed. For example, some coastal and floodplain areas have low relief and do not show a clearly-defined drainage pattern. Whereas the regional direction of surface-water flow may be readily apparent, the direction of localized flow may not be. This also is the case in certain areas covered with glacial sediments, where kettles and potholes, depressions with no outlet, characterize the landscape. Areas of karst topography may have very complex drainage systems, much of which may be dominated by interconnected dissolution cavities beneath the land surface. The appearance of sinkholes on topographic maps and "disappearing" rivers is an indication of karst. To further complicate the picture, surface drainage directions in some of these areas may change, depending on seasonal variations in ground water level. Surface drainage in such areas may be unpredictable without field investigation.

Another complicating factor in interpreting drainage features from topographic maps is caused by drainage engineering. In urban and industrial areas, this mainly takes the form of storm sewers, which may not be apparent on a map. The municipal or county engineer's office, sewer department, or site engineer should be able to provide this information. In addition, urban areas have roadways and parking lots, which nearly always include some form of drainage features (e.g., drainage ditches, culverts, or detention ponds).

Agricultural areas might have buried drainage systems, frequently referred to as drainage tiles. Drainage tiles may be ceramic, plastic, or metal perforated pipes that are buried in parallel rows in fields to divert excess infiltration from crop root zones. Drainage tiles are not marked as such on topographic maps. Locating them requires local knowledge, the scrutiny of aerial photographs of the area, and field checking. In interpreting subsurface agricultural drains it is important to bear in mind that these drains often convey water to a discharge point or ditch, which might be visible on the map. Nevertheless, a field visit still is needed to confirm such observations.

Ground water movement (subsurface drainage) is difficult to interpret from topographic maps, but some clues may be evident nonetheless. For example, recharge and discharge areas may be apparent. In humid regions, lakes, ponds, and wetlands may exist as discharge areas, with recharge coming from the surrounding higher ground. In arid regions, dry washes or ephemeral streams may collect surface drainage during rainstorms, and then lose the water through their beds. These act as recharge areas to underlying aquifers. Active quarries, which sometimes can be spotted on a topographic map, often serve as discharge areas. This is because ground water that seeps into a quarry may be drained to a sump and then pumped out and away from the quarry area so that quarrying operations are not affected.

Beyond topography and drainage, some other features of hydrogeologic significance may appear on topographic maps. Aerial photographs, discussed earlier, may show the same features with greater clarity, and their study is highly recommended for identifying key features of project sites. The following provide examples of such features:

Geologic Features. Floodplains, deltas, eskers, drumlins, moraines, and rock contacts are examples of geologic features that may be observed on a map and may indicate something about the nature and permeability of soils in the area. For instance, esker deposits tend to be well sorted (poorly graded) and show higher values of hydraulic conductivity than drumlin deposits, which are typically poorly sorted (well graded). As another example, consider a slope that changes from steep to gradual to steep again. This change in slope may be due to the relative durability of alternating layers of sedimentary rocks such as shale and sandstone, which may indicate the existence of a complex sedimentary flow system. Similarly, many other features might be interpreted from a map or aerial photograph, which might reveal important hydrogeologic information. A thorough review of these should be performed.

Roads and Roadways. As discussed earlier, these features usually are accompanied by drainage systems. Unpaved roads also may show on the maps, but sometimes these have no drainage systems. The investigator should look for dead-end roads, especially unpaved roads in rural areas, as these sometimes lead to illegal dumping sites. In such cases, maps probably will not show the refuse piles at the road's end. Nevertheless, such areas should be designated for further field scrutiny. In areas with snowy and icy winters, roadways may be salted, which might be a source of ground water contamination.

Airports, Shopping Malls, and Large Parking Lots. These may be hydrogeologically significant in terms of runoff, because they are large areas covered with impermeable materials. This, in turn, may affect the water budget of the area by increasing surface runoff to nearby lakes or streams. In addition, activities at these sites may affect ground water quality because of the runoff of gasoline and oil from parking lots and the runoff of deicing chemicals and jet fuels from airports.

Buried Pipelines. These may be drawn and labeled on topographic maps. The process of digging a trench for a pipeline and then backfilling the trench may result in loose soil, which might have a higher hydraulic conductivity than the surrounding materials. As a result, a filled trench for buried pipelines or cables may act as a preferred pathway for ground water flow.

Cemeteries. These often are shown on topographic maps and may be important for two reasons. First, cemeteries may be a source of contaminants to ground water. Second, in some regions, especially those with cold winters, cemeteries may indicate that soils in the area are sandy. Because water drains easily from sandy soils that are situated above the water table, these soils are not likely to freeze solid in the winter. Consequently, it is easy to dig graves in the dry, sandy soils, even in midwinter. This makes sandy areas more desirable locations for cemeteries than areas with clay-rich soils.

Golf Courses. Large golf courses might be labeled on topographic maps. They may be a significant source of fertilizers and pesticides, both to surface water and to ground water. In addition, golf courses may require extensive irrigation, which might affect the water budget of an area.

Mines and Quarries. These may be shown on topographic maps, especially surface operations. Mines and quarries may require extensive pumping of water to make operations possible, and this may significantly alter the direction of ground water flow in the surrounding area. Inactive mines and quarries might discontinue pumping, so they may fill with water. Sand and gravel mines are sometimes labeled as such on topographic maps; these indicate the presence of coarse-grained sediments.

Precipitation may seep through tailings piles near a quarry or mine, dissolving material as it percolates through the piles. If this water reaches the ground water, it may have a negative impact on water quality. Likewise, springs or streams sometimes issue from a mine opening or drain from a mined area. This water may contain dissolved acid-forming materials (forming "acid mine drainage"), metals, solvents, or other chemicals, such as cyanide, used in the mining and processing of the coal or ore. Any of these can have a profound effect on water quality.

Orchards and Cropland. These may be labeled or may simply appear as patches of green on a topographic map. Agricultural areas may be a source of agricultural chemicals that might affect ground water and surface-water quality. In addition, irrigation might affect ground water levels or surface runoff.

Wells. Most wells do not appear on topographic maps. However, they occasion-ally are noted. Wells have obvious hydrogeological importance, usually indicating the existence of a formation having high hydraulic conductivity. However, not all wells indicated on topographic maps are water supply wells. Wastewater injection wells may appear on topographic maps. In addition, oil and natural gas wells may be simply noted on the maps as "Wells." Further investigation and field checking are necessary to confirm the purpose of the wells.

Water Towers. Some water towers are shown on topographic maps. The presence of a water tower may be an indicator that ground water is being used in the area.

Tanks and Chemical Storage Areas. These areas, including features such as tank "farms", waste lagoons, or impoundments, represent potential spill areas and potential sources of surface-water and ground-water contamination.

Railroad Tracks. Tracks or embankments from old rail lines may be a source of chemical contaminants from track materials, leakage from rail cars, or herbicides used to keep tracks clear. Existing tracks are shown clearly on a topographic map, and some old embankments may be visible as well.

Water-Treatment Operations. Sewage-treatment plants, septic leach fields, or other water-treatment operations may appear on some maps. It may be important to find out where the untreated water comes from and where the treated water goes. These factors may affect water budgets, drainage directions, or water quality.

Soils

Information on the soil types at a site can reveal to hydrogeologists something about the site's vulnerability to ground water contamination, its drainage charac-teristics, the presence of wetlands, its suitability for development, and more. In addition, certain hydrological investigations require information on soils. For example, the Soil Conservation Service (SCS) Curve-Number procedure, used to estimate the amount of runoff that will occur from project areas, requires that information on soil type and land use be used as input data.

Soils at a site may have been examined and mapped by the Soil Conservation Service, a part of the U.S. Department of Agriculture (USDA), or they may have been investigated by the county or state agricultural agency. A soil survey for the county may have been published; if so, this may be a valuable source of information.

The soil surveys published by the USDA include aerial photographs on which have been drawn lines outlining boundaries between soil types. Each outlined area has an alphanumeric code, and each code is referenced to a soil title. Each soil is described in detail with regard to grain size, color, structure, typical soil profile, suitability for various uses, and more.

Using a soil survey takes some practice. To simplify the process, the investigator should turn to the page in the survey that gives directions (often inside the front cover), and follow them. Box 2.1 gives an example.

Not all areas of the Unied States are covered by a soil survey. Those that are covered may have been surveyed more than 20 years ago. Some surveys are out of print or otherwise unavailable. If the county soils agency cannot provide a copy of the survey or other soil information, try an agricultural agency or the soils or agriculture departments of a local university. Some university libraries and public libraries maintain collections of soil surveys.

Information on soil surveys may be obtained by calling the National Soil Survey Center at (402)-437-5423 or by writing to them at the following address:

National Soil Survey Center
USDA Soil Conservation Service
Federal Building, Room 152
100 Centennial Mall, North
Lincoln, NE 68508

The Soil Conservation Service also maintains information on the World Wide Web.

Geology

Information on site geology can be used to predict what hydrogeologic conditions might exist at a site. Information on type and thickness of surficial deposits, types of rocks, and thickness, structure, and fracturing of rock units all may give an idea of the hydrogeologic setting. Geologic reports and maps can provide this information.

Geologic reports or maps of the area may have been published by the U.S. Geological Survey (USGS), the state geological survey, the state water survey, or other agencies such as the U.S. or state environmental protection agencies, Bureau of Land Management, conservation department, or department of natural resources. If the site has been studied before, consulting or engineering reports may be available. If a roadway, railway, bridge, or mine or quarry are near the site, the departments of transportation or mines may have published some valuable data. It may not be possible to find a study that specifically addresses the geology of a particular site, but even a general treatment almost always provides helpful information. If buildings exist in the area, there may be engineering reports that include soil boring or drillhole data used to design the foundations.

In examining geologic reports and maps, try to extract information about surficial deposits, bedrock features, and hydrogeologic features. With regard to surficial deposits, look for the following:

BOX 2.1

Using a Soil Survey

A soil survey is used to determine the types of soil that may be found at a particular site. In general, to use a soil survey, begin at the back of the survey book. An index map included there shows major streams, roadways, towns, and township and range boundaries (Fig. 2.1). The index map also shows the numbers of the photo sheets or photo pages that cover each area in the survey. Use the index map to locate the appropriate photo page for the site of interest.

Next, turn backwards to the photo pages. Note that each photo page is numbered. Go to the correct photo page and locate the site of interest. This can be more difficult than it seems, especially in areas where landmarks are few. Use features such as roads, streams, railroad tracks, and buildings for orientation. Check the scale on the photo. It can be helpful to use a topographic map of the site simultaneously with the photo page, but be aware that the scales may be different (Fig. 2.2). In addition, keep in mind that significant changes in land use and vegetation may have occurred since the survey was published.

After locating the site on the correct photo sheet, make a list of all the soil symbols that appear at the site. In this example, the numbers 152, 154, and 171B, fall within the site boundaries. No other special symbols appear at this example site, but if they had, they should be put on the list as well.

To interpret the numbers and symbols on the photo sheet, turn to the page immediately preceding the photo sheet and read the key given there for the soil names and symbols. Write down each soil name and the meaning of each symbol. For example, in this case, the soil names are as follows:

152 Drummer silty clay loam

154 Flanagan silt loam

171B Catlin silt loam, 2% to 5% slopes. (*Note:* The letter "B" in this soil name indicates the degree of slope. Had the soil had 0% to 2% slopes, it would have been identified as 171A. Steeper slopes are indicated by the letters C, D, E, and so on.)

Next, turn to the list that has a title similar to "Index to Soil Map Units" or "Guide to Mapping Units" (see the table of contents to find the location of this list). This index or guide gives the page in the survey on which each soil unit is described.

Finally, turn to the page that describes an individual soil unit, and read the description. Keep in mind that the information given is of a general nature and may not apply specifically to the project site.

If a particular use of the soil is being considered, it may be helpful to examine the various tables in the soil survey that summarize the characteristics of each soil unit. For example, information on each soil's engineering, physical, and chemical properties may be tabulated, as may be information on each soil's suitability for such uses as building site development, construction materials, septic systems, wildlife habitat, and agricultural uses. ■

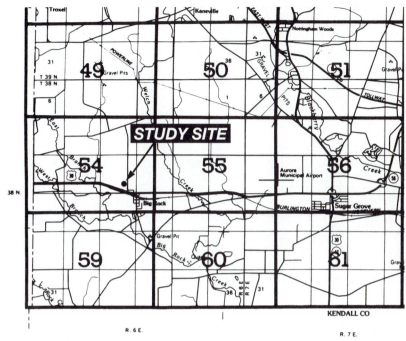

Fig. 2.1   The first step in using a soil survey is to use the index map to determine on which photo sheet the site is located. In this example, the site is located on Photo Sheet 54.

What surficial deposits are present? What type are they? What are their thicknesses? What are their origins and ages?

If the surficial deposits are glacial, fluvial, lacustrine, coastal, marine, or aeolian in origin, try to identify characteristic landforms. This may give hints about sediment type, lateral extent, and depth of the deposits.

Are the surficial deposits colluvial (formed in place) or alluvial (transported to this site from somewhere else)? If alluvial, what was the transport mechanism: water, wind, ice, mass movement, or humans? What erosional processes have affected the deposits, and how, in turn, have these erosional processes affected the hydrogeological properties of the deposits?

Look for the following information about bedrock at the site:

What rock units are present, and of what type and age are they? Where are they exposed, and why? What is the attitude of the rock units? In what structural setting is the area situated? What is the degree and orientation of fracturing? What erosional processes have affected the rocks? Has secondary dissolution influenced discontinuities within the rock mass and had an impact on its secondary permeability features?

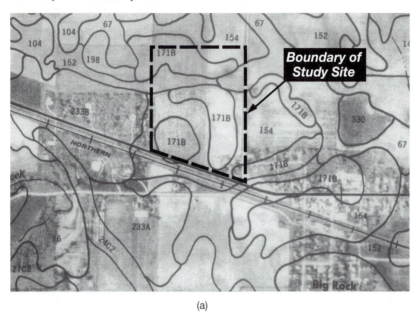

(a)

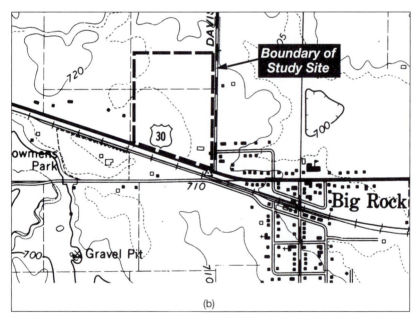

(b)

Fig. 2.2   Locate the site on the photo sheet (a) and read the symbols for the soil types at the site. It may be helpful to use a topographic map (b) as a guide, but be careful to adjust for the difference in scales.

Hydrogeology

Information about site hydrogeology may be gleaned from geologic reports and maps, described in the preceding section. Critical questions to be answered are the following:

What aquifers and aquitards are present?

What is the nature of permeability in the aquifers? Is it related to grain size, sorting, and cementation of the aquifer (primary permeability)? Or is it due to fracturing or dissolution (secondary permeability)?

What are water levels in the various units? How variable is water level within a particular unit?

What are flow directions in the various units? How variable is this?

What units are tapped at pumping centers in the area? What units are receiving injected fluids?

What are potential sources of ground water contamination in the area?

The main sources of information on site hydrogeology usually are the U.S. Geological Survey and the local water management districts. In addition to the sources listed in the "Geology" section, some other sources may be valuable as well. The local health department may keep records on well-water quality. Local well drillers may be able to provide information on subsurface conditions and water levels. And local property owners may know something about depths and water levels of their wells. Well logs and well completion diagrams should be on file with the appropriate agency, but what constitutes the appropriate agency varies from place to place. A local well driller should be able to provide the name of the appropriate agency. It may be the county health department, state water survey, county or municipal engineering department, environmental agency, department of natural resources, or other agency.

Information obtained from these agencies might exist in the form of developed transmissivity, storativity, or hydraulic conductivity values, in ideal situations. But even in less-than-ideal situations, boring logs, well completion reports, water-level records (especially if the well is a public water supply or commercial well), or well abandonment reports might be available. This information might be used to check the types of surficial deposits and rock units, find the number and type of aquifers and aquitards, and determine general water levels. Keep in mind that water levels fluctuate, and levels given in the reports very likely will not reflect the current situation.

Hydrology

Surface-water data, flooding, wetlands, and weather data may be important in some studies. Surface-water data may include records of discharge, stage, and physical and chemical characteristics of streams, lakes, ponds, and wetlands. In some cases, records of erosion, sedimentation, and dredge/fill operations may be available, too. This information may be kept by governmental agencies such as the USGS; state geological surveys or water surveys; state departments of conservation, natural resources, or environmental agencies; the Army Corps of Engineers;

the Federal Emergency Management Agency (FEMA); the U.S. Environmental Protection Agency (USEPA); the Bureau of Land Management (BLM); the U.S. Forest Service (USFS); the local, county, state, or federal transportation departments; or the local or county engineers. Each year the USGS publishes state-by-state compilations of data on daily stream discharge at thousands of gaged sites across the United States. Water-level records can give information on historical high and low levels, mean levels, and daily or seasonal variations. The USGS maintains World Wide Web pages that give real-time data of water level for some streams.

Surface-water data also may include data on flooding. FEMA publishes Flood Insurance Rate Maps (FIRMs) that delineate the boundaries of areas at risk of flooding. This may give information on historical high-water levels and may have a bearing on studies of inundation, wetlands, and vegetation. Local newspaper stories may give anecdotal information about flood levels.

Wetland maps are available through the National Wetlands Inventory, which is administered by the U.S. Fish and Wildlife Service (FWS), a division of the U.S. Department of the Interior. The National Wetlands Inventory delineates the boundaries of various types of wetlands. Although this information obviously might be used in studies involving potential destruction of wetlands, it also might indicate conditions of water levels, types of vegetation, and can provide information related to site access. The FWS maintains information on the World Wide Web.

Weather data might include records of precipitation, snowfall, evaporation, and temperature. This is available from the National Weather Service (NWS), a part of the National Oceanic and Atmospheric Administration (NOAA), which is a division of the U.S. Department of Commerce. NOAA maintains information on the World Wide Web.

Land Use

Information on land use may be particularly valuable in studies of surface-water and ground-water flow and contamination. Land-use data may provide hints as to type of contaminants and approximate age of contaminant release. Planning commissions are a good source of information on land use. Two other useful tools in this type of investigation are aerial photographs, discussed earlier in this chapter, and fire insurance maps.

Fire insurance maps are detailed maps of the floor plans of industrial and commercial properties in the United States. They were developed to identify areas of particular risk for fire or other health or safety problems, and to pinpoint equipment and features of the building important to fire fighting. The most notable of these are the Sanborn fire insurance maps; these have been in publication since the 1860s.

Fire insurance maps identify the type of construction of a building, locations of hydrants and sprinklers, chimneys, boilers, elevators, storage tanks, and other features that might be related to a ground water problem. These maps are particularly useful in environmental property assessments, where historical data may be of great significance. Historical fire insurance maps may identify problem areas that are no longer visible at the site.

Fire insurance maps may be obtained from planning commissions or municipal offices; some libraries, including the Library of Congress, maintain collections of these maps. Hatheway (1992) gave a brief review of the history, uses, and sources of Sanborn maps.

Action Plan

An action plan should be drawn up before going to the field. This plan should consider site access, goals of field activities, schedule, budget, personnel, contractors, and potential problems. This section describes each of these considerations.

Site Access

Site access includes legal, physical, and personal access. For legal access, all necessary permissions should be secured before setting foot on a site. Physical access must be arranged, as well: keys to gates, wells, or other locks must be available, and tools for accessing wells should be taken to the site (e.g., screwdrivers, wrenches, pliers). If a vehicle is to be driven on the site, it must be capable of handling the terrain. Stream crossings, deep ruts, or uneven terrain may dictate the use of a four-wheel drive or other utility vehicle. Personal access must be assured; if a particular type of clothing is needed (e.g., wading boots) or if personal protective equipment is required (see "Health and Safety Plan," which follows), this should be known in advance.

Goals of Field Activities

For maximum efficiency, goals should be clarified before going to the site. Is this a sampling trip? A site walk-through? A remedial project? What should be accomplished by the end of each day?

Schedule

Construct a schedule of all the work that will be accomplished at the site, breaking it down into specific tasks. This may include everything from arrival at the site and setting up instruments to decontaminating, packing, and hauling equipment home. Leave time for record keeping, communication between site personnel, equipment warmup and stabilization, minor problems, and breaks. Be flexible.

Budget

The cost of field operations is only one portion of the entire project budget. However, this portion of the budget must be tied closely to the schedule and list of tasks. Personnel (time), equipment, materials, supplies, travel, contractors, and many other items may need to be estimated and budgeted. In addition, each type of project (e.g., a consulting project for a corporate client, a remediation project funded by the federal government, or a research project funded by a granting agency) will have its own budgetary requirements. For further discussions of business practice and professional ethics, see Fetter (1994) and Watson and Burnett (1995).

In some projects, it may be necessary to hire contractors for certain operations, for example, drilling and excavating. These operations are put out for bid and proposals by the contractors are reviewed carefully before one is chosen. In such situations, it is imperative to ascertain the degree of competence of a contractor to complete the work before hiring. Hiring the lowest bidder may not be the most cost-effective action, if the contractor is not qualified to do the job. If the low bid is a cost estimate based on the use of inadequate equipment, poor-quality materials, insufficiently qualified personnel, or workers who lack the experience, expertise, training, or certifications necessary for a particular task, then cost overruns and schedule delays are likely to occur.

Personnel

Who will do which tasks? Will work be done in teams? Who will lead the teams? Consider the training and experience of field personnel in assigning tasks.

Contractors

If contractors will be used, they should be scheduled and have all of the information they need well in advance of the field day. For example, drilling contractors need to know where the site is, what equipment to bring, and what conditions (geological as well as health and safety) they might encounter at the site. This all should be established clearly and confirmed before a contract is agreed upon. In fact, contractors will need to have this information in order to prepare a reasonable bid for the work.

Problems

Problems in field work usually are related to equipment, contractors, time constraints (often due to inclement weather), or a handful of other difficulties. In preparing a realistic budget and schedule, the investigator should anticipate the problems of an assignment. Typical problems generally arise as a function of the geographic area (e.g., weather) or type of investigation; for example, geophysical surveys tend to be plagued by equipment failures.

Equipment. Problems with equipment may occur for a variety of reasons. *Forgetting equipment* is a problem that usually can be prevented by having a complete checklist established ahead of time. Some pieces of equipment can be improvised in the field or after a trip to the local hardware store. Others cannot be improvised, and forgetting them means that the field work comes to a standstill. This can be costly and is certainly frustrating.

Sometimes *equipment fails to work or gives spurious results.* This can be minimized by putting equipment through careful checks before going to the field. The equipment should be tested to ensure that it functions properly, under the conditions that will exist at the field site. Some consulting companies have equipment managers who thoroughly check all field equipment when it is returned from the field, and prepare it for its next trip out. However, field personnel who are unfamiliar with the operation of the equipment should practice using the equipment before going to the field. (This is described more completely in the section in this chapter on "Equipment List and Check.") If, despite careful checks, the equipment still seems to be malfunctioning, check the following:

> Is the equipment plugged in to a power source? Are batteries charged? Are electrical connections tight and correctly made? Does the equipment need gasoline or other fuel? Is power to the instrument turned on? Has it had sufficient time to warm up?
>
> Are fluid and vacuum lines free of leaks? Are they free of silt, vegetation, or other matter that might cause clogs? Is the instrument properly calibrated? Have buffer or other calibration solutions been used properly? Is the instrument's "range" selector set correctly for the conditions present?

When *equipment is not appropriate for the job,* it is usually because unexpected conditions are encountered. Examples of inappropriate equipment are a 4-inch pump for a 2-inch well; a suction pump (with lift of only 20–25 ft) for a 30-ft depth to water; or a chemical analytical instrument with a sensitivity too great or low to detect the chemical of interest. Careful planning should prevent most of these problems. It often is helpful while planning to envision actually using the equipment in the field. The more information available, the fewer the problems that are likely to arise.

Conflicts about equipment use may arise when two or more field teams need to use the same piece of equipment at the same time. To avoid this, picture or map out the field site. Where will teams be working, and how far is it between them? Can equipment be shared easily? What is the fieldwork schedule? Will two teams need to use the same item at the same time? If there is any doubt, take more of the necessary items.

Problems with *contractors* sometimes occur. These may include such things as failure to show up at the site, failure to communicate clearly with the hydrogeologist or other personnel, failure to perform the work adequately, or other difficulties. Sometimes, conflicts may arise between two contractors; for example, an earthmoving operation may be attempting to dump soil at the precise spot where a drilling rig is performing a soil boring. Careful advance planning and clear communication will prevent most conflicts of this type, but when they occur and cannot be easily resolved, they should be taken to the job supervisor.

Miscellaneous problems may arise in the field. Property owners or neighbors (whether well-meaning or malicious) may interfere with the work. Animals or plants may present problems. These may range from nuisances (mosquitos, biting flies, dive-bombing blackbirds, poison ivy) to dangers (poisonous snakes, angry bulls). Likewise, weather conditions may range from the uncomfortable to the truly hazardous. Rain, snow, hot sun, and cold wind make field work less enjoyable and can cause delays, but usually they can be tolerated. However, lightning storms, dangerous high winds, potential flash floods, severe dust storms, and extreme heat or cold make conditions unsafe, and work should not be done in these conditions unless the field personnel are properly trained and equipped to deal with the dangers.

Other health and safety conditions may present problems and should be anticipated and dealt with in a health and safety plan. This is described in what follows.

Health and Safety Plan

A health and safety plan addresses risks and hazards related to field work. A health and safety plan should cover all aspects of the work, including travel to the site, working on the site, and travel from the site. Such a plan spells out emergency procedures, communication channels, and precautions related to risks and hazards of the work.

In working with contaminated ground water, it may seem at first glance that the main hazards are potential exposure to toxic chemicals through such processes as inhalation of fumes, skin contact with fluids or soil, or accidental ingestion of contaminants. Some hazardous materials may be flammable, explosive, reactive, or radioactive, and obvious hazards are related to these characteristics. But risks and hazards are not limited to these. A health and safety plan should consider the perhaps more mundane but potentially equally dangerous problems such as vehicular traffic, heavy lifting, falls, noise, collapse of trenches or other excavations, falling objects, extreme weather, drowning, accidents with hydrogeological equipment, accidents involving drilling rigs or other heavy equipment, poisonous or dangerous animals and plants, and even problems with unfriendly human visitors to the site, among other risks and hazards. Each site has its own unique risks and hazards, and the health and safety plan should give clear steps for dealing with all of them.

Training field personnel can help make the work safer for everyone. The training should include first aid and CPR (cardiopulmonary resuscitation), but also should include training on how to use potentially dangerous field equipment. Personnel working with hazardous materials may need special safety training. At some sites, including sites that fall under the Resource Conservation and Recovery Act of 1976 (RCRA); the Comprehensive Environmental Response, Compensation, and Liability Act of 1980 (CERCLA), also known as the "Superfund"; the Superfund Amendments and Reauthorization Act of 1986 (SARA), and other state or locally regulated sites, field workers may be required to have completed a 40-hour health and safety training course. This training is commonly referred to in the United States as the "HAZWOPER" (Hazardous Waste Operations and Emergency Response) or the "OSHA 40-hour Course" (after the Occupational Safety and Health Administration, which regulates safety at hazardous materials sites). Some hydrogeological companies offer HAZWOPER training in-house to their employees. It also is offered by commercial environmental training firms, professional hydrogeological associations, and others. HAZWOPER training is offered by a few universities, usually as extension courses.

One of the many aspects of field site activities that should be discussed in the health and safety plan is the appropriate level of protective clothing and breathing protection. OSHA regulations spell out four different levels of this personal protective equipment (PPE) and describe the conditions that warrant each level. These levels are described in what follows, but reviewing this information is not a substitute for taking a health and safety course that trains participants in the use of PPE. The information is included here so that the entry-level hydrogeologist might gain some familiarity with the terminology. Do not attempt to use PPE without proper training.

The four levels of protection described by OSHA are designated by letters (see Fig. 2.3). Level D is the lowest level of protective clothing and equipment. It is used when no respiratory protection is needed. Level C protective clothing and equipment includes air-purifying respirators, but a low level of skin protection is needed. Level B is worn when the highest level of respiratory protection is needed, but a lesser level of skin protection is needed. Level A is worn when the highest level of respiratory, skin, and eye protection is needed.

When some components of PPE are used at a field site, they can have a measurable, even drastic, effect on the project schedule. Donning PPE, doffing it, and decontamination take time that must be factored into the schedule. In addition, work shifts may need to be cut short when extreme weather conditions prevail, because of the risk of dehydration, fatigue, or other difficulty. In very hot weather, for example, workers may need to take a break after working for 30 minutes or less. This should be considered in advance, while the schedule is being developed, in order to keep the project on schedule.

A description of the elements of the basic levels of protective equipment follows.

Level D PPE

At Level D, no PPE is required for respiratory protection. Level D protective clothing provides a low level of skin protection and is used when there is no potential for contact with hazardous levels of chemicals or radiological contamination. It includes coveralls over regular clothing, work gloves, PVC or latex rubber surgical/lightweight gloves when sampling or handling any potentially contaminated surface or item, safety glasses/goggles, steel-toe rubber or leather boots and outer boot covers, and a hard hat.

Level C PPE

Respiratory protection for Level C includes an air-purifying respirator, full-face or half-face mask equipped with a cartridge or canister. Level C protective clothing provides a moderate level of skin protection. It is used when the potential exists for contact with chemicals and/or radiologically contaminated materials, but when protection from liquids is not required and when potential vapors, fumes, gases, or dusts are not suspected of containing levels of chemicals harmful to skin or capable of being absorbed through the skin. Level C protective clothing includes coveralls over regular clothing, rubber/chemical resistant outer gloves, inner gloves of lightweight PVC or latex rubber, safety glasses/goggles, face shield if a splash hazard exists, steel-toe rubber boots, outer disposable booties, a hood if required for radiological work, hard hat, and hearing protection.

Level B PPE

Level B respiratory protection is positive pressure, full face-piece, self-contained breathing apparatus (SCBA), or a positive-pressure, supplied-air respirator. Level B protective clothing provides a high level of skin protection, and it is used when the potential exists for contact with chemicals and/or radiologically contaminated liquids that could saturate or penetrate cloth coveralls. It also must

Level A: SCBA or supplied-air respirator; totally encapsulating nonpermeable suit

Level B: SCBA or supplied-air respirator; suit providing high level of skin protection, but not necessarily vapor tight

Level C: Air-purifying respirator; coveralls, gloves, safety glasses, boots, and hard hat

Level D: No respiratory protection; coveralls, work gloves, safety glasses, boots, and hard hat

Fig. 2.3 The levels of personal protective equipment (PPE) are defined by the type of breathing protection required at each level.

be a situation in which potential vapors, fumes, gases, or dusts containing levels of chemicals harmful to skin or capable of being absorbed through the skin are not anticipated. Level B protective clothing includes a hooded, one-piece, nonpermeable, chemical-resistant outer suit; coveralls as an inner suit; regular clothing under coveralls; outer chemical-resistant work gloves taped to an outer suit; inner gloves of lightweight PVC or latex rubber taped to an inner suit; chemical-resistant, steel-toe boots taped to an inner suit; disposable outer boot covers taped to an outer suit; hard hat; and hearing protection.

Level A PPE

Level A respiratory protection includes positive-pressure, full facepiece, self-contained breathing apparatus (SCBA), or positive-pressure, supplied-air respirator. Level A protective clothing is used when the potential exists for splash or immersion by chemicals and/or radiologically contaminated liquids, or for exposure to vapors, fumes, gases, or particulates that are harmful to skin or capable of being absorbed through the skin. Level A protective clothing includes a totally encapsulating, nonpermeable, chemical-resistant suit; coveralls worn inside the full-encapsulating suit; regular clothing under the coveralls; disposable gloves and boot covers; chemical-resistant, steel-toe and shank boots; hard hat; and hearing protection.

At every level, other protective apparatus that may be used includes a cooling unit, two-way communication radios, cold-weather gear/clothing, and/or protection from biological hazards and pests.

Equipment List and Check

Before field work begins, the field equipment that will be needed should be identified, assembled, tested, and packed. Use the action plan to help identify what will be needed. Include both the equipment itself and any supplies that are required. Pack more supplies than seem necessary. Include supplies and equipment for decontamination, if appropriate. Some companies may have an equipment manager whose responsibility it is to keep track of equipment and ensure that it is working properly. However, the hydrogeologist still must tell the equipment manager which items will be needed.

Make a list of everything that will be needed. Gather the materials and then look carefully at every single item. Check the equipment, even if it has always worked before. Is it clean? Are all the parts present? Does it work? Are the batteries charged? Check the size and type of batteries and take spares. Check all cables and plugs. Are they working? If any instruments include or are attached to computing equipment, does the computer boot? What power source does it require? Are cables needed? A printer? What diskettes or software are required? If the equipment requires calibration, gather any necessary standards or calibrating solutions.

Also gather tools that might be needed in the field, both for normal use of the equipment and for emergency repairs in the field. In addition to these specific tools, it helps to have a tool kit containing all-purpose items. An assortment of screwdrivers; pliers; wrenches; clamps; a few assorted nuts, bolts, screws, and nails; a short-handled shovel or trowel; wire, string, and duct tape; one or two 5-gallon buckets; some disposable gloves; a few plastic beakers; a clean tarp or sheet of plastic; and a flashlight with spare batteries all may come in handy. A geologist's compass, rock hammer, and hand lens should be included as standard equipment. Health and safety equipment, and personal protective equipment (PPE) should be checked out before it is packed. Each field worker should have a timepiece with a second hand or digital watch or clock showing seconds. Finally, take a field notebook and writing implements, a calculator, necessary forms and permissions, and maps.

At The Site

Upon arrival at the field site, the inexperienced field worker should not immediately unload equipment. First, it should be determined that this is, indeed, the correct site. At some sites, it may be necessary to check in with the site supervisor. If this is the first visit to the site, it may be helpful to walk or drive around the site to gain a general idea of conditions, both hydrogeological and health and safety–related. Describe the general conditions in the field notebook. It may also be helpful to sketch a map of the site in the field notebook.

After this field reconnaissance, unload the equipment. If the work will take place at only one location, first unpack and unfold a clean tarp or sheet of plastic; place the equipment on it to keep it clean as it is organized, assembled, and checked. If the project requires that field personnel move about the area, set up a home base. This may be a vehicle, office, shed, or some other location. The home base ideally should be secure, sheltered from rain, wind, sun, and blowing dust, and it should not block roadways or gates.

Take note of health and safety conditions. Are they different from the expected conditions? If so, make appropriate changes to the health and safety plan, and communicate these changes to all personnel. Ensure that all health and safety equipment is accessible and in working order, and that everyone knows where it is.

When the home base is established and safety provisions have been made, begin the field work.

Before Leaving The Site

Before leaving the site, clean and pack the equipment. If anything is to be left at the site overnight, secure it and flag it to provide visibility and identifying information. Load the vehicle. With hands empty, so that any remaining items may be picked up (and nothing missed), check the area one final time to be sure that nothing is being left. Make the final entry in the field notebook, noting the time. If necessary, sign out with the site supervisor. Ensure that the area is as secure upon departure as it was upon arrival: Wells should be locked, gates should be closed and locked, and alarms should be reset.

Keeping Field Notes

Keeping field notes is an essential part of a field hydrogeologist's job. Field notes are a record of what actually takes place in the field. In cases that end up in litigation, a field notebook may be a critical legal evidentiary document. It takes some practice to learn to keep field notes; some guidelines are given here.

Field notebooks are available from geological, forestry, or surveying supply houses. They come in a variety of sizes and colors, and with a variety of patterns of preprinted lines, column dividers, and graph paper. Ideally, field notebooks should be constructed with water resistant covers and paper, to keep writing clear even when the pages are wet from rain or from a dunking in a stream.

Some workers prefer to use larger, looseleaf sheets for certain uses. Well logs and some hydrogeological test results or measurements are most conveniently recorded on preprinted forms. These sheets should be kept in a waterproof case such as a map case. If attached to a clipboard, they should be covered with a sheet of acetate to prevent mud spatters or water damage.

On the inside of the front cover of the field notebook, write your name (or the company name) and give contact information so that the book may be returned if it is lost and found. Include notice of a reward to anyone who finds and returns a lost field notebook. If the field site is far from your home or office, give a local address and phone number as well, so that someone who finds the lost book will not have to make a long distance call to return it.

Number all of the pages in the book consecutively in the upper, outside corners. Then, on the first page of the book, write the heading, "Table of Contents." At the beginning, leave a few pages blank, and add entries to the table as notes are added to the book. Add an entry each time it seems appropriate, for example, when work on a new project, at a new site, or on a new test begins. Ideally, a new notebook should be used for each new project.

Write notes in waterproof ink or pencil. Implements that make a fine, dark line are best, such as a 0.7-mm HB mechanical pencil or fine-tipped black pen. The writing should not smudge on the water-resistant paper; test this before writing notes. Take several pens or pencils into the field in case one is lost or stops working.

Regarding erasures, opinions vary. Some people prefer to write in pencil so that mistakes can be erased. Others believe that erasing anything make it appear as if notations were "doctored" after the field work is over, calling into question the integrity of the notes. These workers simply line out mistakes, instead of erasing them.

Whatever policy on erasures is adopted, it is essential that the notes be neat, clear, and easy to read. In many, if not most, hydrogeological firms, the field notebook is the property of the company, not the individual. As such, the notebook may be used by several different people, each one needing to read the notes of the previous workers. For clarity, notes should be printed, not written in script.

For convenience, some hydrogeologists find it helpful to keep a 6-inch ruler sandwiched in the pages of the field notebook. It also may be helpful to keep a rubber band or two wrapped around some pages and the binding of the book so that it opens quickly and easily to the page that is in use.

Accuracy and precision are extremely important in keeping field notes. To maintain accuracy, write what you observe and only what you observe. Should it be important to record something observed by someone else, identify the source of the information.

To maintain precision, be careful in descriptions of locations, instrument readings, distances, and other measurements. For example, writing that a sample was taken at a well "near the main building" is far too vague. How near? In what direction? Which building is the "main" building? Too many questions may be unanswered by a notation this vague. Use a sketch map with a scale, and couple this with very precisely worded descriptions. On the other hand, do not imply a degree of precision that is not warranted. For example, unless the distance between two wells was measured, it is impossible to say that they are "100 ft apart." Instead, indicate that the distance is "about" or "approximately" or "estimated to be" 100 ft. When a measurement is made, record the method that was used: for example, pacing, a measuring tape, or map distance all may be used to determine the distance between the two wells.

Field notes must be kept *in the field*. Notations with the most integrity are those made at the time of the observations, not after later reflection and second-guessing. If anything is added to the field notes later, it should be clearly identified as a later addition, and should be dated and initialed. Field notes should not be recopied into another field notebook to make them neater, because information may be lost (or added) in the recopying. Notes should be kept neatly in the field and left intact as a permanent record of field activities.

To help keep the notes neat and organized, leave a wide margin at the edge of each page. Use this margin to keep running notations of time and weather, or other brief comments. Begin notes on each new test, new day, or new project on a new page. Use titles or headings to demarcate new tests or projects. Use tables where information lends itself to a tabulated format. (More information on the use of tables is given in what follows.) When multiple instrument readings or other observations are to be taken over a period of a few hours or days, leave space for the later readings.

Begin field notes for each day by writing a heading with the project name and the date, and a precise location description. Describe the weather. Temperature, precipitation, cloud cover, wind, and humidity all may prove to be important in a hydrogeological study. As conditions change throughout the day, use the outer columns to add brief updates on the weather. Name the field personnel and give the company name of any other contractors involved in the work. Give a brief description of the purpose or goal of the field work.

If this is the first field work done at a particular location, give a description of the field area. Topography and relief, geologic and hydrogeologic features, surface-water and ground-water conditions or indicators, vegetation, and human-made features all may be important.

When using an instrument, piece of equipment, or other apparatus, describe it. Give the name, the manufacturer, and the model/serial number. If appropriate, sketch the equipment or field setup, showing connections of hoses, gauges, valves, wiring, and other important components.

Describe the procedures used in the field. For some procedures, this can be done by simply referring to standard methods or manufacturer's instructions. For other procedures, where standard methods do not apply, give a step-by-step listing. Other data and observations taken in the field may include measurements, instrument readings, notations of significant features, tables, graphs, figures, maps, or notes about potential errors.

Maps, cross-sections, and sketched diagrams often are the most useful additions to field notebooks. These drawings convey a great deal of information in a simple and useful format. Every map or diagram must have a scale, whether exact or approximate. Maps must have a North arrow. Give each map or diagram a title or caption to explain what is being illustrated. When the figure is complete, look at it as if you were another hydrogeologist seeing it for the first time. Is it clear and understandable? Does the figure convey the information that it intends to convey? Would another hydrogeologist be able to understand and interpret the salient points of the figure?

Tables can be used to organize a large amount of data in a neat format. Use tables to record repetitive observations, particularly numerical data. Each table must have a title. Of primary importance in constructing a table is defining each column clearly. The column heading should be unambiguous, and if the column contains numerical data, the heading must include the units of measurement. A final column labeled "Comments" or "Notes" can be used for miscellaneous information. Large tables may extend across two facing pages in a field notebook.

Graphs may be highly useful notebook entries. When a graph is constructed in the field notebook, it should have a meaningful title. Axes of the graph must be clearly labeled, and units of measurement must be given for each axis.

At the end of the work, whether at day's end or when leaving a site, make a notation such as "Done for the day" or "Left site." Note the time and initial the entry. This practice makes it clear that any further notes were not written while at the site.

Surface Water

Chapter Overview

The study of hydrogeology focuses on water below the surface. But subsurface and surface waters are inextricably linked, and the study of surface water often provides valuable clues to what is happening underground. This chapter examines the field methods that apply to data collection from springs and seeps, streams, and lakes and ponds. In each case, field observations and notations are described, as are methods for making basic hydrologic measurements.

Springs and Seeps

Springs and seeps are discharge areas, places where ground water moves out of sediments or rocks to the surface. When the rate of flow is rapid, especially if flow is visible, the discharge is called a spring. If flow is slow, it is called a seep. Some springs and seeps discharge from a single point, for example, a joint or a fracture in a rock. Others have the geometry of a line, for example, where a hill slope intersects the water table. Still others might be broad areas.

A spring or seep used as a water supply sometimes must be developed, meaning that some construction must be done to enhance the flow rate and channel the flow. This typically involves digging away soil or other surface matter; placing a high-permeability material such as sand, gravel, or perforated pipe to capture and route flow; installing a section of solid-walled pipe to act as the discharge point; and using concrete, bricks, steel, or mud to seal off all but the discharge point so that water is forced to flow into the pipe (Fig. 3.1). Ideally, a spring used for water supply is located upslope from the user, so that gravity flow can be used to convey the water to the user, thereby eliminating the need for a pump. But if it is downslope, a shed called a "spring house" might be built around the spring in order to house a pump, and sometimes a treatment facility (Fig. 3.2).

The location of a spring or seep may be a valuable piece of information in understanding the hydrogeology of an area. To find certain springs and seeps, streams and their tributaries may be followed upstream to their ultimate source. Intermittent streams may appear at the surface many times throughout their course. Often, springs and seeps are located on hillsides; they may be identified by a point or a line on the slope with damp soil and/or a change in vegetation. A soil color change may indicate a seep, but be careful: This might be unrelated to the presence of water and might simply indicate a change in the underlying sediment or rock type. In addition, the mere presence of damp soil does not mean the water is coming up from below—it may be perched. Careful interpretation of geologic maps

Fig. 3.1 Develop a spring by providing a pathway for seepage to collect and allow the collected water to flow through a pipe. (Courtesy of Tim Andruss.)

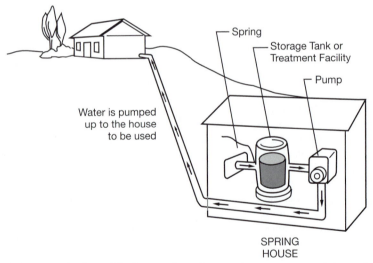

Fig. 3.2 A "spring house" built around a spring may house a pump and storage tank or treatment facility.

may help the user locate likely areas of springs. For example, springs may appear along the contacts of permeable units over impermeable units, or where highly dissolved limestones outcrop. To find a developed spring, look for boxes, culverts, sheds, cement-block housing, pipes sticking out of the ground, or similar structures. If flow rate of a spring is high enough, its presence might be revealed by the sound of trickling water. Check topographic maps of the area: They may show spring locations.

Why Is This Spring Here?

That is a good question to ask any time a spring or seep is encountered in a field study. The answer may illuminate the geologic setting, position of the water table, flow directions, past land-use practices, or other aspects of the hydrogeology of the area.

The location of a spring or seep may be the result of hydrogeologic or human factors, or both. Hydrogeologic factors include those related to the hydrologic regime and to the geologic setting. What is the position of the spring with respect to the water table? Because a surface-water body is the surface expression of the water table, the spring may be located at the water table. This is particularly true if the spring is part of a line of springs on a hillside. They are discharge points, so flow in the water-bearing formation would be toward the spring, at least locally. A spring may indicate a zone of higher permeability than surrounding areas, due perhaps to a change in lithology or to a fracture or joint.

A spring or seep may be instead an expression of a confined layer's potentiometric surface which is at a level above that of the ground surface. These "flowing artesian" springs or seeps might be located in areas where low-permeability materials appear at the surface (Fig. 3.3). Not all saturated, marshy areas with low-permeability materials fit this description, however; some are simply areas where water is perched, delayed in its downward movement.

Geologic factors giving rise to a spring or seep would include a zone of high-permeability near the ground surface, with hydrogeologic conditions such that hydraulic head is higher below ground surface than above it. Fractures, joints, or solution cavities might result in springs, as could a layer of sand or other relatively high-permeability material near ground surface. Leachate springs issue from the side slopes of some landfills. Figure 3.4 shows several geologic settings that may give rise to springs and seeps.

Humans may have made changes to a flow system that result in an artificial spring or seep. What appears to be a spring may be the result of a leaky water main, sewer line, or drain pipe. In farmed areas, drainage tiles, which are

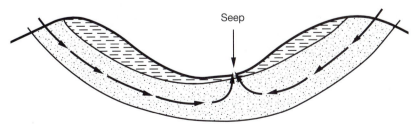

Fig. 3.3 Springs and seeps need not be located where surficial materials have high hydraulic conductivity. Here, a seep results from thinning of a low-conductivity layer at the discharge point.

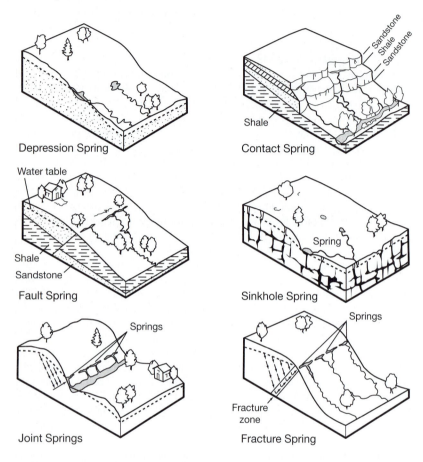

Fig. 3.4 Several geologic settings that may give rise to springs and seeps (after Fetter, 1994).

perforated pipes laid underground to collect and redirect water, may convey water to a "spring" (Fig. 3.5). Outlets from roof drains, roadway drains, and parking lot drains may appear as springs.

Field Observations
If a spring or seep is encountered in the field, appropriate field notations would include the following:

- *General Appearance.* Location; topographic setting; how spring was located; whether it is developed; if spring has point, line, or area configuration; surrounding land use; extent of wetted areas.

Fig. 3.5   Drainage tiles may convey water to an outlet point that appears to be a spring. (Courtesy of Tim Andruss.)

- *Hydrologic Setting.* What clues about the hydrologic setting of the spring are given? Examine soil/rock type, fracturing, presence of other springs or other hydrologic features.
- *Vegetation.* Note presence, condition, and type of vegetation in and near spring.
- *Staining.* Note presence, color, and extent of staining on rocks or sediments, vegetation, and development structures. Red and orange stains may be present if spring water contains abundant iron; black may indicate manganese; other colors may indicate other particular constituents. If the stained area is larger than the area currently being wetted by the flow, it indicates a history of higher flow. It may be obvious from the condition of the stain how recently the higher flow occurred.
- *Geochemical Measurements.* See Chapter 9 on geochemical field measurements and sampling. Sampling from springs may require that the spring first be developed. *Note:* This may not be practical or permissible; check with the landowner before altering anything.
- *Discharge.* Is the spring singular or are there many discharge points? Is it a diffuse seep? If diffuse, it may be necessary to provide it with a preferred path in order to get enough flow to measure. This may be done by developing it, either temporarily or permanently. To develop a spring, follow these steps:

  1. Secure permission of the property owner to modify the spring. In some situations, it may be necessary to obtain a permit from a regulatory agency, as well.
  2. Dig out the surficial matter (vegetation and several centimeters of topsoil) around the area of the seep or spring. The size of the area to be dug out depends on the nature and purpose of the spring development as well as on the characteristics of the spring. Backfill with gravel as needed.

3. Block the flow of the spring or seep using impermeable material such as concrete, clay, boards, cement blocks, large flat rocks, or sheets of metal.
4. Install a pipe through the impermeable layer into the spring or seep, allowing flow to come through the pipe (Fig. 3.6).
5. Use a bucket and timing device to determine flow rate. The bucket should be calibrated with liter or gallon markings; prepare the bucket at the office before going into the field.

If the spring is already developed, use a bucket and timing device to measure flow. Note whether flow or seepage is occurring about the sides of the developed, flowing area. This is difficult, if not impossible, to measure, but is part of the discharge nonetheless.

If the spring is undeveloped, yet flow occurs in a fairly well-defined channel, provide the flow with an improved path or channel by laying down a sheet of plastic. Route the flow into a bucket; use a timing device to determine flow rate.

In all these cases, it is likely that the true value of discharge is greater than the measured value. It is very difficult to ensure that all flow goes into the measuring device; there is likely to be some underflow. If a spring flows upward into a pond or puddled area, it may be impossible to measure discharge of the spring at its source. In this case, it may be useful to measure the discharge of the resulting stream, if there is one. Alternatively, seepage meters may be installed (see the section on seepage meters in this chapter) to determine discharge into the pond. Keep in mind that the discharge of a spring may vary considerably depending on

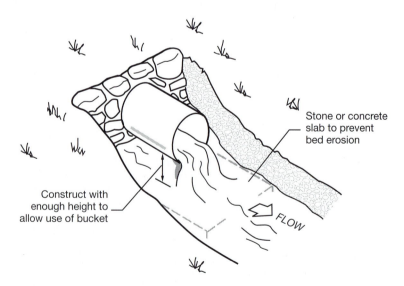

Fig. 3.6 Install a pipe through the impermeable layer into the spring, allowing flow to come through the pipe (after Brassington, 1988).

characteristics of recent precipitation events, season of the year, proximity to irrigated fields, and other similar factors.

Delineating Drainage Basins

A drainage basin or watershed is the area throughout which surface water drains into a particular body of surface water. The body of water might be a stream, lake, wetland, ocean, spring, or pond. Be aware that a surface-water drainage basin is not necessarily the same as a ground water drainage basin. In a ground water drainage basin, we delineate the area throughout which ground water drains into a particular aquifer or out of a particular discharge area. This might or might not coincide with the surface-water drainage basin superimposed above it.

It is important to delineate a surface-water drainage basin in any study of the water budget of an area, the impact of some activity that might affect quality or quantity of water in a basin, or the potential development of water resources in a basin.

Topography determines the boundaries of a surface-water drainage basin. Therefore, a topographic map can be used to make a first approximation of the basin boundaries. (Ideally, any drainage basin map made in this way should be field-checked.) Follow these steps (Fig. 3.7):

1. For a stream's drainage basin, find the point where the stream enters another body of water (its "base level").
2. Put a pencil tip at the base level, and then move it, intersecting contour lines at right angles, to the topographically highest nearby location.
3. Continue in this fashion. The goal is to draw a line surrounding the area on which rainfall would drain into the stream of interest. Rain falling on the other side of the line would flow into a different stream or other body of water.
4. Interpret the topographic contours this way: Set a pencil point down anywhere on the map, and imagine what would happen if a marble hit the ground (as shown on the map) at that point. Would it roll ultimately into the stream of interest? If so, then that point belongs in the drainage basin of that stream. If it would end up in a different stream, then it does not. If that is the case, then move the pencil closer to the stream of interest and repeat the conceptual "marble test".
5. The point where it is unclear which way the marble would roll, because it might roll either way, is situated on the drainage divide, or boundary of the drainage basin.

It can be difficult to make an absolute determination of drainage-basin boundaries from a map alone, especially if there is a broad floodplain or if there are extensive wetlands where the stream meets the next body of water. Flat topography also can be a complicating factor, as can human-made diversions of drainage. For a more precise map, field-check the basin boundaries. Field checking the boundaries of a large basin may be impractical; it may be best simply to check the most problematic areas.

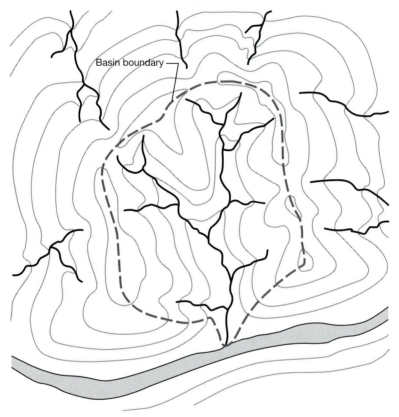

Fig. 3.7 Use a topographic map to make a first approximation of the basin boundaries.

To field-check the map, again begin at the mouth of the stream. Walk uphill, initially walking along the steepest uphill path. Move slowly, at each step checking. "If a drop of rain fell to my left and ran overland to a stream, would it flow into my stream, or somewhere else? What if it fell to my right?" If you are walking along the drainage divide, then the right and left sides should give different answers to this question. Look behind yourself frequently to check yourself. Survey flags may be helpful; position them at the basin boundary and periodically review them as you proceed.

Bear in mind that especially in urbanized areas, drainage pipes and sewers may drain an area that topographically would appear to be part of one basin, while actually conveying the water into a different basin. Roadways and other embankments or structures may divert natural flow paths, rerouting the water to another basin. Local governmental engineers, sewage-treatment plant operators, and highway engineers can provide information on these structures or conveyances. Other con-founding features are sinkholes, especially in karst regions, and glacially formed kettles and "potholes." In regions where these features appear, especially if they make up a significant fraction of a basin, it might be necessary to do a smaller-scale study.

Streams

Why Is This Stream Here?

In a field study of an area that includes a stream, ask the question, "Why is this stream here?" The answer may illuminate flow paths of water through the area. Obviously, streamflow follows topographically low areas. But what feeds the stream? Why is there water in the stream? Why doesn't the water soak into the stream bed? If the stream is intermittent, why is it so? If it is perennial, what keeps the water flowing? Does the stream flow in a single course, or is it a braided stream (i.e., one with many separate channels)? Why?

Sometimes the source of a stream is a single point: a spring, industrial outfall, or drainage pipe. Alternatively, the source might be runoff: water flowing across the ground surface to the stream, or baseflow: ground water discharging into the stream through its bed, all along the length of the stream. An ephemeral stream, which flows only in association with a storm event, probably has only runoff and perhaps some shallow ground water flow as a source. A perennial stream is ground-water-fed and flows even in times of drought.

What keeps the water in the stream? Gaining streams gain more and more water as they flow downstream, both from tributaries and from ground water discharge into the stream. Losing streams have less discharge as they go downstream, because water infiltrates the stream bed and becomes ground water. If water does not infiltrate into the stream bed, it is for one of two reasons: Either the perme-ability of the stream bed is too low to allow it or the ground water gradient is such that water flows in to the stream, not out of it. Intermittent streams contain water along parts of their course, but not along others, as soil and hydrogeologic condi-tions change (Fig. 3.8).

Field observations

Several stream parameters may be measured in the field. While working in and around streams, safety should be of primary importance. Before wading into any stream, test the bottom to determine what the footing is like. If the current is too fast to allow you to keep your balance easily, it is unsafe to work in the stream. Do not even consider using ropes or harnesses unless specialized, advanced training in this specific area has been completed. Such devices in the hands of a novice can be deadly.

Field observations about a stream should include the general topographic set-ting and condition of the floodplain. Note the character of the stream banks and bed. What are the sediments and rocks like? Describe the vegetation.

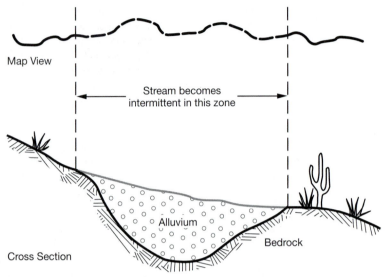

Fig. 3.8 One of many possible situations giving rise to an intermittent stream.

Observations about the following are appropriate in a field study of a stream:

- General topographic setting
- Site-specific topography and relief (a sketch or profile may be helpful)
- Character of the floodplain and floodplain development. (For example, is there a floodplain? How wide is it relative to the stream's width? What is its cross-sectional geometry?)
- Description of the stream banks and bed
- Sediment and rock exposed in cuts and in the stream bed
- Soils on the bank and washover deposits
- Vegetation: plant species, density, and condition
- Evidence of animal activity in the stream on its banks (e.g., tunneling, burrowing, and dam building)
- Field observation of moisture content of the floodplain soils (dry, moist, or saturated)
- Depositional features (sand bars, debris, or high-water marks)
- Erosional features (scouring, exposed tree roots, unstable vegetation, or building foundations that are undercut)
- Human development (buildings, roadways, bridges, railroads, dams, diversion structures, drainage pipes, rip-rap, levees, and spillways)
- Evidence of flooding (high-water marks, debris in trees, muddy or silty coating on vegetation and structures, and dead aquatic animals on banks)
- Bank stability (undercutting, trees with roots exposed hanging over the water, evidence of soil creep, slides or washouts, unvegetated scars of mass movements)

Width

The width of a stream changes as discharge changes. Discharge changes from moment to moment, day to day, and season to season, depending on rainfall, snowmelt, and ground water interactions with the stream. Because width and discharge may change significantly, any stream-width measurement should be correlated with a discharge measurement made at the same point.

The width of an unchannelized stream may vary significantly along its course and as discharge varies. As a result, depending on the objectives of the field project, several measurements may be needed. For small streams, measurements of width may be made easily with a fiberglass or steel tape measure. Measure perpendicular to the flow direction and to the stream banks. If the stream is wide, the tape may sag, resulting in a measurement that is too high, so pull it taut. Also, if the tape drops into the water, the current will pull it, again resulting in an inaccurate measurement. The approximate width of larger streams may be measured on a map. Keep in mind, however, that a change in discharge may significantly affect the width. If a stream is inaccessible by wading, a bridge over the stream may be the only point at which a width measurement may be safely made.

Distance-finding equipment that uses light or laser beams may be used on wider streams. Surveying equipment also may be used. If it is impossible to measure distance in one sighting, a bridge or island may serve as a triangulation station.

For instructions on using triangulation, refer to the *National Handbook of Recommmended Methods for Water-Data Acquisition* (USGS, 1980).

The channels of a braided stream (one with many separate channels) shift rapidly. Because of this, it may be most useful to consider the width of the entire set of channels, depending on the purpose of the study.

Depth

Like width, the depth of a stream varies with varying discharge. In addition, depth typically varies across the stream from bank to bank, and may vary considerably even along short longitudinal distances. Meandering streams are generally deeper on the outside of a curve and shallower on the inside. "Pool" areas generally are deeper than "riffles." Even in a straight stretch of a seemingly regular stream, the bed may be thoroughly irregular and may contain bars and channels. If depth is a critical factor, as it is in any estimation of discharge, sufficient measurements should be taken to characterize it. Stream beds change as deposition, erosion, and flooding occur, so depth measurements may be valid for only a short time, depending on the stream processes.

In small streams, depth may be measured with a plastic or metal graduated rod (plastic meter sticks are particularly handy for this). If a long enough measuring rod is available, it may be lowered into the stream from a bridge or platform. A calibrated, weighted line also may be used for this purpose, but unless the weight is heavy, the current is likely to pull the weight and line downstream, giving an inaccurate reading. A portable crane (a "bridge crane") and a cable reel may make it easier to handle the heavy weight in these locations. Streamlined weights with fins may be used to minimize the effect of the current.

When measuring near bridge abutments, large rocks, or other obstructions, be aware of eddy currents that might distort the water surface. These currents change and, in some cases, even reverse flows. Avoid measuring in these areas; if this is impossible, increase the number of measurements made.

A weighted line also may be used to make depth measurements from a boat. In this case, it is important to know the exact location of the measurement. This may be found by sighting with a compass or surveying instrument to two fixed landmarks or by stretching a graduated cable across the river, and making measurements at a set interval. A surveying instrument may be set up on the banks of the stream, and the boat may be directed (by visual signals or by radio communication) to the precise measuring spot. A global positioning system (GPS) unit may be used in some cases, but this method is not precise enough to warrant its use except in very wide rivers. Depth-sounding equipment may be used, especially if the stream is deep; versions of these instruments are sold in some sporting goods stores as "fish finders."

Because river depth may vary significantly from bank to bank, longitudinally along the river, and at different stages of water elevation, it is important to take many measurements to get an accurate cross-sectional view of the stream. When measuring across the width, especially if the width and depth measurements are to be correlated with discharge measurements at the same points, make at least twenty measurements. Depending on the irregularity of the stream bed and the purpose for which the measurements are to be made, it may be necessary to take more.

Depth may change over time as the stream moves, erodes, and deposits sediments. When discharge changes, depth changes; depths should be correlated with a discharge measurement.

Gradient

The gradient of a stream is the change in elevation of the water surface between two points divided by the horizontal distance traveled by the water. The gradient of a stream changes along the course of the stream, and the overall gradient commonly is different from the gradient measured along discrete reaches of the stream (Fig. 3.9).

The gradient of a relatively short section of a stream may be measured using simple surveying instruments (transit or alidade, meter stick, measuring tape). The gradient measures the change in slope of the water *surface,* and the simplest place to make this measurement is where the water-surface elevation equals the land-surface elevation, or at the water's edge. If the stream has a very rapid current, there may be a "banking" effect on a curve, wherein the water level is higher on the outside of the curve than the inside. When this condition occurs, either avoid the problem by making the measurement at another place in the stream or make two measurements, one on either bank.

Two people are needed to make accurate measurements of gradient. One holds a graduated rod (e.g., a meter stick or stadia rod) so that its bottom end sits precisely at the surface of the water. The easiest way to achieve this is by setting the rod precisely at the water's edge. Attempts to hold the rod level with the stream

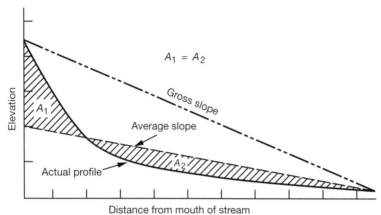

Fig. 3.9 Gradient of a stream changes along the course of the stream. The overall (gross) gradient is different from the average gradient and from the gradient at a discrete section of the stream (after Viessman, Lewis, and Knapp, 1989).

surface while standing in the middle of the stream probably will not be successful. Another person takes a surveying instrument (such as an alidade, transit, or theodolite) some distance upstream or downstream; the direction is a matter of convenience. The height of the instrument above the water surface is measured. A level line is then sighted to the stadia rod, and the reading on the rod is noted. Distance from the instrument to the stadia rod is measured, by tape or accurate map distance. The gradient is calculated as shown in Fig. 3.10.

If the gradient is very low, it may be difficult to measure it in the way described, because change in elevation will be small over a large distance, and accuracy of the method will suffer. Gradient may be estimated from gaging station information collected and published by the USGS by comparing the gage height and datum at two stations, and measuring the map distance along the flow path between them. Alternatively, elevation of the water surface may be surveyed at two points and referenced to benchmarks. However, because stream level may change during the time the measurements are being made, they should be made simultaneously, or they should be repeated several times, alternating between upstream and downstream locations.

Cross-Sectional Area

The cross-sectional area of a stream is the area of the stream perpendicular to flow. It varies along the course of the stream, and it varies as a function of current velocity and discharge. It should be estimated by first constructing an accurate cross section of the stream, based on width and depth measurements. Once the cross section is complete, the area of the stream may be calculated by using any standard method, four of which are described here:

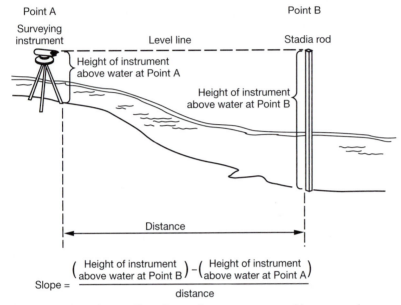

Point A   Point B

Surveying instrument

Level line

Stadia rod

Height of instrument above water at Point A

Height of instrument above water at Point B

Distance

$$\text{Slope} = \frac{\left(\begin{array}{c}\text{Height of instrument}\\\text{above water at Point B}\end{array}\right) - \left(\begin{array}{c}\text{Height of instrument}\\\text{above water at Point A}\end{array}\right)}{\text{distance}}$$

Fig. 3.10   To calculate gradient along a discrete segment, position a surveying instrument at the water's edge at Point A and a stadia rod at the water's edge at Point B. Measure distance between the points. Calculate gradient as the change in elevation over the distance.

1. Trace the cross section onto graph paper and count the number of grid blocks enclosed in the area.
2. Consider the area as being composed of many trapezoids. Each trapezoid is defined by the water surface, the stream bed, and two measurements of depth made along the transect across the stream. Calculate the area of each trapezoid by summing the two depth measurements and dividing by 2 and then multiplying by the width of the section. Sum the areas of the several trapezoids to get the total cross-sectional area of the stream.
3. Use a planimeter, an instrument that measures areas on maps.
4. Use a computer-assisted drafting program that has the capability of measuring areas enclosed within irregular shapes.

Wetted Perimeter

The wetted perimeter (WP) is the length of a line along the part of the stream bed that is under water, in a direction perpendicular to flow. The WP of a stream always is greater than its width. This is especially the case if the stream is narrow and deep. See Fig. 3.11 for examples. WP is best measured by constructing an accurate, to-scale cross section of the stream based on width and depth measurements, then measuring the WP on the cross section. Braided streams may be treated as the sum of many individual channels, or as one channel, depending on the purpose of the study. WP varies with stage and stream width, which vary with discharge; WP should be correlated with a discharge measurement made at the same point and time.

Hydraulic Radius

The hydraulic radius is the cross-sectional area of a stream divided by the wetted perimeter. This quantity is used in estimating stream discharge with the Manning Equation (see the section on the Manning equation later in this chapter).

Stage

The stage of a stream is the elevation of the water surface above a datum. A datum commonly used is mean sea level. However, a datum may be chosen at any local point, selected carefully to be easily visible and permanent, that is, not affected by erosion, scour, or deposition. If a local datum is used it should be referenced to a nearby benchmark to correlate it with mean sea level. Buchanan and Somers (1968) gave a thorough description of the design and use of instruments used by the USGS to measure and record stage. Also see USGS (1980) and Herschy (1995), for more information.

Gages used to measure stage may be of the recording or non-recording type. Nonrecording gages simply show the relative elevation of the water surface; an observer must read and record the elevations manually. A common type of non-recording gage is a staff gage. A staff gage is essentially a large-scale ruler, affixed to a permanent object and referenced to a datum (Fig. 3.12). The observer notes the height of the water surface as indicated by the ruler. Staff gages should be installed at locations unlikely to be affected by scouring or by floating debris. Installation of a staff gage in a stream with a significant current will cause eddy currents about the gage. For this reason, if the measuring side of the gage faces upstream or downstream, the readings will be anomalously high or low. Instead, the staff gage should face neither upstream nor downstream, or it should be placed

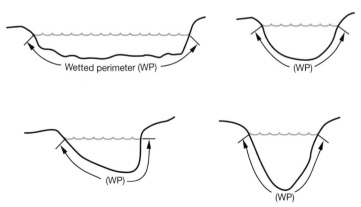

Fig. 3.11   Wetted perimeter of a stream always is larger than stream width. WP may be much larger than the width if the stream is narrow and deep.

Fig. 3.12 A staff gage.

in a location where strong current will not interfere with the reading. Alternatively, the staff gage may be placed in a stilling well. A stilling well is a pipe or other structure that allows free flow of water into and out of the well, but that "stills" the water so that gage readings are accurate. The well must not obscure the observer's view of the staff gage.

Staff gages are relatively inexpensive and durable. If read on a daily basis at the same time of day over a long period, they can give a good indication of flow conditions over that period. However, they are not capable of yielding a good record of brief flow events unless the observer stays nearby to read the gage at short time intervals.

If all that is needed is a record of peak stage during a flow event, a crest stage gage may be constructed and used in tandem with a staff gage. A crest stage gage is essentially a tube, open at bottom and top, installed near or attached to a staff gage or other calibrated marker or surveyed point. Water flows freely into and out of the base of the tube, and air escapes out the top. Pieces of burnt or ground cork are put in the tube; they will float on the water's surface. When water level rises, the cork rises with it (Fig. 3.13). When water level drops, the cork adheres to the sides of the tube, indicating the height of maximum stage. In one design, the tube is clear, so that the cork may be observed through the walls of the tube. In another design, a rod is set within the tube, and is withdrawn to make a reading. See Buchanan and Somers (1968) for more information. To "reset" the gage, water may be poured down the tube and allowed to run along the sides or along the rod.

Although this method is simple and inexpensive, it has a few drawbacks. The gage cannot be read from a distance. It records only the highest peak stage, and it cannot distinguish between peaks. As a result, if there are two or more flow events between readings, it may be unclear which event caused the highest stage. Finally, the gage must be read manually.

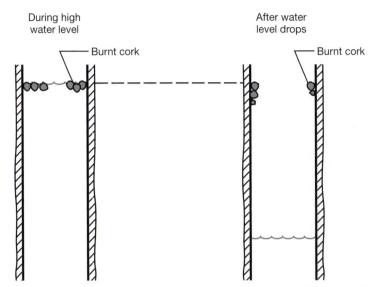

Fig. 3.13 An inexpensive maximum ("crest") stage gage. Pieces of burnt cork floating in a tube adhere to the sides of the tube when water level drops.

Continuous-recording gages are much more expensive than staff gages, but offer significant advantages. They provide continuous water-level measurements; no observer is required in the field. The gage records measurements for periodic retrieval or automatically transmits them to a base office.

Several designs of continuous-recording gages are available. The simplest is the mechanical float recorder, or "Stevens-type" gage. In this design, a pulley and cable are set up, with a float attached to one end of the cable and a counterweight attached to the other end. The float floats on the water surface; when the water level rises and falls, it rises and falls, too (Fig. 3.14). This turns the pulley, which is attached to a recording device. The recording device may be a strip chart recorder, on which charts must be changed daily, weekly, or monthly by the operator. Or it may be a device that digitally encodes that data for later downloading, either by an operator at the site or via remote telemetry.

A stilling well is used to allow measurement of water level unaffected by current, and to protect the float from debris. The well may be located within the stream or on the bank; the latter option may involve excavation and/or drilling (see Fig. 3.15). The well may be made of any material that can be worked and that will protect the float from debris, but will still admit water. For example, a stilling well may be made of corrugated pipe, stainless steel, PVC, or even (under conditions of low velocities and little debris) clothes dryer vent pipe into which holes have been cut to allow water to flow freely in and out.

A stilling well should be constructed such that it functions no matter what the height of the water within the well. Intake pipes or holes in the casing should be

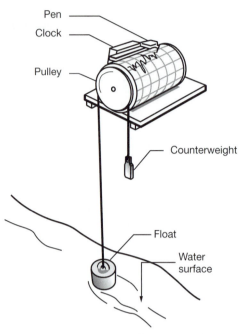

Pen

Clock

Pulley

Counterweight

Float

Water surface

Fig. 3.14 A continuous-recording gage. The float stays at the water's surface, rising and falling with changes in water level. This causes the chart recorder to rotate under a pen. At the same time, the clock drags the pen along the recorder drum at a steady rate, providing a continuous record of water-level changes with time. Measurements also may be recorded digitally.

placed at many elevations so that flow can be admitted to the well at any stream stage. The holes must be large enough to allow water to flow freely in and out, but small enough to protect the gage from debris and to damp out surges or waves.

A continuous recording gage is installed on a platform. The platform may be located in the stream or at its edge. Wherever it is located, it must be accessible for servicing; responsive to changes in stream level; and secure from erosion, flood debris, boaters, and vandals. Construction of the platform must be secure, even in floods, and it must be sturdy enough to allow a person to climb onto the platform to service the gage. To determine how high to make the platform, check the levels of previous high flows in historical records of the U.S. Geological Survey (USGS) or Federal Emergency Management Agency (FEMA), interview local residents, or look for high water marks or other evidence of flooding high above the stream bed.

The entire gage must be elevated high enough on its platform to keep it dry in case of flooding. A "doghouse" of sheet metal or plywood should be constructed around the gage, with hinges to allow for access, and a secure lock to keep out vandals. In areas where the gage may inspire a round of target practice by vandals with guns, a more secure installation may be required. In more expensive and secure

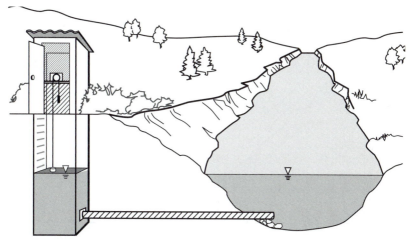

Fig. 3.15   Stilling wells may be located within the stream or on the bank; this may involve drilling or excavating the bank. Intake tubes are subject to siltation and must be maintained regularly (after Manning, 1997).

installations, the gage may be placed in a locked shed constructed of relatively thick metal.

Velocity

Velocity of a stream varies from top to bottom of the flow profile, across the stream from bank to bank, and along the flow course of the stream (Fig. 3.16). Velocity varies as a function of discharge and cross-sectional area, and stream-velocity measurements should be reported with discharge measurements made at the same time and same point in the stream. Buchanan and Somers (1969) provided detailed information on measurement of velocity and discharge at USGS gaging stations. More information is also available in USGS (1980), Dunne and Leopold (1978), and Herschy (1995).

Velocity may be estimated by direct or indirect methods. Direct methods include using a float, current-velocity meter, or tracers.

Float Method

The float method is the simplest and most rapid of the three direct methods, though it has significant disadvantages. In this method, a float is dropped in the stream and timed as it travels a given distance. Velocity is the distance traveled divided by the time it takes to travel that distance. Because the float is at the surface of the water, this method does not give a representative measurement of the average stream velocity.

The float may be anything that floats and is easily visible, but some floats work better than others. It is tempting to simply use sticks of wood or leaves picked up off the ground near the stream. However, it may be difficult to see a stick in the

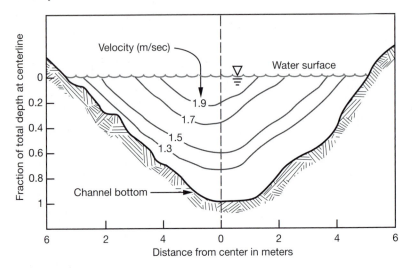

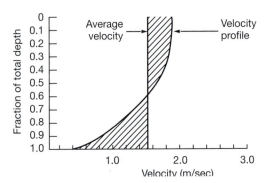

Fig. 3.16   Velocity of a stream varies from top to bottom of the flow profile and across the stream from bank to bank (after Ward and Eliot, 1995).

water, or to be sure it is the "right" stick. Also, sticks and leaves tend to hang up on debris, and they may float so close to the top of the water that they are affected by wind. Leaves, especially, may be severely affected by wind blowing across the water surface. Because velocity in a stream varies from top to bottom, an object that floats too close to the top allows one to measure only the velocity near the top. This is faster than the average velocity. Fishing bobbers are visible, but float too close to the surface and are affected by wind. Oranges constitute a good float. Oranges float somewhat below the surface, are visible, roll around obstructions, and if lost, generally pose little environmental problem.

Depending on the width of the stream, it may be necessary to divide the stream into sections or "channels" and run the test in each (Fig. 3.17). Several repetitions should be conducted in each section. Beware of using too many sections; the float may be drawn into a single flow path no matter where it is released. If this is the case, using multiple sections will not add meaningful information. A few trials should reveal the utility—or lack thereof—of multiple sections.

Length of the segment tested should be chosen based on accessibility, convenience, and precision required. The segment chosen should be relatively straight and of regular gradient, and, ideally, free of debris. The longer the segment tested, the more precise the measurement will be. Length of the segment should be measured by using a measuring tape or a surveying method.

Two operators are needed to run the float test. One should be positioned upstream, and the other, downstream. Distance between them should be measured. The upstream operator releases the float and starts the clock, and the downstream operator catches the float and signals to stop the clock. The float must be released upstream of the starting line, so that it has ample opportunity to become entrained by and move with the current. The operator should stand out of the way of the anticipated path of the float, to avoid obstructing it as it floats downstream.

When the float crosses the starting line, the upstream operator starts the clock, which might be a stopwatch, a digital watch, or a watch with a sweep second hand. The float is allowed to travel downstream toward the other operator, who raises a hand (or communicates via radio) when the float nears the finish line and brings the hand down sharply when the float crosses the line. The upstream operator stops the clock when the signal is given. The downstream operator catches the float and, ideally, returns it to the upstream operator so that the test can be repeated.

The test should be repeated several times (at least three) in every section or "channel" of the stream. Section widths should be measured so that an average

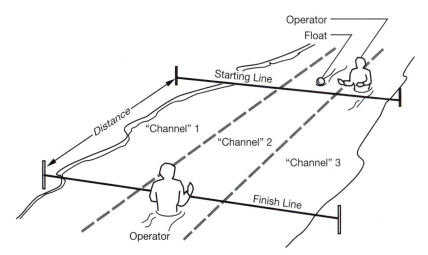

Fig. 3.17 The float test. It may be necessary to divide the stream into sections or "channels" and run the test in each. Here, operators will perform several trials in each of three channels.

velocity may be calculated. If discharge is to be estimated, then depths should be measured as well, so that a good estimate of cross-sectional area in each section can be determined.

A sample of calculations is given in Table 3.1.

Trials should be repeated in each section until at least three trials are made with no errors. Errors should be described in the "Notes" column of the data table, as shown in the table. If a measurement is inexplicably and significantly (more than 10%) higher or lower than the other measurements in that section, another trial should be run.

The float method is generally limited to use in streams that are accessible by wading. It may be used for some other streams by operators standing on the banks. The upstream operator throws a float into the water, waits until it crosses the starting line, starts the clock, and watches for the downstream operator's signal that the float crossed the finish line. However, the inaccuracies in throwing the float and in the angles introduced by observing the start and finish lines from the banks introduce significant error. This method is suitable for approximate estimates of surface velocity.

Current-Velocity Meter

A current-velocity meter is an instrument used to take measurements of stream velocity at a particular point in a stream. A common design features a propeller or set of cups mounted on a shaft. The shaft may be oriented horizontally or vertically. The propeller or cups catch the current and spin on the shaft at a rate proportional to current velocity (Fig. 3.18). These meters are battery-operated; each rotation of the shaft generates an electronic signal that may be digitally displayed or may generate a clicking sound in a set of headphones worn by the operator. The operator reads the digital readout or counts the number of clicks generated per given time period, and translates this into velocity by using the calibration for the particular instrument. Some types can be connected to a datalogger to digitally record continuous readings.

Another flow velocity meter design is based on creating a magnetic field within the sensing device. Changes in the amplitude of voltage correlate with flow velocity. This type of meter is useful for measuring flows with significant debris, which tends to clog the mechanical meters.

Meters may be rated according to the velocity they are capable of measuring. The manufacturer should be able to provide information on each meter's capabilities. Some models may work only in low-velocity streams; others may work well in high-velocity streams, but may not have a high precision at lower velocities. Small units called "wading rods" are mounted on a rod; these are best for relatively slow flows in shallow streams that can be safely waded. In deeper streams, or when it is unsafe to wade in a stream, measurements can be made from a bridge or from suspension stations by using a different type of gage. These larger units can be suspended from a cable that is manipulated by a portable crane and cable reel. In these situations, a streamlined, heavy sounding weight with fins is attached to the cable to hold the gage in place against the pull of the current.

TABLE 3.1
Sample Calculations for the Float Method

Section of Stream	Width[a] (ft)	Trial	Distance[a] (ft)	Time[a] (sec)	Surface Velocity (ft/sec)	Average Surface Velocity in this Section (ft/sec)	Weighted Surface Velocity (ft/sec)	Notes
1	11	1	100	23.5	4.26	4.20	1.32	
		2	100	25.0	4.00			
		3	100	23.0	4.34			
2	10	1	100	22.0	4.55	4.61	1.32	
		2	100	21.3	4.69			
		3	100	21.7	4.61			
		4	100	21.7	4.61			
3	14	1	100	25.1	3.98	3.98	1.32	Float caught in branch
		2	100	—	—			
		3	100	25.2	3.97			
		4	100	25.1	3.98			

Surface Velocity = sum of weighted velocities = 4.23 ft/sec

Average Velocity = (multiply uncorrected velocity by 0.85 to account for variation of velocity with depth in stream) = 3.59 ft/sec

Weighted Surface Velocity is calculated by weighting sectional values according to the width of the section:

$$\text{Weighted velocity} = \frac{\text{width of section}}{\text{total stream width}} \times \text{Velocity measured in that section}$$

[a]Indicates measurements that must be taken in the stream. Others may be calculated on dry land.

Fig. 3.18 A current-velocity meter. The cups catch the current and spin on the shaft at a rate proportional to current velocity. Each rotation causes a clicking sound in the headphones.

Preparing for the Test. In some situations, it may be worthwhile to conduct two trips to the site: one to prepare for the test and one to actually perform the test. If stream conditions are likely to change significantly, the second visit should be conducted soon after the first. For both site visits, take into the field the velocity meter(s) and spare batteries, a depth-measuring device, a measuring tape or some other device for determining stream width, a "tag line" (see what follows) and means to anchor it, tape or surveyor's flags, a calculator, and a field notebook and pencils. If measurements are to be conducted from a bridge or suspension station, take a portable crane, cable and reel, and a heavy, streamlined sounding weight with fins (or "fish").

To prepare for the test, begin by finding an appropriate stretch of stream. The stretch should be straight, relatively unobstructed, and relatively free of vegetation and tree branches or other debris. If the operator intends to wade and carry a rod-mounted meter (a "wading rod"), the depth, current, and footing of the stream bed should be checked to determine if measurements can be made safely. If the measurements are to be made from a bridge, a portable crane with a cable and reel may be set up on the bridge so that a heavy weight may be used to anchor the gage at the correct measuring point.

Measure the width of the stream, and determine the number of intervals and the locations of the measuring points. Because velocity varies across the width of the stream, finding an average velocity requires making several measurements. To determine the appropriate intervals, measure the total width of the stream and divide it into a number of sections. Using 25 to 30 sections may provide results with fairly good precision. If obstructions such as bridge piers or abutments cause complex flow conditions, increase the number of sections. Very narrow streams may require fewer sections. Use your best judgment, based on the conditions at the site and the profile of the stream channel. The sections need not be of equal width, especially if flow conditions are complex. Ideally, each section should carry less than 10 percent of the total stream discharge. A section should not be narrower than the length of the current meter propeller or distance from rod to the end of the propeller, because such a narrow section yields no additional helpful information.

Set up the tag line, which will be used to identify the measuring points during the test. A tag line is a cable stretched across the stream perpendicular to flow, and anchored by using stakes or some other secure means (Fig. 3.19). Some tag lines are heavy cable with steel beads at regular intervals; when anchored securely, it may be possible to use this type of tagline to steady a boat while a measurement is being made. Depending on conditions, it may be possible to use a length of sturdy rope or a steel tape as a tag line. Once the width of the measurement intervals is determined, mark the line with tape or surveying flags at the locations where meter readings will be taken.

Measure the depth and velocity of the stream at a few points. Based on these measurements, choose the appropriate meter for the test (see the manufacturer's recommendations). If rotating-element meters are used, such as the "Price type" meters, use the smaller or "Pygmy" model if stream depth is less than 1.5 ft, and the larger model if the stream is deeper (USGS, 1980). Check to be sure the meter is working: Spin the propeller or cups, and watch the display or listen to the headphones to make sure that the display device is accurately indicating the rotation. (Ideally, this test should be performed before going to the field.)

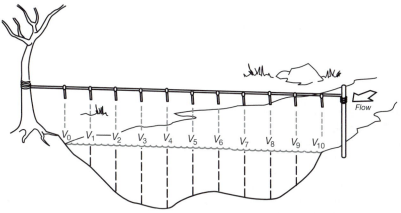

Fig. 3.19 Stretch a tag line across the stream and secure it at both ends. Mark the points at which velocity will be measured. Measure velocity at the edges of each section, not the center.

If vegetation is growing in areas where measurements are to be taken, the area should be raked to remove enough vegetation upstream and downstream of the measuring point to provide unobstructed flow and to keep vegetation from interfering with the rotation of the propeller or cups.

Before starting the measurements, construct a data table. This will facilitate recording the measurements during the test. A sample data table appears in Fig. 3.20.

Running the Test. To make a velocity measurement, go to the first measurement point in the stream, which has been previously identified on the tag line. Record the distance of this point from the bank. Measure and record the depth of the stream at the point.

Next, determine the depth at which velocity will be measured. Stream velocity varies from bottom to top of the stream. When depth is less than 2.5 ft (0.76 m), average velocity at a particular point may be approximated as the velocity at six-tenths of the depth (as measured from the top surface of the stream). Some meters automatically adjust to measure velocity at this depth; others use a system of dual scales to make the setting fairly simple.

When depth is more than 2.5 ft (0.76 m), the USGS suggests another procedure. At these points, a more accurate estimate of average velocity may be obtained by measuring velocity at two-tenths and at eight-tenths of the depth, and averaging these two measurements.

To make a measurement, adjust the height of the meter so that it measures at the appropriate depth. Ensure that vegetation and other debris will not interfere with the reading. Place the meter at the tag line. Stand to the side and slightly downstream of the meter, well away from it; stand sideways, facing the streambank. Hold the meter far enough away from yourself that you will not affect the current velocity at the meter. Keeping the rod vertical, orient the propeller or cups on the meter so that they face upstream. Make the measurement by counting the number of rotations in a certain time period. Depending on the type of meter being used, measurements may need to be of 1-minute duration or longer. Some meters may require less time. Record the value; then move to the next measuring point.

Calculations. A sample of stream measurements is shown in Fig. 3.21. If measurements were made at six-tenths of the depth, they approximate the average velocity at the measuring point. If measurements were made at two-tenths and eight-tenths of the depth, average them to find the vertically-averaged velocity at the measuring point.

To determine the total cross-sectional area of the stream, sum the cross-sectional areas of each section. To find the average velocity, divide the total discharge by the total area. Calculations for this method are summarized in Table 3.2. An example calculation is given in Table 3.3.

These measurements of average velocity also may be used to find average discharge; see the later section "Discharge," for more information.

Data Collection Form for Measuring Average Stream Velocity and Discharge Using a Current Meter

Date _____ Field Personnel _____ Job No. _____

Stream name/number _____

Location _____

Meter type, manufacturer, and model number _____

Meter coefficient (velocity per revolution) _____

Meter mounted on rod / cable (circle one) _____

Comments:

Observation point number	Trial No.	Position on tag line (or distance from edge)	Total depth at observation point	Depth at which measurement is made	Number of revolutions	Elapsed time during trial	Notes

Fig. 3.20 A sample data table for stream-velocity measurements.

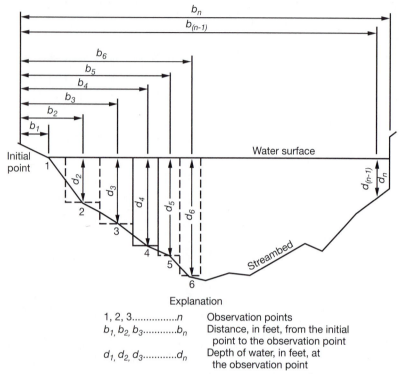

Explanation

1, 2, 3.................n	Observation points
b_1, b_2, b_3............b_n	Distance, in feet, from the initial point to the observation point
d_1, d_2, d_3............d_n	Depth of water, in feet, at the observation point

Fig. 3.21   A sample of stream measurements (after Buchanan and Somers, 1969).

Tracers

Tracers may be used to determine "time of travel", which is one method of finding velocity. The method is too complex to describe here, but several good sources (Wilson, 1968; Kilpatrick, 1970; Smart and Laidlaw, 1977; Hubbard and colleagues, 1982; Kilpatrick and Cobb, 1985; Herschy, 1995) describe it in detail. Briefly, the method involves releasing some material (usually a dye or a salt) at a point upstream, and then measuring the concentration of the material downstream. The "plume" of tracer disperses, diffuses, and is diluted as it moves downstream. Time of travel is generally taken to be the length of time it takes the peak concentration of the plume to reach the downstream point divided by the distance the tracer traveled. Critical considerations include the choice of tracer, measuring method, and measuring points. In addition, the appropriate regulatory agencies should be consulted to determine if a permit is needed before adding the tracer to the stream.

Indirect Methods

Some velocity measurements are *indirect,* that is, factors related to velocity are measured in the field, and the velocity is estimated from them. This is particularly

TABLE 3.2

Calculations of Average Stream Velocity and Discharge Using a Current Meter

Column A Observation Point	Column B Distance from Observation Point to Initial Point	Column C Width of Section Encompassing this Observation Point	Column D Depth at Observation Point	Column E Area of Section	Column F Vertically-Averaged Velocity at this Observation Point	Column G Discharge Through this Section
Initial Point (a well-anchored pt. located on stream bank)	0	—	—	—	—	—
1 (edge of stream)	b_1	$\dfrac{b_2 - b_1}{2}$	d_1	Col. C × Col. D	v_1	Col. E × Col. F
2	b_2	$\dfrac{b_2 - b_1}{2} + \dfrac{b_3 - b_2}{2}$	d_2	Col. C × Col. D	v_2	Col. E × Col. F
3	b_3	$\dfrac{b_3 - b_2}{2} + \dfrac{b_4 - b_3}{2}$	d_3	Col. C × Col. D	v_3	Col. E × Col. F
4	b_4	$\dfrac{b_4 - b_3}{2} + \dfrac{b_5 - b_4}{2}$	d_4	Col. C × Col. D	v_4	Col. E × Col. F
5	b_5	$\dfrac{b_5 - b_4}{2} + \dfrac{b_6 - b_5}{2}$	d_5	Col. C × Col. D	v_5	Col. E × Col. F
etc.						
n (edge of stream)	b_n	$\dfrac{b_n - b_{n-1}}{2}$	d_n	Col. C × Col. D	v_n	Col. E × Col. F

TOTALS: Cross-sectional area = _____   Total discharge = _____

Average velocity = Total discharge/cross-sectional area

TABLE 3.3

Sample Calculation of Average Stream Velocity and Discharge

Column A Observation Point	Column B Distance from Observation Point to Initial Point (ft)	Column C Width of Section Encompassing this Observation Point (ft)	Column D Depth at Observation Point (ft)	Column E Area of Section (ft²)	Column F Vertically-Averaged Velocity at this Observation Point (ft/sec)	Column G Discharge Through this Section (cfs)
Initial Point	0	—	—	—	—	—
1	2.0	2.5	0.00	0.0	0.00	0.00
2	7.0	5.0	0.91	4.5	0.90	4.05
3	12.0	5.0	0.93	4.5	1.00	4.50
4	17.0	5.0	1.81	9.0	1.09	9.81
5	22.0	3.75	2.67	9.75	1.21	11.80
6	24.5	2.5	2.69	6.5	1.19	7.74
7	27.0	2.5	2.56	6.25	1.24	7.75
8	29.5	2.5	2.54	6.25	1.46	9.13
9	32.0	3.75	2.61	9.75	1.73	16.87
10	37.0	5.0	2.47	12.0	2.67	32.04
11	42.0	5.0	2.47	12.0	2.51	30.12
12	47.0	5.0	2.46	12.0	2.55	30.60
13	52.0	5.0	2.47	12.0	2.43	24.30
14	57.0	4.0	2.59	10.0	2.41	24.10
15	60.0	3.0	2.63	7.8	2.45	19.11
16	63.0	3.0	2.74	8.1	2.52	20.41
17	66.0	3.5	2.68	9.1	2.45	23.11
18	70.0	4.5	2.66	11.7	2.32	27.14
19	75.0	5.0	2.47	12.0	2.49	29.88
20	80.0	5.0	2.48	11.5	2.37	28.44
21	85.0	5.0	2.31	9.5	2.22	25.53
22	90.0	5.0	1.90	9.0	2.05	19.48
23	95.0	5.0	1.81	9.0	1.89	17.04
24	100.0	5.0	1.80	9.0	1.52	13.68
25	105.0	2.5	0.00	0.0	0.00	0.00

TOTALS: Cross-sectional area = 214.2 ft²   Total discharge = 413.49 cfs

Average velocity = Total discharge/cross-sectional area = 413.49 cfs/214.2 ft² = 1.93 ft/sec useful when the current of the stream is too rapid to allow for safe direct measurement. In some streams, this is particularly relevant when the stream is in flood. Two methods are commonly used: the Manning equation and back-calculation from rating curves.

Manning Equation. The Manning equation uses four factors to estimate velocity: slope, wetted perimeter (WP), cross-sectional area, and a roughness factor (the Manning n value). The formula is as follows, when length is measured in meters:

$$V = \frac{R^{2/3} \, S^{1/2}}{n}$$

where V = average velocity, m/sec
          S = slope of the water surface, m/m
          R = hydraulic radius (R = area/WP), m²/m
          n = roughness coefficient (Manning n) (see Table 3.4)
or where length is measured in feet:

$$V = \frac{1.49 R^{2/3} \, S^{1/2}}{n}$$

where V = average velocity, ft/sec
          S = slope of the water surface, ft/ft
          R = hydraulic radius (R = area/WP), ft²/ft
          n = Manning n (see Table 3.4)

Note that the units indicated must be used with the appropriate formula.

For normal flows, the slope used in the Manning equation is the same as the stream gradient and may be measured as described earlier. The slope of flood flows may be more difficult, or even dangerous, to measure, or the flood may recede before it is possible to measure it. In this case, it may be possible to use high-water marks to measure slope. Alternatively, the slope of the floodplain may be assumed to equal the slope of the water surface during a flood; relying on this assumption may introduce some error, depending on local topography.

The wetted perimeter used in the Manning equation is found as described earlier in this chapter. For flood flows, when it is too dangerous to make a direct reading, or when the flood recedes before the measurement can be made, the best procedure is to construct a cross-sectional profile of the channel and floodplain, up to the highest stage of the flood, and measure the WP from this profile. This should be done before the flood or after it has passed, to avoid injury. If both cross sections are measured and drawn both before and after floods, changes in channel morphology may be evaluated.

TABLE 3.4

Values of Manning's *n* for Channels of Various Types

Type of Channel and Description	n		
	Minimum	Normal	Maximum
Minor streams (top width at flood stage <100 ft)			
Streams on plain			
1. Clean, straight, full stage, no riffles or deep pools	0.025	0.030	0.033
2. Same as above, but more stones and weeds	0.030	0.035	0.040
3. Clean, winding, some pools and shoals	0.033	0.040	0.045
4. Same as above, but some weeds and stones	0.035	0.045	0.050
5. Same as above, but lower stages, more ineffective slopes and sections	0.040	0.048	0.055
6. Same as 4, but more stones	0.045	0.050	0.060
7. Sluggish reaches, weedy, deep pools	0.050	0.070	0.080
8. Very weedy reaches, deep pools, or flood-ways with heavy stand of timber and underbrush	0.075	0.100	0.150
Mountain streams, no vegetation in channel, banks usually steep, trees and brush along banks submerged at high stages			
1. Bottom: gravels, cobbles, and few boulders	0.030	0.040	0.050
2. Bottom: cobbles with large boulders	0.040	0.050	0.070
Floodplains			
Pasture, no brush			
1. Short grass	0.025	0.030	0.035
2. High grass	0.030	0.035	0.050
Cultivated areas			
1. No crop	0.020	0.030	0.040
2. Mature row crops	0.025	0.035	0.045
3. Mature field crops	0.030	0.040	0.050
Brush			
1. Scattered brush	0.035	0.050	0.070
2. Light brush and trees, in winter	0.035	0.050	0.060
3. Light brush and trees, in summer	0.040	0.060	0.080
4. Medium to dense brush, in winter	0.045	0.070	0.110
5. Medium to dense brush, in summer	0.070	0.100	0.160
Trees			
1. Dense willows, summer, straight	0.110	0.150	0.200
2. Cleared land with tree stumps, no sprouts	0.030	0.040	0.050
3. Same as above, but with heavy growth of sprouts	0.050	0.060	0.080
4. Heavy stand of timber, a few down trees, little undergrowth, flood stage below branches	0.080	0.100	0.120
5. Same as above, but with flood stage reaching branches	0.100	0.120	0.160
Major streams (top width at flood stage >100ft)			
Regular section with no boulders or brush	0.025	—	0.060
Irregular and rough section	0.035	—	0.100

Source: Chow (1959)

Cross-sectional area is determined by constructing a graphical cross section of the channel and measuring area by one of the methods described earlier. For flood flows, be sure the cross section includes the floodplain, up to the highest reach of the flood.

The Manning roughness coefficient, or Manning n, is chosen from the table of values. The description best fitting the channel's characteristics is determined, and then a value is read for that description.

The Manning equation gives an approximation, and should be relied on only when other methods cannot be used, such as when flood waters make direct methods unsafe.

Back-Calculation from Rating Curve A rating curve is a graph showing the relationship established between discharge and stage of a stream at a given point (Fig. 3.22). Once the rating curve has been established, if the stage is known, the discharge can be estimated from the graph. Discharge divided by cross-sectional area equals velocity. Cross-sectional area can be determined by constructing a profile of the channel, as described earlier, for a given stream stage, and measuring the area from the profile. Then discharge at that stage is divided by the cross-sectional area at that stage to yield velocity at that stage.

Discharge

Discharge (Q) of a stream is its volumetric flow rate, in units of volume per time. Discharge may be determined by calculations based on velocity and area

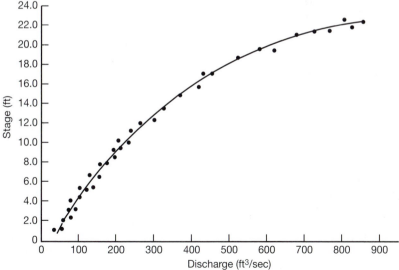

Fig. 3.22 A rating curve.

measurements, measurements with a weir, or estimations from a rating curve and stage data. For more information, see Buchanan and Somers (1969), Dunne and Leopold (1978), USGS (1980) and Herschy (1995).

Calculating Discharge from Velocity

Discharge is velocity of the stream flow multiplied by the cross-sectional area through which the flow occurs. When velocity is measured with the float method or current-velocity meter method, the cross-sectional area is determined as a part of the method. The cross-sectional area of the entire stream multiplied by the average velocity gives the discharge.

It also is possible to find the discharge within each section by multiplying area times average velocity in each section. These discharges may be summed to give the discharge of the entire stream. The result should be the same as that from the method described in the preceding paragraph.

Any discharge measurement is a single data point that represents conditions at a single point in time and space. Discharge varies over time and along the length of the stream. No single measurement should be considered to be representative of the stream discharge, because discharge varies with time and distance.

If velocity was estimated using the Manning equation, then cross-sectional area has already been determined. Multiply the average velocity by the area to get the discharge. Be cautious in using this number; recall that the results of the Manning equation are based on estimation of the *n* value, and this introduces imprecision.

Weirs

Weirs are structures built into the stream that force the water to flow through an opening of a known size and shape. Measuring the height of the water flowing through the opening allows one to calculate the discharge. The following discussion focuses on sharp-crested weirs. Sharp-crested weirs are constructed of a thin plate with a notch through which water flows. Weirs designed specifically for discharge estimation have openings that may be rectangular, V-notched at 90° (or, less commonly, 60°), or trapezoidal (Fig. 3.23).

Building a weir in a stream involves, essentially, damming it. Obviously, this is a labor-intensive and time-consuming undertaking, and it may be a cost-effective method only if discharge is to be monitored at a site for an extended period of time. In addition, the stream must be small enough to make weir-building feasible. It may be necessary to obtain a permit from the local or state conservation or environmental authorities before building the weir. This may be especially relevant if the stream holds trout or other fish species that might be disturbed by construction of the weir. Permission to build the weir must be obtained from the land owners. Existing dams, whether across small or large streams, may be used as weirs if the opening through which water flows is of an appropriate shape (for more information, see USGS, 1977).

Weirs may be constructed of any material that blocks flow. They may be small, temporary installations made of plywood, or large, permanent concrete dams, or anything in between. The important aspects are the following: Water must flow only

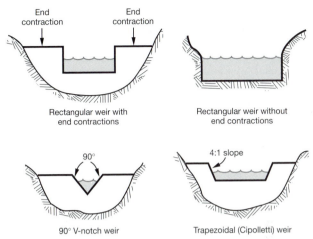

Fig. 3.23 Weirs have openings of varying shapes.

through the notch, the height of flow through the notch must be measurable, and leakage around the sides and bottom of the weir must be eliminated or minimized.

A temporary weir across a small stream can be constructed in a few hours, if the materials are at hand. The project requires several people. Planning must take place before construction begins. Measure the width and depth of the stream. Examine the stream bed. Is it made of sediments? What type? How far is it to bedrock? How high does the water rise when the stream is in flood? Remember that even a small stream, even a dry stream bed, may become a raging torrent in a sudden storm. The weir should withstand even these extreme flows.

How will water height in the notch of the weir be measured? A staff gage may be affixed to it and read from the bank, but only if the weir is accessible. A recording gage may be installed in the pool of water behind the weir. This, too, must be accessible and safe from floods. Of what material will the weir be made? If plywood, sheet metal should be fastened across the lip of the plywood. This will make a smooth surface for water to flow over, protect the weir from current-carried debris, and slow the deterioration of the plywood. To further smooth the surface, the notch should be beveled, typically at a 45° angle.

How big should the notch be? It must accommodate the flow, even in small floods. Estimate the discharge using the float, current meter, or Manning methods; estimate the largest and smallest expected flow rates; and then back-calculate from the weir equations (shown in what follows) to ensure that the notch is sized appropriately. If quantifying small variations in flow is important, do not make the notch too large, because then it will take huge differences in discharge to make small, difficult-to-measure variations in height of water. Finally, consider whether the weir will have "contractions." A contraction is the length from the edge of the notch of the weir to the point where the weir meets the stream bank. A contraction forces the water to flow away from the banks and toward the center of the weir in order to pass through the opening. If the weir has contractions, contraction length should be greater than three times the head of water above the weir crest.

In some studies, measuring normal flows with great accuracy is important, but measuring the largest and smallest flows is not important. The weir design should take this into account. Weir notches may be rectangular or V-shaped. V-notches are better for quantifying small changes in flow. A mixed design, with a smaller V-notch set in a larger rectangular notch, yields greater accuracy at low flows, and lesser accuracy at high flows.

How will the weir stay in place? It may be possible to dig out a portion of the stream bed, put the weir in place, and then shovel the bed material back to secure the weir. The edges of the weir should extend into the stream bed and banks, to discourage seepage around the edges. It probably also will be useful to further secure the weir by staking it or building a frame to support it. Will the character of the stream bed interfere with this process? If the stream bed is sediment, the task may be simple; but what if it is bedrock? Will it be necessary to divert the stream while constructing the weir? How will this be accomplished?

No matter what method of weir building is to be used, it would be wise to be prepared for anything. Take into the field shovels, several hammers and 16d nails, 2×2's, 2×4's, plywood, a long carpenter's level, sheet metal, metal shears, sledge hammers, quickset concrete, a bag or two of bentonite chips, hatchet, keyhole saw, rip saw, exterior-grade waterproof caulk, cordless drill, and a few bolts and nuts. Scrap welding materials (e.g., lengths of angle iron) might also come in handy. Depending on the scale of the project, a portable auger or posthole digger might be useful as well. To conserve materials and make the weir more structurally sound, build it across a narrow portion of the stream. If appropriate and possible, take advantage of natural features (e.g., large tree trunks or boulders) to support the weir.

Using the Weir. Measure the height of water that backs up behind the weir. The critical measurement is the height of the water above the lowest point in the notch of the weir. This measurement should be made several feet upstream of the weir itself, to avoid the possible interference of surges. If the height of the water is to be measured by a recording gage, the gage height must be referenced to the elevation of the notch. For rectangular weirs, measure the width of the notch.

If using a large dam as a weir, first determine if it can be approximated to be a sharp-crested weir. If not, consult the *National Handbook of Recommended Methods for Water-Data Acquisition* (1980) for further information and calculations. If the dam can be used to approximate a sharp-crested weir, survey the opening, measure it from an accurate map, or ask the agency (e.g., Corps of Engineers) that maintains it for the length of the crest.

Calculate the discharge from the weir measurements as follows:

Rectangular Notch, Without End Contractions.
Using metric units:

$$Q = 1.84(L - 0.2H) \, H^{3/2}$$

where Q = discharge, m³/sec
       L = length of the rectangular weir opening, m
       H = head of water above the weir crest, m
   Using British units (Fetter, 1994):

$$Q = 3.33(L - 0.2H)H^{3/2}$$

where Q = discharge, ft³/sec
       L = length of the rectangular weir opening, ft
       H = head of water above the weir crest, ft

Rectangular Notch, with End Contractions.
   Using metric units (Fetter, 1994):

$$Q = 1.84(L - 0.2H)H^{3/2}$$

where Q = discharge, m³/sec
       L = length of the weir crest, m
       H = head of the water above the weir crest, m
   Using British units:

$$Q = 3.33(L - 0.2H)H^{3/2}$$

where Q = discharge, ft³/sec
       L = length of the weir crest, ft
       H = head of the water above the weir crest, ft

Trapezoidal (Cipolletti) Weir. Trapezoidal (or Cipolletti) weirs have notches with steep sides that slope at a ratio of 4 (vertical) to 1 (horizontal). Because the sides are so steep, the formula for finding discharge through a Cipolletti weir is the same as that for a rectangular weir.

V-Notch (90°).
   Using metric units (USGS, 1980):

$$Q = 1.343H^{5/2}$$

where Q = discharge, m³/sec

H = head of the water above the apex of the V-notch, m

Using British units (Driscoll, 1986):

$$Q = 2.4381H^{5/2}$$

where Q = discharge, ft³/sec
H = head of the water above the apex of the V-notch, ft

V-Notch (60°)
Using metric units (Watson and Burnett, 1995):

$$Q = 0.778H^{5/2}$$

where Q = discharge, m³/sec
H = head of the water above the apex of the V-notch, m

Using British units (Driscoll, 1986):

$$Q = 1.4076H^{5/2}$$

where Q = discharge, ft³/sec
H = head of the water above the apex of the V-notch, ft

Note that the units that are indicated in the formulas must be used.

Flumes
A flume is similar to a weir, except that instead of water flowing through an essentially two-dimensional opening, it flows through a channel of known dimensions. A Parshall flume is an example (Fig. 3.24). A Parshall flume may be constructed in a similar fashion as a weir, but the dimensions and leveling are more critical and exacting. Measurement of the height of water in the flume allows one to find discharge. For more information, see Driscoll (1986), USGS (1980), or Kilpatrick and Schneider (1983).

Seepage Through Stream Beds
Streams may gain or lose water through their beds. Because perennial streams are a surface expression of the water table, direction and rate of flow depend on the ground water gradient adjacent to the stream. Direction and rate of seepage vary with time and from point to point along the stream. A stream that gains water as it flows downstream is called a gaining stream; one that loses water is called a losing stream (Fig. 3.25).

Fig. 3.24 A Parshall flume.

Whereas a stream may be considered a gaining or a losing stream under normal flow conditions, flow direction may completely reverse if meteorologic conditions change. For example, after an intense rain storm or sudden snow melt, a stream that normally gains water as it flows may rise to such a high stage that the normal direction of flow is reversed, and the stream begins recharging ground water.

Measuring Rate and Direction of Seepage

Rate and direction of seepage through a stream bed may be measured directly, using a seepage meter. Direction of flow may also be measured indirectly, using one of a number of methods that are described in what follows.

Seepage Meters

Seepage meters are essentially upside-down buckets or drums pushed into a stream or lake bed. A bag containing a measured amount of water is attached to the end of a tube, and the other end of the tube is inserted through the wall of the bucket. After some time, the amount of water in the bag is measured again. A gain or loss indicates that the stream is gaining or losing water, respectively (Fig. 3.26).

Constructing a Seepage Meter.

Materials: Five-gallon plastic bucket or a steel drum (for streams with hard beds), #7 or #7.5 rubber stopper, 5-in. length of tubing, 1-qt plastic bag, or IV bag from a medical supply house, small rubber bands or quick-connectors

Tools: Drill, 1/4-in bit, 1-in spade bit, hacksaw, vise grips or clamp, pliers, knife

Cut off and discard the top portion of the bucket or drum, leaving the bottom with about 8-in. of the sides intact. Invert the bucket or drum and use the spade bit to drill a 1-in. hole about 1 to 2 in. from the edge of the bucket. Holding the stopper with vise grips or clamp, drill a 1/4-in. hole through the stopper from top to bottom or use a cork borer.

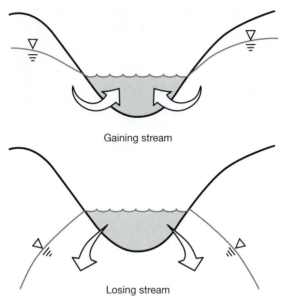

Gaining stream

Losing stream

Fig. 3.25 Flow is toward the stream in a gaining stream, and away from the stream in a losing stream.

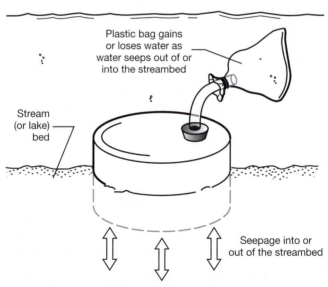

Plastic bag gains or loses water as water seeps out of or into the streambed

Stream (or lake) bed

Seepage into or out of the streambed

Fig. 3.26 A gain or loss of water in the bag attached to a seepage meter indicates that the stream is gaining or losing water, respectively.

Pull the tubing through the hole in the stopper. To do this, first trim the end of the tubing at an angle. Insert the thinner part of the tubing through the hole, and use the pliers to pull it through farther. Check to be sure that water can flow freely through the tube.

Insert the stopper into the hole in the bucket. Partly fill the bucket with water to test the watertightness of the stopper.

More information on seepage meters may be found in a paper by Lee and Cherry (1978), who suggested using a 55-gallon drum instead of a 5-gallon bucket. Plastic buckets are lighter and have no dangerous sharp edges, but may not work if the stream bed is hard.

Using a Seepage Meter

Materials: Small rubber bands (or tubing quick-connectors), seepage meter, 1-qt plastic bags or medical IV bags

Tools: Watch, 100-ml graduated cylinder (plastic) with 1-ml gradations, meter stick or measuring tape

Choose a location for the meter. The stream bed should be free of large cobbles, debris, and vegetation. Note that because direction and rate of flow may vary significantly along a stream bed, it would be wise to install several meters at various locations along the stream.

Measure the cross-sectional area of the opening in the meter. With the stopper removed, install the meter by inverting it and pressing it into the stream bed, leaving about 3 in. of it protruding from the bed. Tilt the meter slightly while installing it to allow air to escape out the hole, which should be positioned at the top of the tilt.

Measure about 250 ml of water into the plastic bag or IV bag. Wrap the rubber band tightly around the top (or connect the quick-connector), and then insert the top portion of the tubing in the tubing/stopper assembly into the bag's rubber band–wrapped opening. Do not push the tubing in very far, so that when the bag is held above the stopper, all the water can flow freely out of the tubing.

Holding the stopper above the bag, slowly lower the assembly into the water. Allow the water to press all of the air out of the bag. Continue lowering the stopper slowly, until water inside the tube rises to the point where it almost spills out (Fig. 3.27). Put a fingertip over the end of the tube, invert the stopper, and insert it into the hole in the meter. You must press hard enough to make a watertight seal, but not so hard as to jet water into the bag. When in place, the entire assembly should be completely submerged, with no air bubbles in the bag. Note and record the time.

After some time has passed (1 to 2 hours may be enough, depending on the gradient and the permeability of the stream bed), retrieve the stopper and bag. Do this by reversing the installation procedure: Gently loosen the stopper. Lift it slowly, just enough to allow you to put a finger over the end. Invert the stopper and bring it to the surface. Pull the tubing out of the bag, and pour the contents of the bag into a graduated cylinder. Do this very carefully, as the bag may now hold more than the cylinder can hold. Do not overtop the cylinder, or the test will be ruined.

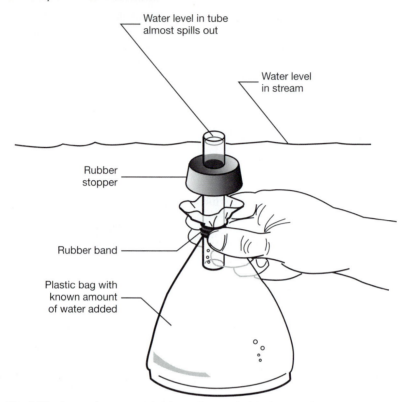

Fig. 3.27 Lower the stopper slowly, until water inside the tube rises to the point where it almost spills out. Then put a fingertip over the end of the tube, invert it, and press the stopper into place in the bucket.

The difference between the volume of water in the bag before and after the test is the volume that seeped into or out of the stream bed through the area of influence of the meter.

Calculation. Based on Darcy's law,

$$Q/A = K(dh/dl)$$

where Q = discharge

           A = cross-sectional area through which flow occurs

           K = hydraulic conductivity of the material through which flow occurs

       dh/dl = hydraulic gradient

calculate *Q/A* as follows:

$$\frac{Q}{A} = \frac{\text{volume that seeped into or out of bag}}{\text{elapsed time} \times \text{cross sectional area of meter}}$$

Be careful to indicate the direction of seepage, that is, is seepage into or out of the stream bed?

To determine the gradient (and thereby make a calculation of hydraulic conductivity, *K* of the stream bed possible) a minipiezometer may be used. This is described in the following section.

Minipiezometers

Lee and Cherry (1978) described a simple and inexpensive method of determining direction of seepage through a stream or lake bed using minipiezometers. A minipiezometer is a tube inserted into the sediment of the stream or lake bed and allowed to fill with water. By comparing water level in the tube and in the stream or lake, one can determine which direction water is flowing (Fig. 3.28).

Constructing a Minipiezometer.

Materials: Clear vinyl or tygon tubing (9/16-in. or 1.43-cm ID, approximately 5 ft long), "T" connector to fit vinyl tubing, hand vacuum pump (may be purchased from a scientific supply house), translucent polyethylene tubing (5/16-in. or 0.31-cm ID, approximately 10 ft long), small screw-type hose clamps, electrical conduit (5 ft long), carriage or lag bolts (5/8-in. or 1.59-cm diameter)

Tools: Hammer, ruler, screwdriver to fit hose clamp, and knife

Cut a length of polyethylene tubing about 5.5 or 6 ft long. Make a 1- 2-in-long "screen" in one end of the tubing by cutting tiny holes or slots in it or by stretching a piece of fiberglass cloth or a nylon stocking over the end and securing it with a rubber band, wire, or a clamp. Set this aside. Cut the vinyl tubing into three pieces: 1, 3, and 5 ft long. Attach the longer two pieces to the crossbar of the T connector, and the shorter piece to the upright of the T connector. Using the hose clamp, attach the longer piece of vinyl tubing to the intact end of the polyethylene tubing. Tighten the clamp until the connection is airtight (Fig. 3.29).

Using the Minipiezometer. Loosely fit a carriage bolt into the end of a piece of electrical conduit. Holding the bolt in place, set the bolt and conduit in the stream bed. Pound the conduit into the stream bed to a depth of anywhere from several inches to a foot. The deeper it is inserted, the more precise the measurement will be, but the more difficult it will be to remove it. Also, in streams with a bed of hard or compacted clay, it may be impossible to get a reading if the minipiezometer is inserted too deeply into the clay. Once the conduit is in place, thread the polyethylene tubing, screen end first, into the conduit. Remove the vinyl tubing by loosening the clamp.

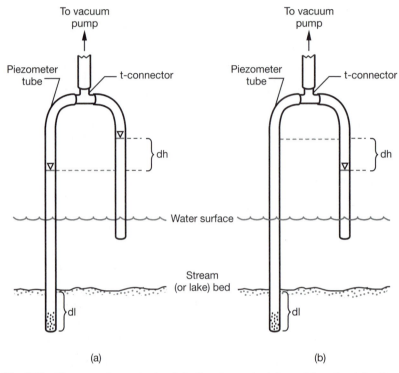

Fig. 3.28 By comparing water levels in the piezometer tube and the other tube, the vertical gradient can be determined. In both diagrams, gradient is *dh/dl*. In (a) Flow is downward, and in (b) flow is upward.

Holding the polyethylene tubing in place, remove the conduit. The tubing should now be partly stuck in the stream bed. Being careful not to dislodge it, reattach the vinyl tubing. Tighten the clamp to ensure an airtight connection.

Submerge the other end of the vinyl tubing in the water. It may be helpful to attach a weight to it to ensure that it does not float. Compare water levels in the two pieces of vinyl tubing. If the water level is visible in the vinyl tube attached to the polyethylene tubing, the stream is gaining water. If the water level is not visible in the piece attached to the polyethylene tubing, the stream is losing water. Attach the vacuum pump to the short vinyl tubing. Gently draw a slight vacuum in the tubes, just enough to bring both water levels up to a point where they are visible. Measure the difference in height between them. This is the Δh, or change in hydraulic head. Across what distance does it occur? To determine this, reach into the water and grasp the polyethylene tubing at the level of the stream bed. Hold tightly and pull out the tubing. Without letting go, measure the length of the polyethylene tubing that was buried. This is the Δl.

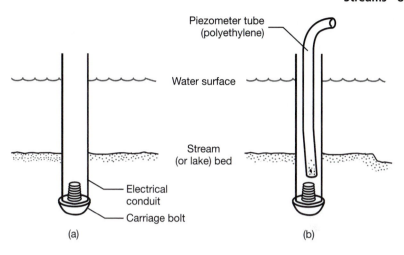

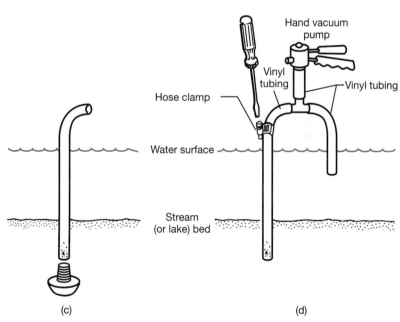

Fig. 3.29 Steps in installing a minipiezometer: (a) Set the carriage bolt inside the conduit and pound it into the stream or lake bed. (b) Thread the polyethylene tube into the conduit. (c) Keeping the polyethylene piezometer tube in place, remove the conduit. The carriage bolt is no longer needed, but remains in place. (d) Attach vinyl tubing to the polyethylene tubing, and tighten the hose clamp to make the connection airtight.

Calculation. Calculate the hydraulic gradient as follows:

$$\frac{\Delta h}{\Delta l} = \frac{\text{difference in water levels in tubes}}{\text{length of polyethylene tubing buried in stream bed}}$$

If a seepage meter was used to determine discharge per area at the same location as the minipiezometer, hydraulic conductivity of the sediments may be calculated as follows:

$$K = \frac{Q}{A(\Delta h/\Delta l)}$$

where Q/A is taken from the seepage meter measurement.

Indirect Methods

Although seepage meters can indicate seepage conditions at a point, if conditions vary greatly, or if they are not conducive to using a seepage meter, it may be more useful to use indirect methods of determining seepage direction.

Comparison of Discharge at Two Points. A simple, but accurate, indirect method to measure seepage into (or out of) a stream involves gaging stream discharge at two points along the course of the stream. In a gaining stream, discharge at the upstream point will be less than discharge at the downstream point. In a losing stream, the opposite will be true. In selecting locations to conduct discharge measurements, be aware that tributaries or other overland flow entering the stream will increase the flow. This increased flow is not due to seepage through the bed, although it is an important component of the water budget of the stream. Points for discharge measurement therefore must be chosen carefully to avoid the effects of tributaries.

Discharge measurements in this method must be made at the same time; otherwise, changes in water level with time will render the data useless. Similarly, if significant evaporation is taking place from the surface of the stream, flow may decrease, even though the stream is not losing water through the bed. This may interfere with the analysis.

Finally, this method should not be used when discharge conditions are changing rapidly, such as during or after a precipitation or snowmelt event.

If two USGS gaging stations are located on the stream, it may be possible to compare readings for the same date and time. However, these stations are frequently located so far apart that any of these confounding factors may be present.

Comparison to Water Levels in Nearby Wells. Water levels in wells near the stream may be a useful indicator of direction of water seepage through the bed. If the water table is at a lower elevation than the water level in the stream, the stream is losing. If the water table is higher, the stream is gaining. To determine the difference in elevations, it is necessary to survey the water levels.

For this method to be valid, water in the well must be in hydraulic contact with water in the stream. This is most likely to be the case if the well is screened through the same rock or sediments that comprise the stream bed. One way to check for hydraulic connection is to monitor changes in well-water level as water level in the stream changes. Likewise, if pumping water out of the well (and not returning it to the stream) causes a decline in water level in the stream, the two probably are hydraulically connected.

Sampling Stream Water

Stream water may be sampled if chemical, biological, or physical analysis of the water is to be performed. Typical samplers are a simple water jug or sample bottle, a "remote sampler" (weighted sampling bottle attached to a long rope, used for sampling streams inaccessible by wading), and Kemmerer-type samplers (for depth-specific sampling). Sample containers should be chosen on the basis of sampler type, analytical procedures, and container material. These considerations are more fully discussed in Chapter 9, "Geochemical Measurements."

Sampling location and timing of sampling events are important considerations that may affect the analytical result. It is easiest to sample from the edge of the stream at a point where it is most accessible, at a time convenient for the researcher. However, chemical, physical, and biological factors all may vary with sampling location. Anyone who has waded in a stream knows that temperature varies with depth and location; other parameters vary as well. Time of day may determine activity of microorganisms, which may affect dissolved oxygen and other parameters. The season of the year may affect suspended sediment or organic matter loads, as well as discharge. Discharge is a critical factor, as higher discharge may correlate with lower concentrations of dissolved species due to a dilution effect.

Sampling protocol is described in Chapter 9, "Geochemical Measurements."

Lakes and Ponds

Why Is this Lake Here?

Lakes and ponds may occur where the water table intersects the ground surface; they occur in local topographic low spots. Large lakes occur in regional topographic low spots. In addition, in regions where hydraulic conductivity of the near-surface materials is low, some lakes or ponds may occur perched above the water table.

Lakes and ponds may have obvious sources and outlets if they are stream-fed and stream-drained. But they may also occur where no inflow or outflow is apparent. Some of these lakes and ponds may be fed by ground water. This type of lake is particularly common in glaciated areas. They may occur as kettle lakes or potholes. Ground water-fed lakes also are common in karst areas, for example, where sinkhole lakes occur. Other lakes or ponds may be fed only by precipitation, if conductivity of the lake bed is low. In some places, lakes and ponds may be the result of dammed streams, whether dammed by natural or human causes. Human-made ponds may also be simple pits dug by backhoe or bulldozer.

Whatever the nature of the investigation, it is wise to determine why a lake exists. It is equally important to determine where water in the lake comes from and where it goes. The water budget of a lake may be constructed to evaluate the inflows and outflows. Streams, precipitation, and evaporation may be obvious sources of inflow to and outflow from a lake, but ground water may also account for inflow and outflow. Ground water flow directions may be determined by a water-level map or by comparison of water temperatures in the lake and in nearby wells. Human activity may result in inflows (e.g., drainage pipes, drainage tiles) or outflows (e.g., spillways).

Limnology, the study of lakes, is a complex field of study with its own rich and extensive literature. This literature should be consulted in detailed hydrologic studies of lakes.

Field Observations

A general description of the field setting of a lake or pond might include the following factors:

- Topography and relief
- Approximate dimensions and shape
- Character of the bank and bed: topography, relief, location of marshy or wetlands areas, inflowing and outflowing streams
- Sediments and rocks: type, color, and thickness of soils and sediments; exposure, type, orientation, and weathering characteristics of rocks
- Vegetation: for vegetation on the banks and in the lake, presence or absence of plants, dominant species, abundance
- Animal activity: on the banks and in the lake, presence of burrows, signs of beaver activity, visible organisms in the water or on the surface
- Degree of saturation of the soil: Are the soils surrounding the lake saturated, moist, or dry?
- Human development: houses, buildings, roads, bridges, causeways, parking lots, sewers, drainage pipes, fences, dams, spillways, culverts, penstocks, rip rap on banks, boating activity, docks, piers, fueling stations, etc.
- Erosional features: undercut banks, indicators of soil creep on banks, mass movement scars on banks, rip rap
- Depositional features: infilling of lake channels, siltation
- Evidence of lake level changes: changes in coloration of banks, flooded structures, flooded forested area (may result in die-off of trees)

Depth

Depth of a lake may be estimated from bathymetric maps, if available. Depth also may be determined directly, by depth sounding with a weighted line or with an acoustic sounder such as a "fish finder." More information on depth sounding is described in the earlier section on stream depth.

Inflow and Outflow

A water budget for a lake or pond should account for all of the inflows and out-flows. Stream inflow to and outflow from a lake may be determined by measuring discharge of all streams entering or draining the lake. Seepage of ground water through the bed of a shallow lake may be estimated with seepage meters and minipiezometers. Ground water-level maps may be used to determine flow conditions around and through the lake. Precipitation and evaporation may be measured by using the appropriate meteorological instruments.

Soil Water and Ground Water

Chapter Overview

Soil water and ground water are the main objects of a hydrogeologist's investigations. Under saturated conditions as water infiltrates the ground and percolates downward, it passes through the unsaturated (or vadose) zone. Within this zone, water does not completely fill the pore spaces. The water is under tension, and is held tightly to mineral grains and organic matter through the forces of adhesion and cohesion (Fig. 4.1).

As water moves downward through the vadose zone, it eventually reaches a point where the pore spaces are filled, yet water is still under tension. This is the capillary fringe, or tension-saturated zone (Fig. 4.2). Water in this zone is still held by forces of adhesion and cohesion. The capillary fringe is directly above the water table. Its thickness is a function of the diameter of pore spaces between the grains. When pore diameters are smaller, a thicker capillary zone forms. When pore diameters are larger, a thinner capillary zone forms.

The water table is the surface at which the pore pressure of water is zero; it is neither under tension nor positive pressure. Below the water table, the pore pressure is greater than zero. At and below the water table, pore spaces are saturated.

Above the water table, water is called soil water. Once it reaches or moves below the water table, it is called ground water. This chapter explores methods for studying both soil water and ground water.

Soil Water

Soil water is not strictly speaking ground water, but it is important in hydrogeologic investigations nonetheless. For example, measurements of infiltration and soil moisture content are important in agriculture and in monitoring drought conditions. Infiltration rate may tell how rapidly a spilled contaminant will seep into the ground. The hydraulic conductivity of the unsaturated zone is a measure of how rapidly that zone will transmit fluid.

This section outlines methods for investigating infiltration rate, soil moisture content, and unsaturated hydraulic conductivity of the vadose zone.

Infiltration

Infiltration is the process by which water moves from the surface into the subsurface. The rate at which this occurs is called the infiltration rate, and it is governed by the type and textural features of the soil as well as by antecedent moisture

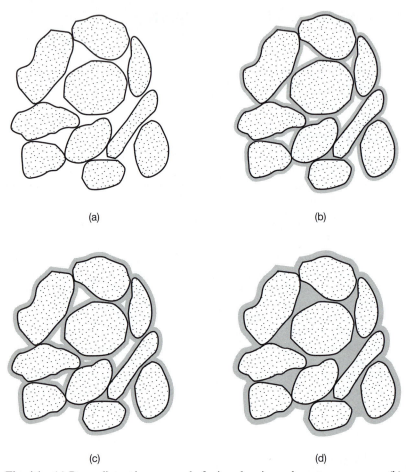

Fig. 4.1 (a) Dry sediment is composed of mineral grains and empty pore spaces. (b) On addition of a small amount of water, the force of adhesion causes water to form a thin coating around the grains. (c) With addition of more water, the force of cohesion causes capillary spaces to fill, but some pore space is still empty. (d) When the sediment is saturated, all pore spaces are full of water.

content. For example, soil in a dry condition immediately prior to a precipitation event will have a higher infiltration rate than the same soil in a saturated condition. As infiltration proceeds, the soil may become more and more saturated, and the infiltration rate will decline until it reaches a steady value. Because of this, we are most often interested in this final, steady value, or ultimate infiltration rate (Fig. 4.3).

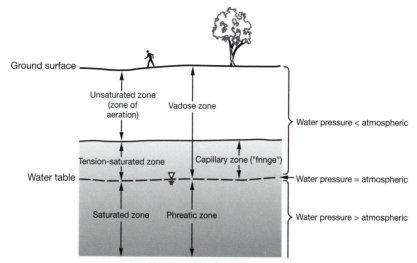

Fig. 4.2   In the unsaturated zone, water does not completely fill the pore spaces. In the tension-saturated zone, water fills the pore spaces, but is held under tension. In the saturated zone, water fills the pore spaces and is able to move freely.

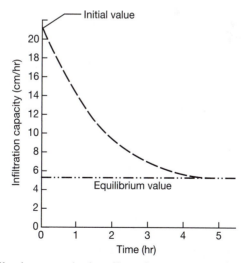

Fig. 4.3   As infiltration proceeds, the soil may become more and more saturated, and the infiltration rate declines until it reaches a steady value, the ultimate or equilibrium infiltration capacity.

Measurement of Infiltration Rate

Infiltration rate is best measured in the field, rather than in the laboratory. Important controlling factors such as vegetation, root channels, animal burrows, and soil structure are not easily preserved in a sample which is extracted, transported, and then tested in a laboratory. Infiltration rate may be measured in the field by two methods. Infiltrometers are fairly small in size, ranging from a few centimeters to about a meter in diameter. Because they measure infiltration rate across a small area, they give limited information, particularly in the case of inhomogeneous materials. Areal infiltration tests, as the term suggests, measure infiltration over a larger area. They are more suitable than ring infiltrometers when sediment conditions vary significantly at a site.

Infiltrometers. Infiltrometers are of two types: single-ring and double-ring. Single-ring infiltrometers are simpler in design and require fewer measurements, but may not give as accurate a measure as double-ring infiltrometers. Procedures for using both are described here.

Single-Ring Infiltrometers

A single-ring infiltrometer is an open cylinder or ring, usually made of metal, pressed a few inches into the ground. Water is added to the ring and kept at a constant height. The rate at which water must be replenished to maintain that height is converted to inches or centimeters per hour (calculations are explained in what follows); this is the infiltration rate.

Depending on soil type and hardness and the level of sophistication of the test, the ring may be made of a machined stainless steel cylinder or a coffee can, 5-gallon bucket, or a 55-gallon drum with the top and bottom removed. The ring is pressed into the soil to a suitable depth of at least 3 inches. This may be accomplished by simply pressing it by hand into soft soils, by using a hydraulic jack to press it in the case of stiff clays or dense granular soils, or by covering it with a board or thick metal plate and pounding it in with a sledge hammer in the case of very hard soils. Using the last method is likely to cause some disturbance to the soil, and should be used only when the other methods are not feasible. For purposes of installing the ring by hammering, plywood works better than does dimension lumber (e.g., 2 × 4's) or logs, because plywood does not split apart easily.

While inserting the ring, take care not to disturb the structure of the soil, and recompact it where it is disturbed. Vegetation may be left undisturbed or may be removed, depending on the purpose of the test. The ring should be level; use a bubble level to check this.

Select a height at which to maintain water level during the test. The higher the level, the greater the hydraulic head, and therefore the faster the flow into the soil. Higher heads will therefore give faster results, which might be of particular importance when testing low-permeability soils.

Before beginning the test, ensure that an ample supply of clean water is available. To begin the test, pour water into the ring. Take care not to erode or alter the surficial materials by pouring the water vigorously. Fill the ring to the desired height. Start the clock. Using a graduated cylinder, note and record the volume of water added and the time at which it is added. A sample data form is shown in Fig. 4.4. Maintain water level at the preselected height. If the test goes on for a long time, or in dry, windy conditions, cover the ring to minimize evaporation.

End the test when the infiltration rate or rate of water added per time is constant. In some soils, this may take a few hours; in others, it may take a few days. Alternatively, end the test at some predetermined time.

Single-ring Infiltrometer Test Field Data Form

Contract/Job _____ Date _____

Field personnel _____

Location _____

Weather conditions _____

Description of area _____

_____

Description of soil and vegetation_____

_____

Depth to water table _____

Ring diameter_____ Height of water during test _____

Source of water_____ Water pH _____

Water temperature _____ Was ring covered during test? _____

Time	Volume of water added (units: _____)

Fig. 4.4   A sample single-ring infiltration test data form.

Calculations

$$\text{Depth of water added} = \frac{\text{volume of water added}}{\text{area of ring}}$$

$$\text{Infiltration rate} = \frac{\text{depth of water added}}{\text{elapsed time between additions of water}}$$

Depending on how rapid infiltration is, a value of the infiltration rate may be calculated each time water is added, or it may be calculated for given time intervals, e.g., every 5 minutes at the beginning of the test, then every 10–30 minutes, then every hour. Graphing infiltration rate versus time may help to illustrate the change in rates and to determine when to end the test.

A source of error in single-ring infiltrometer tests is that water may escape laterally out from the base of the ring (Fig. 4.5). This makes the infiltration rate seem higher than it actually is. To address this problem, a double-ring infiltrometer may be used.

Double-Ring Infiltrometers

A double-ring infiltrometer consists of a single-ring infiltrometer and a larger, outer ring that encircles the smaller, inner ring (Fig. 4.6). Water is added both to the inner ring and the annulus, or space between the inner and outer rings. Theoretically, water infiltrating from the annulus into the ground should create a curtain about the water infiltrating from the inner ring. This should force the water in the inner ring to migrate downward without lateral spreading, resulting in a

Fig. 4.5 A source of error in single-ring infiltrometer tests is that water may escape laterally outward from the base of the ring.

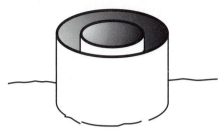

Fig. 4.6 A double-ring infiltrometer consists of a single-ring infiltrometer plus a larger outer ring that encircles the smaller inner ring.

more accurate measurement. Large, sealed infiltrometers may be used in the case of low permeability soils, such as might be found in landfill liners and covers.

A double-ring test is conducted in a manner similar to that of a single-ring test. However, in a double-ring test water is added to both rings, and separate records must be kept for each. The outer ring should be pressed into the soil first, to an appropriate depth (6 inches or more). Next, the inner ring should be centered in the larger ring and pressed in, to an appropriate depth (2 inches or more).

Height of water within the two rings should be approximately the same. As suggested by Beverly L. Herzog of the Illinois State Geological Survey, this can be accomplished by using an IV bag from a medical supply house. The bag should be partially filled with water and connected to tubing which is run into the inner ring and secured with its end below water level. All air should be expelled from the bag and tube. The bag then is placed into the outer ring, where it will float. If all connections are watertight and no air remains in the system, water levels will be equalized by a siphoning action into or out of the bag.

A sample data form is shown in Fig. 4.7.

Calculations. Calculate the infiltration rate for both the inner ring and annulus; a significant difference between the values indicates that lateral spreading is a significant factor, and only the inner ring value should be used. Calculations for the inner ring are made as for a single-ring infiltrometer. Calculations for the annulus must consider the area of the annulus, which is the area of the larger ring minus the area of the inner ring. Infiltration rate for the annulus is then calculated as for the inner ring.

Equipment List

Infiltrometer (single- or double-ring)

Meter stick or steel tape

Means of pressing ring into soil:

   Hydraulic jack or

   Sledge hammer, plywood large enough to cover the infiltrometer (two or three pieces), and ear plugs

Two graduated cylinders (clear plastic, 1000 ml)

Double-ring Infiltrometer Test Field Data Form

Contract/Job _____ Date _____

Field personnel _____

Location _____

Weather conditions _____

Description of area _____

_____

Description of soil and vegetation_____

_____

Depth to water table _____

Ring diameters: Outer ring _____ Inner ring _____

Height of water during test (use same height for outer and inner rings) _____

Source of water_____ Water pH _____

Water temperature _____ Were rings covered during test? _____

Time	Volume of water added to inner ring (units: _____)	Volume of water added to outer ring (units: _____)

Fig. 4.7 A sample data form for a double-ring infiltrometer test.

Two 5-gallon buckets
Source of clean water
Spade (to dig out infiltrometer at end of test)
Watch

Calculator
Field notebook and pencils
Waterproof marking pen
Bubble level
Cover for infiltrometer (to prevent evaporation), if needed

Areal Infiltration Tests. Infiltrometers can give an estimate of infiltration at a single point. But if estimates over a larger area are required, an areal infiltration test may be more useful. In this test, walls are constructed around an area, and water is kept ponded to a constant height within the walls (Fig. 4.8). Time and volume of water added are recorded, and calculations are as for single-ring infiltrometers. As sample data form is shown in Fig. 4.9.

Monitor evaporation as the test proceeds. Water evaporated should be subtracted from the volume added before the infiltration rate is calculated. To measure evaporation, a second "walled" area should be constructed, similar to the test area. But this second area should have a waterproof liner at the base, so that water cannot leak into the ground. During the test, it should be filled with water to the same depth as the test area. The amount of water that must be added to maintain this height should be recorded and subtracted from the uncorrected infiltration rate. In addition, precipitation should be monitored during the test, using a standard rain gage. This should be included in the figures for the amount of water added.

Soil Moisture

Soil moisture is water contained within the pore spaces of soil. It varies according to available water and soil characteristics. Some types of soil can hold a great deal more water than other types, depending on their grain size and sorting, mineralogy, organic matter content, and degree of compaction and cementation. Moisture content is generally given as a percentage:

Fig. 4.8   An areal infiltration test. (Courtesy of Kurt O. Thomsen.)

Areal Infiltrometer Test Field Data Form

Contract/Job _____ Date _____

Field personnel _____

Location _____

Description of area _____

_____

_____

Depth to water table _____

Description of soil and vegetation _____

_____

Water source & application method _____

Depth of water (if water is ponded)_____

Water pH _____ Water temperature _____

Time	Rate of water application (units: _____)	Weather conditions (precipitation, evaporation, wind, temperature, cloud cover)

Fig. 4.9 A sample data form for an areal infiltration test.

$$\text{Moisture content} = \frac{\text{mass of water contained within soil}}{\text{mass of wet soil}} \times 100\%$$

Measuring Soil Moisture

Soil moisture is typically measured in hydrogeologic studies by one of four methods: (1) the gravimetric method, (2) extrapolation from tensiometer data, (3) electrical resistance, and (4) neutron probe. Each of these methods is described here, and equipment lists are provided. Although the gravimetric method is not strictly speaking a field method, it is described here because it is used to calibrate the other methods.

Emerging technology allows soil moisture to be measured in yet another way, by measuring the apparent dielectric constant of the soil. These instruments apply an electrical signal along a transmission line and then monitor the changes in impedance along the line. The changes are related to changes in soil moisture content. Used primarily in plant biological work, current designs work well at the surface, but are of limited use in examining a depth profile of soil moisture.

Gravimetric Method. The gravimetric method involves taking a sample of moist soil, weighing it, drying it, and then weighing it again to determine how much water weight was lost. For more information, ASTM Standard Test Method D2216 may be consulted. ASTM Standard Test Method D4643 is an alternate method using a microwave oven, and may be used when rapid results are required.

Steps in using the conventional gravimetric method are as follows:

1. Sample the soil and place it in a waterproof container. Seal and label the container. Keep the sample cool while transporting it to the laboratory. (Methods for sampling soil are described in Chapter 6.)
2. Label a clean, dry moisture tin and lid or a watch glass and cover. Weigh the container and lid. *Note:* Because some marking pens contain ink that melts or evaporates when heated, do not use them in this test. Instead, use a pencil or etching tool.
3. Place about 25 grams of soil in the container, and cover or cap it.
4. Weigh the closed container and soil.
5. Uncover the container, place the lid under the container, and place the container, lid, and soil in a drying oven that has been preheated to 104.5°C. Allow the soil to dry for 24 hours.
6. Remove the container from the oven and cover or cap it. Place the capped tin in a desiccating vessel until it is cool.
7. Weigh the capped container, lid, and soil.

Calculations

$$\text{Moisture Content} = \frac{(\text{mass of wet soil} + \text{tin}) - (\text{mass of dry soil} + \text{tin})}{(\text{mass of wet soil} + \text{tin}) - (\text{mass of tin})} \times 100\%$$

Advantages of this method are that it is simple, easy, requires no specialized equipment, and gives moisture content directly, with no need for calibration. Also, if some other sample measurements, such as volume, are known, results of the test may be used along with these data to estimate other soil parameters, such as porosity. Disadvantages of the method are that it requires a laboratory, and takes 24 hours to produce results. In addition, it requires destructive sampling of the soil, that is, the sample must be removed from its surroundings and its structure must be altered. The microwave method is more rapid, but should not be considered a replacement for the conventional method. Rather, it is to be used when it is necessary to expedite other tests.

Equipment List

Field

Soil sampling device (e.g., auger, shovel, scoop)

Sample containers (e.g., zipper-closure plastic bags)

Marking pen to label sample containers

Cooler and ice to keep samples cool while transporting them

Lab

Soil moisture tins and lids or watchglasses with covers

Balance with at least 0.01-g precision

Drying oven, preheated to 104.5°C

Desiccating vessel (bell jar with calcium chloride crystals)

Tensiometers. A tensiometer is a tube filled with water and inserted into a borehole. The tube is sealed at the top, but has a porous tip, often ceramic, at the bottom. Water flows out the tip into the soil, which causes a vacuum to form at the top. The drier the soil, the more water will flow out of the tip, and the greater the vacuum will be at the top. A vacuum gauge, manometer, or pressure transducer fitted at the top gives readings in units of soil suction. A tensiometer may be used to give relative measures of soil moisture. However, if absolute values of soil moisture are needed, the characteristic moisture curve of the soil must be developed by using the gravimetric method in combination with tensiometer readings.

Commercially available tensiometers should be used in accordance with the manufacturer's instructions. The general procedure for using a tensiometer follows.

Fill the tensiometer with degassed water, and allow the tip to become saturated. Use a hand vacuum pump to draw gas out of the water by pulling a vacuum at the top of the tube. Auger or bore a hole, and reserve the soil removed from the hole. Install the tensiometer in the borehole. If the soil has low permeability, pack silica flour or very fine sand about the porous tip. Backfill the borehole annulus, if there is one, with the soil removed from the hole. (Some tensiometers come with a coring device that produces a hole exactly the right size for the tensiometer.) It may be necessary to make the soil into a slurry to provide a good seal in the annulus and to ensure good contact between the soil and porous tip. If a slurry is used, response time of the tensiometer will be slower.

Install a vacuum gauge or other reading device (manometer or pressure transducer) at the top of the tube. Record the time, date, and vacuum reading. The vacuum should increase until equilibrium is reached, and then will remain steady until soil moisture conditions change.

Advantages of this method are that once installed, the tensiometer may be left in place for a long time (an algicide may be added to reduce algae growth), trends with respect to time can be monitored easily, and no destructive sampling is needed. Disadvantages are that the readings must be considered simply as relative readings, unless the characteristic soil moisture–soil tension curve has been developed by using the gravimetric method. Tensiometers filled with water may not be used in freezing conditions, although some workers use a 50–50 mix of water and ethylene glycol to prevent freezing. See ASTM Standard 3404 for more information on tensiometers.

Equipment List

Auger (and extensions, if needed) or coring tube

Tensiometer with vacuum gauge or other vacuum measurement device (manometer or pressure transducer)

Hand vacuum pump

Two 5-gallon buckets (one to hold clean water; one in which to mix slurry)

Source of clean water (enough to fill tensiometer and mix slurry)

Algicide, if needed (for long-term installations)

Watch

Field notebook and pencils

Electrical Resistance Method. The electrical resistance method, sometimes referred to as the "gypsum block" method, is based on the principle that electricity will pass through wet soil easier than it will through dry soil. In this method, electrodes (embedded in a porous plate, ceramic block, or "gypsum block") are buried in the soil. A current is passed between the electrodes, and resistance is measured with an ohmmeter (Fig. 4.10). Because soil type, moisture content, and degree of saturation also affect resistance, this method must be used in combination with the gravimetric method to develop the characteristic soil moisture— electrical resistance curve. Otherwise, the method can be used only to measure relative differences in soil moisture.

To use the electrical resistance method, first develop the soil moisture— electrical resistance curve. This step is critical, and some workers recommend replicating these measurements as a quality check. Take a sample of the soil to be monitored, and split it into at least five subsamples. Add water in varying amounts to some subsamples, and allow others to partially dry, to varying degrees. The objective of this step is to produce subsamples of varying moisture contents. Water must be dispersed evenly throughout each soil subsample. Bury the electrode in a subsample, read and record the resistance, and then immediately, and with a minimum of handling, place the subsample in a preweighed moisture tin for moisture content determination using the gravimetric method. Repeat for each subsample. Dry and weigh the samples, determine the moisture content, and plot moisture content versus resistance on a graph.

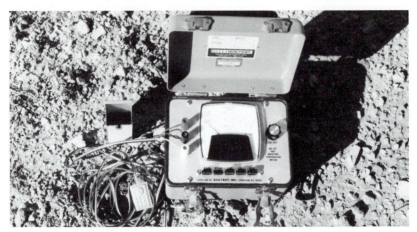

Fig. 4.10 A soil moisture meter. The flattened electrodes and their cables are buried in the soil. A current is passed between the electrodes, and resistance is measured with an ohmmeter.

Once the calibration curve is constructed, make readings at the field site. Bury the electrode at the desired depth and location, and read the resistance. Use the curve to determine moisture content. If a new soil type is encountered, another curve must be developed.

Advantages of this method are that the electrodes can be left in place for long periods of time, simplifying later measurements, and that if several sets of electrodes are used, profiles can be constructed easily. Disadvantages are that the calibration process takes more than 24 hours, and that installing the electrodes disturbs soil structure. If gypsum blocks are used, they may dissolve over time.

Equipment List

 Sampling device (e.g., auger, shovel, scoop)

 Sample containers and labels

 Marking pen to label sample containers

 Electrodes (may be purchased from soil-testing supply companies)

 Ohmmeter or soil resistivity meter

 Equipment for performing gravimetric method

 Graph paper to plot calibration curve

 Field notebook, pencils, and pens

Neutron Probe. Neutron probes are used to determine moisture content of soil into which a borehole is drilled. Adaptations are also available for near-surface (nonborehole) probing. A sending unit gives off fast neutrons, which strike water molecules and slow down to become slow neutrons. A counting device counts slow neutrons. The wetter the soil, the more slow neutrons are produced and can be counted (Fig. 4.11). By raising and lowering the device in the borehole, a soil moisture profile may be made. Use of neutron probes requires specialized training, as they contain radioactive materials.

Sampling Soil Water

Sampling soil water may be desired for purposes of chemical, physical, or biological analysis. Because soil water is held as capillary and hygroscopic water, retrieving it necessarily involves applying a suction to the soil. Soil water samplers and pressure-vacuum lysimeters act on that principle, but differ somewhat in design.

A soil water sampler is an empty tube with an airtight cap on top and a porous tip at the bottom. It is installed in a borehole, and a vacuum is pulled inside. The

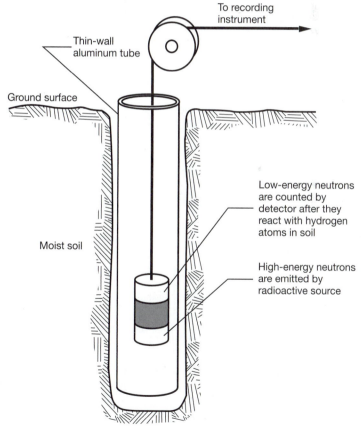

To recording instrument

Thin-wall aluminum tube

Ground surface

Low-energy neutrons are counted by detector after they react with hydrogen atoms in soil

High-energy neutrons are emitted by radioactive source

Moist soil

Fig. 4.11   Neutron probes contain a sending unit, which gives off fast neutrons, and a counting device, which counts slow neutrons. The wetter the soil, the more slow neutrons are produced and can be counted (after Fetter, 1994).

vacuum draws water from the soil in through the porous tip (Fig. 4.12). A sample may then be taken from the tube.

A pressure-vacuum lysimeter also is an empty tube with an airtight cap and porous tip. However, unlike a soil water sampler, a lysimeter is buried completely in the soil. The cap has two holes, and a tube is inserted through each hole and extends to the surface. One of the tubes is longer, reaching all the way to the bottom of the lysimeter, and is used for retrieving a sample. The other tube is shorter, and is used for applying vacuum or pressure.

One problem related to the use of soil water samplers and pressure-vacuum lysimeters is that standard lysimeters do not work at great depths because greater suction is needed. Specialized lysimeters may be used for deep installations. Other problems include contamination of the sample by components leached from the porous cup, as well as changes in redox conditions and pH. To help prevent this, samplers and lysimeters with ceramic porous cups should be cleaned before using by passing a solution of 10% hydrochloric acid through them, followed by distilled, deionized water. Those with stainless steel membranes should be cleaned with 1 to 10% nitric acid, followed by distilled, deionized water. For more

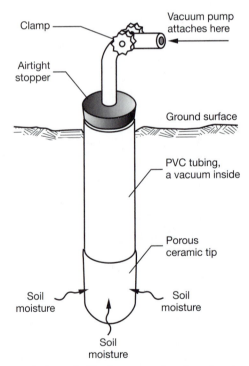

Fig. 4.12 A vacuum is drawn inside the empty tube of a soil water sampler, drawing soil moisture through the porous tip and into the sampler.

information, see Everett and others (1984), Ballestero and others (1991), Wilson (1995), and ASTM Standard Guide D4696.

To install a soil water sampler or pressure-vacuum lysimeter, auger a hole, reserving the soil extracted from the hole. Insert the sampler. Silica flour or very find sand may be placed about the porous tip to help ensure good contact with the soil. If the fit is loose, it may be necessary to fill the annulus with a slurry. The slurry can be made of water and either the cuttings from the hole, or bentonite chips or pellets. If the slurry is used, the first sample or samples extracted will consist of water from the slurry. Use a vacuum pump to draw a vacuum in the sampler or lysimeter, and clamp the tube, maintaining the vacuum within. Note the time and date of installation.

To retrieve a sample from a soil water sampler, loosen the clamp and allow the vacuum to escape. Remove the cap. Put one end of a clean tube down into the sampler, and attach the other end to one outlet of a rubber stopper with two outlets. Place the stopper tightly into a sample container. Draw a vacuum on the other outlet of the stopper; the vacuum in the sample container should draw water out of the sampler and into the container (Fig. 4.13). This may not work at more than very shallow depths (a few feet). If the sampler is installed at greater depths, a bicycle-type hand pump may be used to force water up the tube. At even greater depths, use a bailer or pump to retrieve water from the sampler.

To retrieve a sample from a pressure-vacuum lysimeter, open the clamps on the two tubes. Attach a bicycle pump to the short tube, and direct the end of the long tube into a sample container. Apply pressure with the pump. The sample should be forced up into the container.

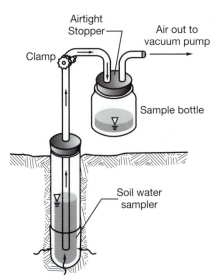

Fig. 4.13   A sample may be retrieved from a shallow sampler by drawing a vacuum in the sample bottle, pulling the sample up through the tubing.

Equipment List

Installing Soil Water Sampler or Pressure-Vacuum Lysimeter
Auger (and extensions, if needed)

Sampler or pressure-vacuum lysimeter and tubing

Vacuum pump

Materials for sealing cap and tubes (e.g., clamps, tape)

Two 5-gallon buckets (one for clean water, one for mixing slurry)

Source of clean water for slurry, if needed

Silica flour or very fine sand

Field notebook, pencils, and pens

Retrieving Sample

For both methods:
Sample container and labels

Marking pen to mark labels

Cooler and ice, to keep sample cool during transport to the lab

Field notebook, pencils, and pens

For soil water sampler method:
Tubing long enough to reach bottom of sampler, plus about 2 feet

Rubber stopper with 2 holes, tightly fit to sample container

Vacuum pump and/or bicycle pump

Means of attaching tubing and vacuum pump to outlets in stopper

For pressure-vacuum lysimeter method:
Bicycle pump and means of attaching it to tubing

Hydraulic Conductivity of the Unsaturated Zone
Hydraulic conductivity (K) is a measure of the ease with which a porous medium can transmit a fluid (such as water). The term hydraulic conductivity is sometimes used interchangeably with the term permeability, but strictly speaking they are not the same. Permeability (also called intrinsic permeability) is a characteristic only of the porous medium, whereas hydraulic conductivity is a characteristic of both the porous medium and the fluid.

Hydraulic conductivity of the unsaturated zone (K_{unsat}) is generally substantially less than that of the saturated zone (K_{sat}), as water passing through a dry pore space must overcome the adhesive and cohesive forces in such a space in order to pass through. In a saturated pore space, the presence of more water makes flow easier because there is enough water to satisfy adhesion and cohesion while still allowing flow.

Unsaturated hydraulic conductivity increases as water content increases, and the water content of unsaturated soils may fluctuate frequently. In a contamination "worst-case scenario," such as a spill from a train car or leaking tank, the unsaturated zone is flooded by water or some other fluid. Because of this, it often is desirable to know the *saturated* hydraulic conductivity K_{sat} of the unsaturated zone. The K_{sat} is the maximum hydraulic conductivity, under "worst-case" conditions.

Field tests conducted to determine K_{sat} of the unsaturated zone generally involve "field saturating," or introducing water into an auger hole in an effort to saturate the soil or sediments. Because this means of saturating the materials is rapid and somewhat artificial, it may not replicate exactly the conditions of true saturation. The following are methods for determining the "field-saturated" hydraulic conductivity (K_{fs}) of the unsaturated zone.

Measuring K_{fs} of the Unsaturated Zone

Several methods can be used to determine K_{fs} of the unsaturated zone. The simplest of these, the constant-head well permeameter test, involves augering or boring a hole, filling it with water, and then watching to see how rapidly water moves into the ground (Fig. 4.14). Alternatively, a Guelph Permeameter or a Compact Constant-Head Permeameter (CCHP) may be used to automate certain aspects of the test.

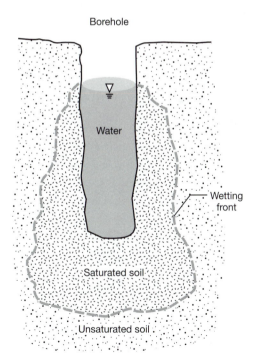

Borehole

Water

Wetting front

Saturated soil

Unsaturated soil

Fig. 4.14 In the constant-head well permeameter test, a hole is augered and filled with water, and then the water level is monitored to see how rapidly water moves into the ground.

Constant-Head Well Permeameter Test. The Constant-Head Well Permeameter Test is similar to the so-called percolation or "perk" tests that often are performed to assess a soil's suitability for use in a septic system. Various jurisdictions (local, county, or state) have various prescribed protocols for percolation tests. Following is a description of a method for determining the field-saturated hydraulic conductivity of the unsaturated zone using a generic version of this test.

Procedure. Auger or bore a hole of the radius r to the desired depth. Before beginning the test, ensure that an ample supply of water is at hand. Determine the height of water to be maintained in the hole. Determine how water level will be maintained. Can the level be found visually? If not, it may be necessary to use a water-level indicator. Determine how water volume will be measured before it is added to the hole. A large graduated cylinder may be useful for this purpose. If flow is rapid, it may be helpful to use two of these: one may be filled while the other is being used to fill the hole. Alternatively, precalibrated jugs or buckets may be used.

To begin the test, fill the hole to the predetermined level, and record the time. Maintain the water level at a constant height (H) above the bottom of the hole, recording the amount of water added and time. Continue the test until the amount of water added per unit time reaches a constant value (Q).

Calculations. Amoozegar (1989) proposed the use of the "Glover solution" to calculate K_{fs} of the unsaturated zone. This calculation is as follows:

$$K_{fs} = Q \frac{[\sinh^{-1}(H/r) - (r^2/H^2 + 1)^{1/2} + r/H]}{2\pi H^2}$$

where K_{fs} = field-saturated hydraulic conductivity of the unsaturated zone
Q = volume of water added to the hole per unit time
H = height of the water level above the bottom of the hole
r = radius of the hole
$\sinh^{-1}$ = inverse hyperbolic sine function

Equipment List
Auger/extensions
Meter stick, steel tape, or water-level indicator
Source of water
Buckets
A few 1-liter plastic graduated cylinders
Watch
Water-level indicator
Field notebook, pencils, and pens

Guelph Permeameter and CCHP. Two instruments may be used to aid in measuring K_{fs} of the unsaturated zone (Fig. 4.15). The Guelph permeameter is a commercially available instrument that uses the Marriote principle to maintain a constant water level in the hole. The Guelph permeameter has a calibrated set of reservoirs that allow for easy measurement of water added to the hole.

Like the Guelph permeameter, the Compact Constant-Head Permeameter (CCHP) is based on the Marriote principle (Amoozegar, 1989). It requires less water and is lighter in weight than a Guelph permeameter.

For either of these instruments, and as with all of the soil water tests described thus far, heterogeneities in soil characteristics can cause wide variations in results. Therefore, the results should be taken to be point-specific, and multiple measurements at various locations should be made.

Ground Water

Ground water is water below the water table. Most of this book deals with ground water investigations. This section introduces the basic concepts of ground water and surface-water interactions, the meaning of water levels in piezometers and wells, and the measurement of water level and discharge.

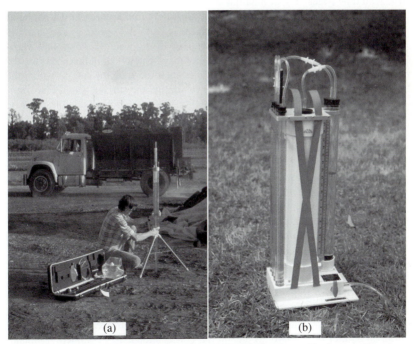

Fig. 4.15   (a) A Guelph permeameter (courtesy of Soilmoisture Equipment Corp.), and (b) a compact, constant-head permeameter (courtesy of Ksat, Inc.).

Surface Expressions of Ground Water

Whether streams, ponds, lakes, or oceans, any surface-water bodies are likely to be surface expressions of ground water. Some, like puddles after a rainstorm, are ephemeral, or short-lived. Others, like the oceans, are perennial, or long-lasting. In either case, we may gain valuable information about subsurface conditions by examining surface-water/ground water interactions.

In some cases and under some conditions, surface water is moving toward ground water. This is the case with losing streams, and it also may be true for losing ponds, lakes, wetlands, or puddles. The opposite condition might also exist: Water might be moving from ground water toward the surface "gaining" stream, pond, wetland, or other water body.

To determine the direction of movement, consider water levels in wells or piezometers near the surface-water body. The wells or piezometers must be open to the same body of rock or sediment that holds the surface water, and they must be hydraulically connected. A lower level in the wells indicates that the surface-water body is losing, whereas a higher level in the wells indicates that it is gaining (Fig. 4.16). Temperature may be another indicator. In temperate regions, ground water tends to be colder than surface water during the summer and warmer than surface water during the winter. Some environments in which this would not be true would be zones of hydrothermal activity (e.g., hot springs), or areas where the surface water originates as meltwater from glaciers. Differences in water chemistry might indicate flow directions, as well.

Any time a surface-water body appears to be losing water to the ground water, consider the possibility that the surface water may be perched. In this situation, the rate at which water is added to the surface water exceeds the rate at which the underlying rock or sediment can transmit or "drain" the surface water. For example, a heavy rain or sudden snowmelt might cause a temporary pond to form on clay-rich soil. A clay lens near the surface in some glacial materials might cause the formation of a perpetual wetland at the surface, whereas the regional water table lies much farther below.

Subsurface Expressions of Ground Water

Geologists look for outcrops of rock to provide clues about subsurface geology. But where no outcrops exist, their information may come from drillhole data. Likewise, hydrogeologists can learn a great deal from "outcrops" of ground water: springs, seeps, and some other surface-water bodies. But where no hydrogeologic outcrop exists, hydrogeologists must rely on data from holes drilled to puncture the ground water's surface and allow us to examine its nature.

Why Is This the Water Level?

Ask that question any time you encounter a water level, whether surficial or subsurficial. If surface water, is it gaining or losing? If subsurface, is the water level from an aquifer or aquitard? If an aquifer, is it confined, semiconfined, or unconfined? Is the water level the result of flow from fractures? Is it from a cave, mine, conduit, tunnel, pipeline, or drainage tile? Was the water level measured in a piezometer? A well? What is the screened or open interval in the well?

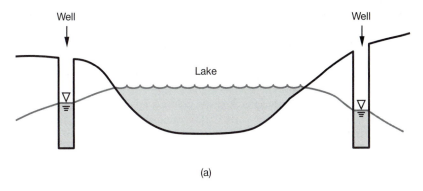

(a)

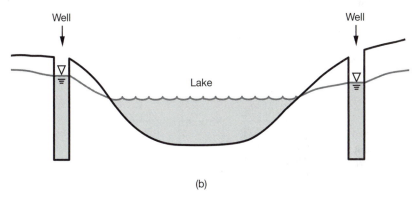

(b)

Fig. 4.16 (a) If the water levels in the nearby wells are lower than the water level in the lake, the lake is losing. (b) If water levels in the wells are higher, the lake is gaining.

One fairly unusual subsurface expression of ground water is water in caves or caverns. Some of these hold subsurface springs, streams, and ponds. Of course, these features occur only where large underground openings exist, and they are particularly likely to be in karst regions. Other underground conduits might not be natural. For example, such human-made features as tunnels, underground mines, or buried conduits or pipes might be affected by ground water. In these cases, consider how permeable the walls of the opening are, and particularly with mines or pipes, consider whether water is actively being pumped or drained from the opening.

Perhaps more common than these features are holes we put in the ground to find the water level. Piezometers and wells are the cornerstones of hydrogeologic data.

Water Levels in Piezometers

A piezometer is a pipe installed in the ground, and it has a very short (less than 1 ft) intake, or screened or open interval. It may even have no screen at all. A piezometer is installed specifically for the purpose of determining hydraulic head at a specific point within an aquifer or aquitard. The bottom opening of the piezometer, or its intake, is the point at which it makes the measurement.

The water level in a piezometer generally does not give directly the position of the water table or potentiometric surface. (Potentiometric surface is referred to as "piezometric" surface in some texts.) A piezometer gives the location of the water table only if its intake is positioned such that it barely penetrates the water table (Fig. 4.17). The likelihood that this will occur probably is not great. If it does occur, it probably will be coincidental, and so piezometers are most useful when their screens are below the water table. However, hydraulic head varies vertically. As a result, if a piezometer is installed with its intake more than a few feet below the water table, vertical components of flow will prevent it from giving the water table location directly.

When enough hydraulic head values are available, taken from multiple piezometers, a flow net may be constructed from the data, and the location of the water table or potentiometric surface may be determined from the flow field. For more information on flow nets, see Chapter 10, "Hydrogeologic Mapping."

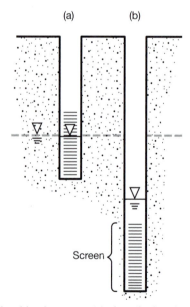

Fig. 4.17 The water level in piezometer (a) gives the location of the water table, but only because it is positioned such that it barely penetrates the water table. The water level in piezometer (b) does not give location of the water table, because its screened interval is too far below the water table.

Piezometer nests are groups of three or more piezometers installed side by side, but measuring hydraulic heads at various depths (Fig. 4.18). Reading the water levels in nested piezometers allows one to determine the direction and magnitude of a vertical gradient. Begin by finding the difference in head (Δh) between two of the piezometers; usually, this is the difference in elevation (above a datum) of water levels in the piezometers. Next, find the distance (Δl) between the points at which the piezometers are open to the ground water. The gradient is $\Delta h/\Delta l$, and the direction of the gradient is from the opening in the piezometer with the higher water level to the opening in the piezometer with the lower level (Fig. 4.19).

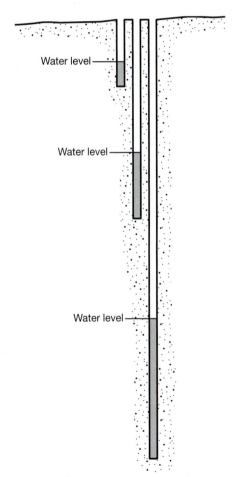

Fig. 4.18   Piezometer nests are groups of three or more piezometers installed side by side, to different depths. They are used to measure vertical gradients. Water levels in this piezometer nest indicate a downward vertical gradient.

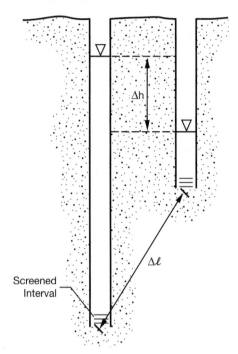

Fig. 4.19 When using piezometers to measure gradient, $\Delta h / \Delta l$, the value of Δh is calculated as the difference in elevations of water in the piezometers. The distance between the openings or screened intervals of the piezometers is Δl.

Piezometers may be regarded as small-diameter wells, and as such, they may not be well-suited for sampling for chemical analysis, because their intakes usually are too short to allow for easy flow of water into and out of the piezometer. For the same reason, piezometers are not appropriate for performing pumping tests or other tasks requiring pumping large volumes of fluid from an aquifer, such as contaminant recovery. They may, however, be suitable for slug or bail tests.

Water Levels in Wells

A well is a pipe installed in the ground, and it has a longer intake than does a piezometer. The optimal length of a well intake depends on desired yield and formation characteristics. Wells are installed for the purpose of extracting ground water, whether for domestic or municipal use, or irrigation; for pumping tests; for extracting samples for chemical analyses; for product recovery; dewatering an area for construction or engineering purposes; or for injecting fluids, as in a recharge well or waste-injection well.

The water level in a well indicates the conditions through the screened or open interval of the well. The water level in the well might be above, at, or below the ground surface. When the well water level is above the ground surface, it indicates that the screened or open interval is under confining conditions. When the well water level is at or below ground surface, it may indicate one of two things. First, it might indicate that the open or screened interval is not under confining conditions; the aquifer is unconfined. Or, second, it might indicate that the aquifer is confined, but that the confining pressure does not create enough head to drive the well water above ground level. Drill hole data, geologic maps and cross-sections, or hydraulic testing of the well might be used to determine which of these conditions is present.

In unconfined aquifers, when the intake is short (no more than a few meters) *and* it intersects the water table, the level of water in the well indicates that of the water table. But if the screened or open interval is longer, or if the top of that interval is positioned below the water table, the water level in the well does not give the level of the water table. Instead, the water level reflects conditions only for that screened or open interval (Fig. 4.20). To locate the position of the water table, a piezometer nest or a well of appropriate design must be used.

Finding Wells

To find already-existing wells in the field, the easiest place to start is to ask the property owner where the wells are. Another source of information is maps of locations given on site plans or in drilling logs. Government agencies might have logs that indicate location. In the field, look for pipes sticking up out of the ground, or plates mounted flush with ground surface. Look for stone- or concrete-lined access pits, possibly covered by wood or metal sheeting. Follow water supply lines leading out of houses or other buildings. Be aware that wells may be inside buildings. Look for stone or clay tile drainage ways. Look for windmills or windlasses that may have been used to pump water out of the ground; be aware that windmills also might be used for generating electricity or grinding grain. Look for storage tanks, water towers, or pumping facilities; any of these might lead you to a well.

Well Depth

The depth of a well is the vertical distance from the bottom of the well to the surface. In order to keep measurements consistent, establish some reference point at the surface. The reference point might be the top of the well casing (TOC), or it might be ground level (grade).

The depth of a well should be recorded on the well construction diagram or log. However, each time the water level is measured, the well depth should be determined by sounding, if possible. This should be done because silt may have settled in the bottom of a well, frost heave may have damaged it, or well construction diagrams may be unavailable or inaccurate. In some wells, however, the presence of pumps and discharge lines may make depth sounding impossible.

To sound the bottom of a well, first check the well construction diagram, if available. This will indicate the approximate depth. Then lower a weighted cable or tape down the hole until it goes slack, and measure the cable length. It may be convenient to use a water-level indicator for this purpose.

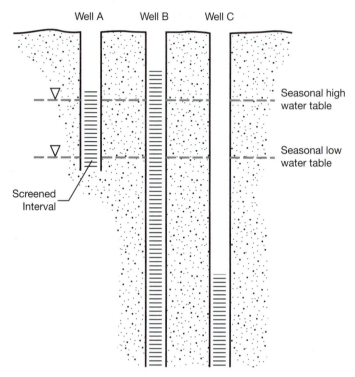

Fig. 4.20 The screen in well A intersects the water table and is short enough to give a good representation of the water table. In well B, the screen is so long that the water level in the well may not accurately reflect the location of the water table. The screen in well C is too far below the water table to be of use in locating the water table.

If the well has a pump installed in it, watch out for the pump lines. The line may accidentally become wrapped around one of them and tangle. Do not play out too much line; it may have caught on something, and as you continue unreeling it, it will just become more and more tangled.

Water-Level Measurements

Water-level measurements provide a basic piece of information used in hydrogeologic studies. Water level may be a "static" level or a "pumping" level; a static level is measured when the well is not being pumped, and has not been pumped for some time. A pumping level is measured when the well is being pumped or has been pumped recently.

The method for making a water-level measurement depends on whether the well is flowing or nonflowing. This section describes some of these methods.

Measuring Water Level in Flowing Wells

In some wells, called flowing wells, water flows out the top of the well naturally and without any pumping. The potentiometric surface associated with these wells is above the ground surface. If a well is flowing, and it is necessary to know the elevation of the potentiometric surface in the aquifer tapped by the well, the water level must be measured either directly or by taking a pressure measurement.

Direct Measurement. Direct measurement involves allowing the water to rise to the level of the potentiometric surface and then measuring its height above ground surface. To use this method, a conduit for the water to rise in must be provided. This method works only if it is physically possible to measure the height above the ground surface. If it is more than about 8–10 ft above the ground, this will be difficult.

Attach the conduit, which might be a pipe or piece of tubing, to the top of the well. It is critical that there be no leaks at the joint or anywhere along the conduit. The pipe or tubing may be of any diameter; it is perfectly acceptable for the tubing to be the small diameter, clear vinyl tubing commonly found in laboratories, even if the well has a much larger diameter. Clear tubing has the added advantage of making it easy to see the water level. Alternatively, a piece of PVC or metal pipe might be attached to the top of the well; in this case, you will need to use some nonvisual method of finding water level. No matter what type of conduit is used, it is not important that it be straight. The critical measurement is the elevation of the water level above the ground, and the path it takes is not significant.

Measure the vertical distance between the ground surface and the water level in the conduit. Record the water level as the distance in feet or meters above the ground surface. Based on survey data, this may be referenced to a known datum, for example, mean sea level.

Equipment List

Conduit and tools and materials for attaching it to the wellhead

Tape measure

Keys and tools to open well and locked gates

Water-level indicator, if conduit is not transparent

Well construction diagram and boring log

Field notebook, pens, pencils, calculator

Air monitoring devices, as needed

Personal protective equipment, as needed

Pressure Measurement. To measure water level in a flowing well by pressure measurement, attach a pressure gauge to the top of the well casing (Fig. 4.21). Stop the flow of water by capping the well; ensure that there are no leaks. Because the well is capped off, pressure or head measurements made in this way are sometimes referred to as "shut-in pressures" or "shut-in heads."

Fig. 4.21 One way to measure the water level in a flowing well is to attach a pressure gauge to the top of the well casing.

Read water pressure in units of length (feet or meters). Record this value as height of the water level above the elevation of the pressure gauge. Note that the pressure gauge is not at ground level; measure and record its elevation relative to ground level.

If the gauge reads in units of pressure, such as psi (pounds per square inch), this pressure measurement can be converted easily to a head measurement. To do this, use the specific weight of water, 62.4 pounds per cubic foot (pcf). An example follows:

$$\text{Head above measuring point gauge} = \frac{\text{gauge reading}}{\text{specific weight of water}}$$

Be sure to account for differences in units. For example, if the pressure gauge reads 10 psi, the head can be calculated as follows:

$$\text{head} = \frac{10 \text{ psi}}{62.4 \text{ pcf}} = \frac{10 \text{ lb}}{\text{in.}^2} \times \frac{1 \text{ ft}^3}{62.4 \text{ lb}} \times \frac{12 \text{ in.}}{1 \text{ ft}} \times \frac{12 \text{ in.}}{1 \text{ ft}} = 23 \text{ ft}$$

Equipment List

   Well cap with fitting to attach pressure gauge

   Pressure gauge

   Wrenches to attach well cap

   Keys and tools to open well and locked gates

   Well construction diagram and boring log

   Field notebook, pens, pencils, calculator

   Air-monitoring devices, as needed

   Personal protective equipment, as needed

Measuring Water Level in Nonflowing Wells
In a nonflowing well, it usually is necessary to lower a probe into the well to find the water level. Probes range from very simple to complex in their construction.

Calibrated Lines. A weighted line is a simple means of finding depth to water. If chemical composition or contamination of the well water is of no concern, the line may be coated with chalk at its lower end. Ordinary classroom chalk works well for this purpose. Lower the line into the well until it has reached the water. Mark the place on the line where it touches the reference point of your choice. The top of casing (TOC) and the ground surface are two common choices; record which point is used, and use the same point every time. Measure the length of the line from that point to the top of the wetted portion, which is shown by the washed-off or wet chalk (Fig. 4.22). That measurement is the depth to water from the reference point.

When using this method, it is helpful to use a line that has been calibrated to the nearest foot or meter or less. Lower the line until the end is wetted and a whole number of feet or meters touches the measuring point. Then simply subtract the length of the wetted portion.

Using a measuring tape as a chalk line eliminates the need for calibrating the line. A steel or fiberglass tape with raised markings may be used, and the end may be chalked or the wetted portion may be measured, without chalking. If using a fiberglass tape, be aware that the tape will stretch somewhat as it hangs down in the well. The tape manufacturer should be able to tell how much stretch to expect; in some cases, this information is printed right on the tape or housing. Be aware that wet fiberglass tape can stick to the side of a well, making measurements difficult. Also, the weight on the end of the line will displace water, thus raising the water level. This is particularly significant in small-diameter wells.

Fig. 4.22 Using a chalk line to measure the depth to water.

To avoid using chalk, it is simpler, cleaner, and more efficient to use a "popper" at the end of the line to indicate water surface. Poppers are cheap and easy to construct; see Fig. 4.23 for an example. Attach the popper to the end of a calibrated line or measuring tape. Lower it until you hear the "pop" it makes when it reaches the water surface. It may be difficult to hear this noise if a pump is running, so this method may not be appropriate in all situations. Don't forget to add the length of the popper to the depth measurement.

Electric cable water-level indicators work on the principle that an electrical current will pass through water fairly easily. These indicators consist of a current source; two wires attached to the source at one end and weighted at the other; and an indicator of current, commonly a light, buzzer, or ammeter. The weighted ends of the wires are dropped down the well, and the current source is activated. When the wire ends touch the water surface, it completes the circuit, and the current indicator reflects that fact (Fig. 4.24).

An inexpensive indicator of this type may be built fairly easily by using light gauge two-strand wire with the ends exposed and a weight taped to one end. The other ends of the wire strands should be connected individually to the leads of a continuity tester, or multitester or ammeter wired to a battery. The disadvantage to this type of homemade instrument is that the line is not calibrated, which will make

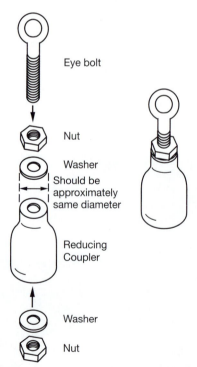

Fig. 4.23 A popper is inexpensive and easy to construct.

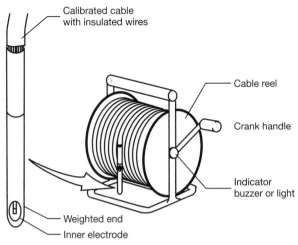

Fig. 4.24   An electric cable water-level indicator.

measurements inconvenient. But especially for shallow water levels, it will work adequately for studies not requiring a high degree of precision.

This type of indicator will not work if a free organic liquid (e.g., gasoline or oil) is floating on top of the water in the well, because organic liquids usually are not good conductors of electricity. The same may be true of water with a very low dissolved-salt content. This type of indicator may give a false reading if the wire ends touch water that has condensed on the inside of the well casing, above the true water surface. If the well is constructed with metal casing, and if the wires contact the casing, it may conduct the current and give a false reading. Finally, in using an electric cable water-level indicator, it is better to rely on the depth-to-water reading obtained while lowering the cable rather than that obtained while raising it. As the probe is raised, water may temporarily adhere to it, giving the impression that it is still submerged, when in fact, it may not be.

In using any of these methods, be certain when reading depth to water on a calibrated line that you are reading up from below the reference point, and not down from above it (Fig. 4.25). In addition, be aware that some measuring tapes are calibrated in tenths of feet rather than inches.

Acoustic Indicators.   Acoustic water-level probes direct sound waves toward the bottom of the well. The sound bounces off the water and returns to the surface, where a detector determines how long it took the sound to "make the trip." A long time interval indicates a deeper water level. These probes are more expensive than electric cable indicators, and they require periodic calibration.

Interface Probes.   Interface probes are particularly useful in ground water contamination studies. These probes detect the level at which there is an interface between water and oil, gasoline, or solvent. They may have varying designs.

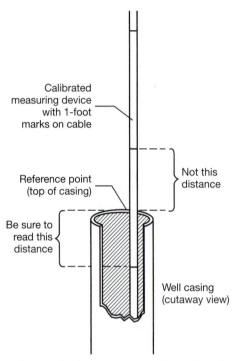

Fig. 4.25 When reading the depth to water on a calibrated line, be sure to read up from below the reference point, not down from above it.

Perhaps the least complex design is simply two electric cable indicators in one probe, with a tiny difference between the height of indicators on the probe. The interface is the level at which the upper indicator is situated within floating "product," or organic liquid, while the lower indicator is in water. At this level, no current will pass through the upper indicator, but current will pass through the lower (Fig. 4.26).

When using an interface probe, remember that the water level in the well will be depressed somewhat because of the presence of the overlying product. The corrected depth to water may be found by using the following formula.

Corrected Measurement of Depth to Water

$$= \text{Field Measurement of Depth to Water}$$
$$- \left(\text{Thickness of LNAPL Layer} \times \frac{\text{Density of LNAPL}}{\text{Density of Water}} \right)$$

Note that the fluid densities may be strongly temperature-dependent.

Float Recorders. When continuous measurements must be recorded, a float recorder may be used for convenience. Design of a float recorder for a well is the same as that of a float recorder for a stream gage. A float attached to a cable floats

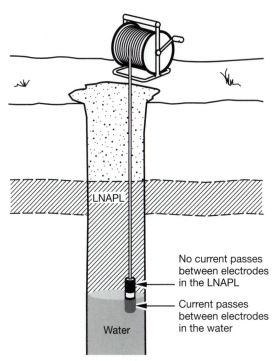

Fig. 4.26   Design of a simple interface probe.

at the water surface. The cable is wrapped around a pulley, with a counterweight at the other end. As water level rises and falls, the float and pulley adjust, and the change is recorded either digitally or by a chart-and-pen system. The major difference between this setup and that of a stream gage (see Chapter 3) is that the float, cable, and counterweight must fit within the diameter of the well, and so are smaller than those used for a stream gage.

Pressure Transducers. Pressure transducers measure a particular kind of water level. They do not measure depth to water in a well. Rather, these instruments measure height of the column of water situated above the probe (Fig. 4.27). This makes them useful for applications such as pumping tests, slug tests, or other situations in which water level changes with time. They are particularly useful in tests in which water level changes rapidly. Pumping and slug tests are described in Chapter 5, "Aquifer and Aquitard Testing."

The design principle of pressure transducers is that the pressure on the tip of the probe is related to the height of the water column covering the probe. An electrical current passes through the tip of the probe, and the current is proportional to the pressure. A digital readout device attached to the transducer cable translates the current into head measurements. These may read out directly or may be stored digitally (in "dataloggers") for later retrieval or downloading.

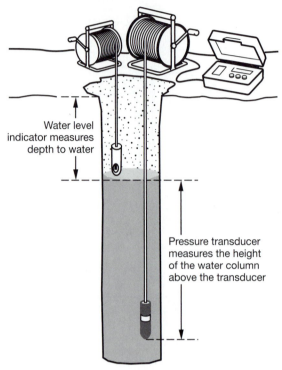

Fig. 4.27 Pressure transducers do not measure the depth to water. They measure the height of the water column above the transducer.

Because the pressure reading must be corrected for barometric pressure, the transducer cable carries a tiny air line that allows the device to compensate for air pressure. Some may require temperature corrections as well.

Equipment List

Tape measure or ruler

Keys and tools to open well and locked gates

Well construction diagram and boring log

Field notebook, calculator, pens, pencils,

Air-monitoring devices, as needed

Personal protective equipment, as needed

Decontamination equipment, as needed

Water-level indicator (one or more of these)
1. Calibrated line
    Tape measure (steel preferred)
    Carpenter's chalk *or* popper and materials to affix it to the tape measure

2. Electric cable water-level indicator
   Water-level indicator
   Spare batteries
3. Interface probe
   Interface probe
   Spare batteries
Other probes as needed (e.g., acoustic probe, pressure transducer)

Discharge

Discharge is the rate of flow of water out of a well, as measured in units of volume per time, for example, gallons per minute (gpm), cubic feet per second (cfs), or cubic meters per day (cmd). Discharge of a well is a fundamental quantity in hydrogeologic studies, because it tells how much water the well is producing. Several methods for measuring discharge are described in this section.

Measuring discharge involves extracting water from the well. This section describes how to measure discharge from a well. If the measurement is made as part of a pumping test (see Chapter 5, "Aquifer and Aquitard Testing"), release the pumped water far from the pumping well, to avoid recharging the aquifer during the test. If the discharged water might be contaminated, regulations may require that the pumpage be contained so it can be tested or hauled away for treatment by a licensed operator.

Measuring Discharge from Flowing Wells

Flowing wells discharge spontaneously, without pumping. To measure discharge from a flowing well, cap the well and direct the discharge through a hose or pipe. Then use the same methods for measuring discharge as described in what follows for nonflowing wells.

Measuring Discharge from Nonflowing Wells

Nonflowing wells have no spontaneous discharge. But if they are being pumped or bailed, several methods might be used to find the discharge rate.

Calibrated Bucket and Watch. The simplest method of determining discharge uses a calibrated bucket and a watch. An ordinary 5-gallon bucket may work well for this; it is most convenient to have two or three of them on hand. If precision is essential, the buckets should be carefully calibrated ahead of time. Direct the flow into the bucket and time how long it takes to fill. This method works only for relatively low flows. Five-gallon buckets work for flows of up to 25–30 gallons per minute (gpm), and 55-gallon drums work up to 200–300 gpm. Have as many buckets or drums on hand as the number of discharge measurements needed. In addition, if using a 55-gallon drum, be sure to have a means of emptying it. A 55-gallon drum full of water weighs well over 450 pounds, and is not easily tipped. Use smaller buckets to empty it. A simple siphon might drain most of the water from the drum, but this is likely to be a slow process.

Equipment List

Buckets or drums

Materials for emptying drums, as needed

Watch (with second hand) or timer

Hose, tubing, or pipe to direct flow into buckets (or away from site, if conducting a pumping test)

Materials and tools to attach hose or pipe to the wellhead

Tank for storage of discharged water, as needed

Keys and tools to open well and locked gates

Well construction diagram and boring log

Field notebook, pens, pencils, calculator

Air-monitoring devices, as needed

Personal protective equipment, as needed

Decontamination equipment, as needed

Flow Meters. Flow meters attached in the discharge line may give flow rate or total flow (Fig. 4.28). To find flow rate from a total flow meter, make two readings, 1 minute apart. The difference is the volume per minute. Of course, any convenient time interval may be used. Flow meters may have limited precision in various ranges; be sure to use one appropriate to the particular situation.

Equipment List

Flow meter

Materials and tools to attach the flow meter to the discharge line

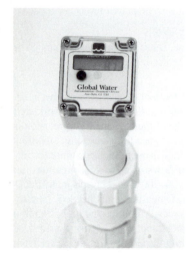

Fig. 4.28 A flow meter attached in the discharge line gives the flow rate or total flow. (Courtesy of Global Water.)

Watch (with second hand) or timer

Hose, tubing, or pipe to act as the discharge line (or to direct discharge away from site, if conducting a pumping test)

Materials and tools to attach hose or pipe to wellhead

Tank for storage of discharged water, as needed

Keys and tools to open well and locked gates

Well construction diagram and boring log

Field notebook, pens, pencils, calculator

Air-monitoring devices, as needed

Personal protective equipment, as needed

Decontamination equipment, as needed

Orifice Weirs. An orifice weir is a hole of precise dimensions through which discharged water is directed. Water pressure is measured and used to determine discharge rate.

Orifice weirs are used to measure very high flows. They operate on the principle that the pressure of water discharging from a given orifice will relate to the discharge. Discharge through a given cross-sectional area determines velocity of flow, and velocity in turn determines pressure of the water at the orifice.

To construct an orifice weir, begin with a discharge pipe through which flow is directed. The discharge pipe must be at least 6 feet or 2 meters long (longer if possible), oriented perfectly horizontally (use a level), and cut exactly squarely at the end (Fig. 4.29).

At the end of the discharge pipe, confine the flow to a known cross-sectional area by using an orifice plate. An orifice plate is a circular steel plate, 1/16-inch thick, fastened over the end of the discharge pipe. The orifice plate must have a cleanly cut perfectly circular hole in its center, through which the flow passes. The diameter of the hole should be less than 80% of the inside diameter of the discharge pipe.

For this method to work, there must be a means of measuring water pressure at the orifice. This is commonly done with a manometer (or piezometer) inserted into the discharge pipe exactly 24 inches "upstream" from the orifice plate. The manometer can be constructed by tapping the discharge pipe for 1/8-inch pipe, filing off any burrs inside the pipe, inserting a 1/8 pipe nipple so that it is flush with the inside of the pipe, and attaching a 4–5-foot-long piece of flexible tubing to the nipple. The tubing should be oriented vertically; at least the uppermost section of it should be clear so that the water level in the manometer is visible. Usually, the tubing is fastened to a meter stick or other calibrated scale to make measurement easier. More information on construction of an orifice weir may be found in Driscoll (1986).

To use an orifice weir, first clear the manometer of air bubbles by lowering it far enough so that water flows out of it. Then raise it and measure the height of the water level in the manometer above the center line of the discharge pipe.

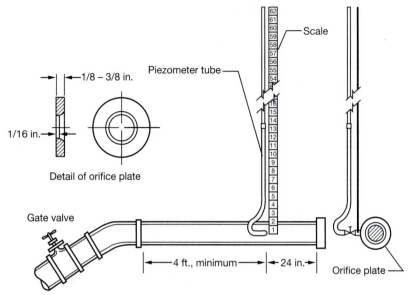

Fig. 4.29 Design of an orifice weir (after Driscoll, 1986).

Calculate the ratio of orifice diameter to pipe diameter, and read the value of coefficient *C* from Fig. 4.30. Then calculate the discharge using the following equation:

$$Q = 8.025CAh^{1/2}$$

where Q = discharge, gallons per minute
               C = coefficient from Fig. 4.30
               A = area of the orifice, square inches
               h = height of the water in the manometer above the center line of the discharge tube, inches.

Note that the units of measurement specified here must be used to facilitate the conversion factor built into the equation.

Equipment List

    Orifice weir
    Materials and tools to attach the orifice weir to the discharge line
    Discharge line
    Materials and tools to attach the discharge line to the wellhead
    Manometer and fittings to attach it to the discharge line

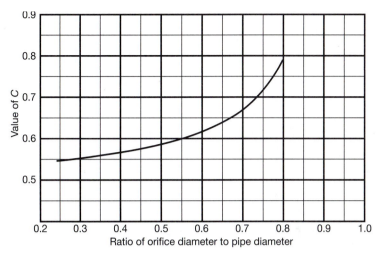

Fig. 4.30   Value of the coefficient C for using an orifice weir (after Driscoll, 1986).

Ruler or meter stick

Pole or support to hold the manometer vertical

Tank for storage of discharged water, as needed

Keys and tools to open well and locked gates

Well construction diagram and boring log

Field notebook, calculator, pens, pencils

Air-monitoring devices, as needed

Personal protective equipment, as needed

Decontamination equipment, as needed

Aquifer and Aquitard Testing

Chapter Overview

Hydrogeologic parameters of aquifers and aquitards are best determined *in situ*. Although hydraulic conductivity, transmissivity, and storativity may be estimated from charts of common values or from laboratory testing, field tests are much more likely to give results that more closely predict the aquifer or aquitard's actual behavior.

This chapter describes principles and procedures for planning and carrying out slug tests and pumping tests, and describes the rudiments of data analysis for these tests as well. Additionally it offers a few considerations on tracer tests.

Slug Tests

What Is a Slug Test?

The purpose of a slug test is to obtain a preliminary estimate of the hydraulic conductivity of aquifer or aquitard material *in situ*. In some situations, storativity may also be obtained from slug test data.

In a slug test, an instantaneous change in water level is caused in a well, and then the rate at which water level returns to the initial level is measured (Fig. 5.1). A faster return to initial head correlates with a higher value of hydraulic conductivity.

Slug tests are of two types: falling-head tests and rising-head tests. When a slug test is performed on a well, one of each type of test may be conducted in sequence. A falling-head test involves causing an instantaneous rise in water level, then observing the head as it falls back to the initial level. A rising-head test involves causing an instantaneous drop in water level, then observing the head as it rises back to the initial level.

Slug tests may be relatively inexpensive, require little equipment, are frequently rapid, and can be used to test the characteristics of either aquifers or aquitards. In addition, slug testing involves extracting little or no water from a well; this is particularly important if the water is contaminated and its disposal is regulated. The major disadvantage of slug testing is that it gives a measure of hydraulic conductivity only for the zone immediately adjacent to the well. As a result, measurements from slug tests cannot incorporate or account for the effects of large-scale heterogeneities or fractures which occur more than a short distance from the well intake. When fractures intersect or occur close to the well intake, however, the results of the slug test may be controlled by the fractures, and may not reveal the nature of the matrix permeability.

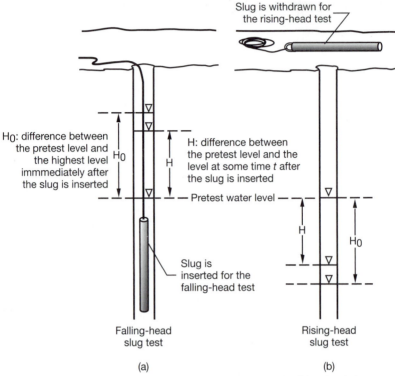

Fig. 5.1 Slug test geometry: (a) falling-head slug test and (b) rising-head slug test.

Slug test field and analytical procedures are described in numerous other sources, among them Freeze and Cherry (1979), Domenico and Schwartz (1998), Kruseman and de Ridder (1990), Dawson and Istok (1991), and Watson and Burnett (1995). The ASTM has developed several standard methods pertaining to slug tests. ASTM Standard Test Method D4044 gives the field procedures, while a half-dozen others describe analytical procedures.

Slug Test Procedures
This section covers slug test planning, field procedures, data processing, and data analysis.

Planning a Slug Test
The first step in a slug test is to assemble the necessary tools and instruments. An equipment list is given at the end of this section. As that list shows, in addition to the normal equipment for opening a well and measuring depth to water, a few special items are required. These are described in more detail here.

An instrument for measuring water level in the well is needed. More specifically, the instrument should be suited to measuring changes in water level. For materials of low to medium hydraulic conductivity, such as clays, silts, silty sands, and most rocks, any water-level indicator may work. For materials of high hydraulic conductivity, such as clean sands, gravels, and fractured rocks, using a pressure transducer in addition to a water-level indicator is recommended, and may be absolutely necessary, because measurements must be taken very rapidly. A data logger will simplify the job, because it can take literally hundreds of measurements per second and store them in digital form for later downloading and processing Lby computer. These instruments are described in Chapter 4, "Soil Water and Ground Water."

A means of instantaneously raising and lowering water in the well is needed. This "slug" gives the test its name. Possibilities are a solid slug, a bailer, a slug of water, and simple air pressure or vacuum.

A solid slug is a cylindrical object, metal or plastic, with a loop or hole at its end where a line can be attached. Slugs may be 2–3 feet long or longer. In higher-conductivity materials, longer slugs are better, because they cause a greater change in water level, which in turn takes longer to recover to the original level. Because the test is longer, readings are more manageable, as they need not be taken as rapidly. In low-hydraulic-conductivity materials, however, tests may take hours or days, so shorter slugs are more convenient. The advantages of using a slug are that, if clean, it does not introduce contaminants into the well, and that the rise and fall of water levels are virtually instantaneous. If a slug is to be homemade, the critical factors are that it should have a sturdy loop for attaching a line, it should be narrow enough to fit down the well and still admit a transducer cable or other water-level measuring device, and it must be heavy or dense enough to sink once it hits the water.

A bailer is similar to a slug except that it is hollow and can be filled with water. It also has a valve at the bottom into which water flows as the bailer is lowered into a well. (See Chapter 7, "Sediment and Rock Sampling and Drilling Methods," for a more complete description.) To cause a rise in water level, fill the bailer with water at the ground surface, and rapidly lower it into the well, below the water surface. Do not do this if the well is to be sampled soon after for chemical analysis. To cause a drop in water level, rapidly withdraw the bailer. Use care in this portion of the test, as the bailer may leak water out the bottom, which would cause the subsequently measured rising head data to be inaccurate.

A slug of clean water—a volume of water poured quickly down the well—may be introduced into the well to cause a rise in head. The advantage of this method is that it costs nothing. However, it has three drawbacks. First, because of the time it takes to pour the water, the change in water level is not instantaneous. Second, only a falling-head test can be performed this way, not a rising-head test. For a rising-head test, which is more accurate in certain situations, some means of withdrawing water still will be needed. And, third, using this method will alter water chemistry within the well, so this method should not be used if the well is to be sampled for chemical analysis.

Air pressure or suction may be used to conduct a "pneumatic" slug test. These tests have the advantage of not introducing any materials into the well at all, which minimizes the risk of contamination. In addition, the change in water level is virtually instantaneous, making it possible to use slug tests even in wells screened in aquifers of very high hydraulic conductivity. For more information, see Leap (1984) and the report by Orient and colleagues (1987).

Other requirements of a slug test deal more with the test well than the equipment. For a slug test to be successful, the column of water in the well must be long enough to accommodate the transducer (if one is to be used) as well as the slug. Information on length of the water column may be available from the boring log or well completion diagram, or from previous data-collection efforts. The diameter of the well must be large enough to accommodate the slug and the water-level measuring device simultaneously. The entire well screen should be submerged below water level in the well throughout the test. For reasons explained under the subsection "Interpret the Results," which follows, if the screen is not submerged through the whole test, the falling-head test will not be reliable. In this situation, only the rising-head test results should be considered credible. For most test conditions, it still will be necessary to insert a slug and allow the water level to fall to the original elevation before conducting the rising-head test; thus, data on the rate of fall of water may still be kept. However, this may not constitute a tenable slug test. Finally, for the slug test to be successful, inserting or withdrawing the slug must cause a significant change in water level. The amount of change will depend on the diameter of the well and the length and diameter of the slug. A short narrow-diameter slug may cause little change in a large-diameter well. For example, a 3-foot-long, 1-inch-diameter slug will cause only a 2.25-inch rise in water level in a 4-inch-diameter well. This small change in water level will not lead to precise measurements during the test; a larger or longer slug should be used.

Equipment List

Watch (with second hand), or stop watch, or timer

Tools and keys to gain access to the well

Slug (solid, bailer, or water) or device to apply air pressure or vacuum

Water-level indicator and spare batteries

Field notebook, pencils, and pens

Calculator

Pressure transducer, as needed, with spare batteries

Data logger with cables to attach to the transducer, as needed

Air-monitoring equipment, as needed

Personal protective equipment, as needed

Field Procedures

Prepare to Run the Test. Before beginning the test, measure the depth-to-water level in the well. The water level must be constant and steady for the test to be valid. Also measure depth to the bottom of the well to be sure that the well can accommodate the slug and water-level measuring device.

If using a transducer, lower it into the water to the appropriate depth. If using a slug or bailer, set the transducer deep enough in the well to avoid damage from the submerged slug or bailer. However, keep the transducer sufficiently far enough above the bottom of the well to avoid damaging it or clogging it with silt.

Check the water level, and wait for it to return to normal. Introducing the transducer and cable into the well will displace some water, and the water level must recover before the test begins. If the well is in a layer of low hydraulic conductivity, this may take a long time. If it takes an hour or more, then expect a very long slug test, perhaps lasting days or more. Under these conditions, it is wise to use a small slug. In addition, using a transducer may not be wise, unless it can be "dedicated" to the well (left undisturbed in the well for the duration of the test, which in such cases may be a period of days or weeks). Measurements will not need to be taken rapidly in a situation like this, and because the transducer measurements must be checked against depth-to-water measurements anyway, the transducer offers no real advantage.

Just before starting the test, measure and record the water level. If using a transducer, the proper level to record is the height of the column of water above the transducer. If using any other instrument, the proper level to record is depth-to-water level from some reference point. Be sure to note what instrument is used, as well as the reference point.

Conduct a Falling-Head Test

Introduce the Slug into the Well. Begin the test by introducing the slug into the water in the well. This should be virtually instantaneous, but it also is important to avoid creating splashes or turbulence in the well. If using a solid slug, it might be wise to lower the slug into the well before the test such that it is just above the water surface. Then, when the test begins, you can lower it quickly into the water. Be careful not to damage the transducer, if you are using one, by dropping the slug onto it.

If using a bailer, consider using the same procedure as for a solid slug. However, be careful that water does not leak out of the bailer before the test begins, causing a premature rise in water level.

If pouring a slug of water into the well, pour quickly. This is particularly important when the formation has high hydraulic conductivity, because water level will change rapidly.

No matter what type of slug is used, once it is introduced into the water, it should not be moved for the duration of the test.

Take Measurements of Water Level. Begin measuring the water level immediately after introducing the slug into the well. The first objective is to measure the highest water level. This level will occur within the first second or so after the slug is lowered, and the level will begin to drop immediately. Be prepared for this before lowering the slug. Record both the measurement and the time of measurement, to the nearest fraction of a second, if possible.

The next objective is to read and record the water level as it returns to normal, and also to record the time each water-level measurement is made. The sample data form (Fig. 5.2) shows readings that should be made in the field. Continue

Slug Test Field Data Collection Form

Job _____ Field Personnel _____

Well Number _____ Date _____ Time _____

Well Location _____

Well Data: Casing Diameter_____ Total Depth of Well_____

            Screen Diameter_____ Length of Screen _____

            Depth of Screened Interval: from _____ to_____

Measurement Point: ____ Top of Casing (TOC) ____Grade ____Other(_____)

Water Level Measurement Devices:

         For measuring depth to water_____

         For measuring water levels during test _____

Depth to Water   **Any trend/change in water**
Before test start _____ **level before test? Explain**_____

Slug: description of slug or injection/withdrawal method _____

         Slug volume (or pressure change)_____

Type of test (circle one):   **Falling head**   **Rising head**

Time	Elapsed Time	Water level (units: ____)	Time	Elapsed Time	Water level (units: ____)	Time	Elapsed Time	Water level (units: ____)

Fig. 5.2 A sample slug test data form.

making and recording measurements of the water level and time, according to a schedule. The schedule will depend on how fast the water level falls, which depends on the hydraulic conductivity (K) of the formation.

For materials of moderate hydraulic conductivity, measure the water level every second for the first 10 seconds, then every 10 seconds for the first minute, every minute for the first 5 minutes, and every 5 minutes until the water level returns to the static pretest level. In materials of lower hydraulic conductivity, begin on the same schedule, but after three constant readings in a row, skip to the next higher time interval. Because a well in materials of low hydraulic conductivity may take days or even months to recover, it may be appropriate to make readings at intervals of 1 to 6 hours or longer, even weeks. Do not disturb the slug or transducer during the test, as this will change the water level and compromise the results.

If the well must be capped while the water level recovers, be sure to use a vented cap so that air pressure does not build up in the well, potentially affecting water level.

In materials of high hydraulic conductivity, measurements must be taken very rapidly. As a result, using a transducer may be a virtual necessity. Using a data-logger in this situation simplifies the test considerably. But if neither of these is available, it still may be possible to take the rapid measurements that are required by using the following method. This method requires two people to take measurements, a china marker (grease pencil), and a water level indicator. An electric cable water-level indicator with an audible signal such as a beeper or buzzer works best in this situation. When the test begins, one person uses a stopwatch and calls out "Read" at 5-second intervals. The other person uses the water-level indicator to find the depth to water. This person should not actually determine the depth, but should simply lower the probe to the water surface, then use the china marker to mark the position on the cable. The timer should indicate intervals of 30 seconds, and the cable marker should make a different mark (perhaps thicker, or doubled, or of a different color) to indicate the 30-second reading. As long as water level keeps changing rapidly, continue taking the measurements at 5-second intervals. Later, change to 10- or 30-second intervals. When the test is over, marks on the cable should be reviewed and translated into actual depth measurements. This method is not easy, and it requires advance practice.

Decide When to End the Test.   How long should the test continue? If a rising-head test will be conducted after the falling-head test, then wait until the water level has completely recovered before starting the rising-head test. Meanwhile, continue recording the time and water level until the water level has recovered at least 80% of the way. Depending on which analytical method will be used to process the data, the values between about 60% and 80% of recovery may or may not be needed, but it will not hurt to have recorded them. If the well must be capped while the water level recovers, be sure to use a vented cap so that pressure conditions in the well are atmospheric.

An example of a completed slug test data form is given in Fig. 5.3.

Slug Test Field Data Collection Form

Job _Bigco, Inc._____ **Field Personnel** _LLS & SAB_____

Well Number _13D_____ **Date** _4/11/97_____ **Time** _9:30 a.m.____

Well Location __25 ft north of NW corner of machine shop_____

Well Data: Casing Diameter_2" ID____Total Depth of Well_72 ft_____

Screen Diameter_2" ID____Length of Screen _10 ft_____

Depth of Screened Interval: from ____62 ft____ **to**_____72 ft_____

Measurement Point: ____ Top of Casing (TOC) ____Grade ____Other(_____)

Water Level Measurement Devices:

For measuring depth to water__electric cable meter S400A_____

For measuring water levels during test _____pressure transducer &_____ display IS89

Depth to Water   **Any trend/change in water**
Before test start ____52.39 ft_____   **level before test? Explain**__No_____

Slug: description of slug or injection/withdrawal method Steel slug 4' long, 1" diam.

Slug volume (or pressure change)___0.0218 ft^3_____

Type of test (circle one):   ⟨Falling head⟩   Rising head

Time	Elapsed Time (sec)	Water level (units: ft)	Time	Elapsed Time (sec)	Water level (units: ft)	Time	Elapsed Time	Water level (units: ___)
10:10 am	Before test	9.25		30	9.49			
10:11	0	10.65		40	9.40			
	1	10.16		50	9.34			
	2	10.22		60	9.32			
	3	10.08		70	9.32			
	4	10.06		80	9.31			
	5	10.03		90	9.31			
	6	9.97	Ended test					
	7	9.95	10:12:30	a.m.				
	8	9.94						
	9	9.89						
	10	9.86						
	20	9.63						

Fig. 5.3 A completed slug test data form.

Conduct a Rising-Head Test

Prepare to Start the Rising-Head Test.   Wait until the water level has fully recovered from the falling-head test. Just before starting the test, measure and record the water level and time.

Begin the Rising-Head Test.   Begin the test by instantaneously removing the slug from the well water. If using a solid slug, withdraw it quickly. Depending on the depth of the well, either pull it to the surface or simply leave it suspended in the well. In pulling it to the surface, there is a risk of bumping the transducer cable or tangling other downhole probes. If using a bailer, it might leak and empty itself if left suspended in the well; this will cause the water level in the well to rise too quickly and yield an erroneous test result. Weigh this risk against the risk of disturbing the transducer or other downhole probes.

Begin measuring and recording the water level immediately after withdrawing the slug. The first object is to measure the lowest water level, which will occur within the first second after the slug is removed. Record both the measurement and the time of measurement, to the nearest fraction of a second, if possible. The next objective is to read and record the water level as it returns to normal, along with time of the measurements. Continue making measurements of water level and time, according to the appropriate schedule, as for the falling-head test.

If the well must be capped before the test is complete, use a vented cap. In the rising-head test, this is particularly important if hydraulic conductivity in the tested formation is high and if the water level in the well is above the screened interval. When this is the case, the water level can rise quickly enough to compress the air in the casing, which can blow the cap off with explosive force.

End the Test.   Stop the test after 60% to 80% of the initial head has been recovered.

Processing the Data

Before the data can be analyzed, they must be processed to yield the appropriate values for analysis. The following steps for processing apply to both falling- and rising-head tests, but the data from each test should be processed separately.

First, find H_0. H_0 is the maximum level above (or below) the pretest level to which the water level rises (or falls) immediately after the slug is inserted (or removed). Find H_0 by subtracting the depth to water *before* the test from the depth to water immediately *after* the slug is inserted or removed.

The choice of H_0 is somewhat arbitrary, but is critical to the success of the method. As the slug drops into the water, a great deal of splashing and turbulence tends to occur. The result is that water levels in the first second or two may vary widely. When anything but a transducer and datalogger are used to make water-level measurements, the choice may be simple, because usually by the time the first measurement is made, the water has stopped splashing.

However, the choice of H_0 is particularly confusing when a transducer and datalogger are used, because the datalogger may record a hundred or more readings in the first second of the test. How can one choose an H_0 value from such an alarmingly large and widely diverging set of points? The answer is to use common sense. Remember that the maximum change in water level (H_0) cannot exceed what is physically possible, given a slug of a certain volume in a well of a certain diameter—even though the transducer readings might indicate so. It is a good idea to calculate the highest water level that might possibly be reached, and not to select an H_0 value which exceeds this. Knowing the volume of the slug, calculate the maximum change it could possibly have caused in the water level by using the following formula:

$$\text{Maximum possible change in water level} = \frac{\text{volume of slug}}{\pi(\text{well radius})^2}$$

Don't forget to use the *radius* of the well, not its diameter, and to convert all measurements to consistent units.

Once the maximum possible change in water level is known, eliminate from consideration all data points that are greater than that value. Choose the recorded value that is closest to, but does not exceed, the maximum possible value. Why not just use the calculated value? The calculated value is a theoretical value that might not ever be achieved, because water instantaneously moves into or out of the well screen as the slug is inserted or withdrawn.

Once the value of H_0 has been chosen, make a table or use a computer spreadsheet program to record and process the water level and time data. If a depth-to-water indicator was used to measure water level, the columns should be as given in Fig. 5.4(a).

If a transducer was used to measure water level, the transducer measurements will give the height of the water column above the transducer. In this case, H_0 is the highest value recorded immediately after the slug is inserted (or the lowest value recorded immediately after it is removed) *minus* the height before the slug was inserted or withdrawn. The data table will look slightly different, too, as is shown in Fig. 5.4(b).

Analyzing the Data

The analysis given here follows the Hvorslev (1951) method. This method is applicable in a wide variety of field settings, and graphing and calculations are relatively simple. The Hvorslev method only gives hydraulic conductivity of the formation, not transmissivity or storativity. Other methods of interpreting the data are available; see Freeze and Cherry (1979), ASTM Standards 4044 and 4104, Dawson and Istok (1991), Fetter (1994), or other texts for details.

Graph the Data. Plot elapsed time versus head ratio (H/H_0) on a one-cycle semilogarithmic graph with the head ratio on the logarithmic axis and time on the arithmetic axis. Ignoring the first few data points, which are likely to be spurious, draw a straight line through the points (Fig. 5.5).

Perform the Necessary Calculations. Read the value of T_0, the basic time lag, from the graph. T_0 is the time at which the head ratio equals 0.37. To find it, find 0.37 on the scale of the head ratio, draw a horizontal line from this point to the line

Slug Test Data Table for Use With Depth to Water Readings

(1) Time after test began (units: _____)	(2) Depth to water (units: _____)	(3) Depth to water minus Depth to water before the test began	(4) Head Ratio (H/H_0)*
Before test began		0	
0		$= H_0$	
1			
2			
3			
etc.			

*H/H_0 is the value in Column 3 divided by H_0. H_0 may be chosen as the highest value in Column 3 and usually occurs at or near time = 0. (See text for explanation of how to choose H_0)

Note that in a falling head test, the values in Column 3 will be negative. Disregard the negative signs, and treat the numbers as if they are positive.

(a)

Slug Test Data Table for use With Pressure Transducer Readings

(1) Time after test started (units: _____)	(2) Height of water above transducer (transducer reading) (units: _____)	(3) Transducer reading minus transducer reading before test started (units: _____)	(4) Head Ratio (H/H_0)*
Before test began		0	
0	$= H_0$		
0.10			
0.15			
0.20			
0.25			
0.30			
etc.			

*H/H_0 is the value in Column 3 divided by H_0. H_0 may be chosen as the highest value in Column 2 and usually occurs at or near time = 0. (See text for explanation of how to choose H_0)

Some data loggers allow the user to select an option which causes the data logger to record the difference between water level before the test and water level during the test. In other words, the data logger calculates and stores the values in Column 3 of the table above. If this option is used, Column 2 may be deleted.

(b)

Fig. 5.4   (a) If depth-to-water measurements were taken, the columns should appear as given here. (b) If a pressure transducer was used to measure the height of water above the transducer during the test, the columns should appear as given here.

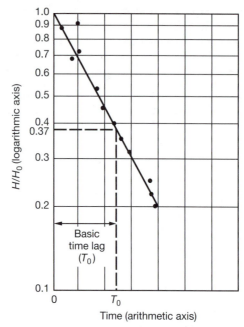

Fig. 5.5   Plot of head ratio (H/H_0) vs. time for data collected during a slug test (Hvorslev method).

connecting the data points, and then drop a vertical line down to the time axis. Read the value; this is T_0.

Calculate K.   Calculate hydraulic conductivity (K) of the formation using the following formula. This formula applies only if the length of the well or piezometer casing is more than 8 times the radius of the screen (or $L/R > 8$):

$$K = \frac{r^2 \ln(L/R)}{2LT_0}$$

where r = radius of the well or piezometer casing
   L = length of the saturated portion of the screen or filter pack (see what follows)
   R = radius of the screen or screen plus filter pack (see what follows)
   T_0 = basic time lag, which is read from the graph, as described earlier
   K = hydraulic conductivity of the formation

If the well or piezometer taps an aquitard, then R is the radius of the screen plus the filter pack (gravel pack), and L is the length of the filter pack (gravel pack).

Interpret the Results.   Check the results with a table of common values of hydraulic conductivity for the rock or sediment type. One such listing is given in Chapter 6, "Describing Sediments and Rocks".

Check the results of the falling- and rising-head tests against each other. (Both types of test should be performed whenever possible.) They should give results that are fairly close to each other, so one can be used to check the other. However, occasionally the two tests will not agree. This is most likely to occur in a well installed in an unconfined aquifer, with a screen that extends above the water table. During a falling-head test in such a well, water might leak out of the screen above the water table into the unsaturated zone. The test result will then give an erroneously high value for K. If this is suspected, the rising-head test should be considered more reliable.

In both the falling- and rising-head tests, the line may be curved at either end. These portions should be ignored during curve fitting.

Finally, recall that the volume of water entering or leaving the formation during the test is very small. If results seem too high or too low, it may be that only the sand or gravel pack surrounding the screen actually was tested, and not the formation itself. To determine if this is likely, use information from the boring log and well construction diagram, along with geological judgment and common sense.

Pumping Tests

What Is a Pumping Test?

In a pumping test, water is pumped out of a well at a known rate over a period of several hours or days. As pumping is going on, water level is monitored in one or more observation wells some distance from the pumping well, and in the pumping well, too. There is a relationship between the test parameters (water levels during the test, pumping rate, time since pumping began, and distance between pumping and observation wells) and the aquifer parameters (transmissivity, T, and storativity, S). For best results, a pumping test should be followed by a recovery test, in which water level is monitored after the pump has been shut off. Data recorded during both phases of the test are used to calculate values of T and S.

Pumping tests have both advantages and disadvantages when compared with slug tests. Because pumping tests involve extracting more water from the aquifer than do slug tests, they do a better job of estimating aquifer characteristics close to and at some distance from the well. Pumping tests can measure characteristics of large-scale heterogeneities and anisotropy. They give a more realistic estimate of how the aquifer and any confining layer actually respond to pumping. However, pumping tests take longer, require the use of a pump, may require the presence of an observation well in addition to the pumping (test) well, and are therefore considerably more expensive than slug tests. In addition, the problem of disposing of the pumped water must be addressed. Finally, pumping tests are less well-suited than slug tests for estimating the characteristics of aquitards.

This section describes how to plan a pumping test, carry it out, and analyze the results. Pumping test field and analytical procedures are described in numerous other sources, among them Freeze and Cherry (1979), Kruseman and de Ridder (1990), Dawson and Istok (1991), Fetter (1994), Watson and Burnett (1995), and Domenico and Schwartz (1998). The ASTM has developed several standard methods pertaining to pumping tests. ASTM Standard Test Method D4050 gives the field procedures, while a few others describe analytical procedures.

Planning a Pumping Test

Planning a pumping test begins with studying the geology of the site. Then, an analytical method is considered, an equipment list is put together, and the pumping and observation wells are designed and installed. Finally, the test parameters must be chosen, including pumping rate and length of the test. The steps are described in more detail in what follows.

Study the Geology

An understanding of subsurface geology is essential to planning a successful test. Answering the following questions will provide a good beginning. When the test is to be performed in an unexplored area, answers to these questions may not be known, in which case informed guesses should be made. These guesses should be based on the study of regional geology, by utilizing existing geologic maps and reports, field observations of outcrop patterns, and topography.

What aquifers, partial aquifers, and aquitards are present, and what are their depths, orientations, and thicknesses?

What is the degree and nature of heterogeneity of the geologic materials and/or formations in the area?

What is the degree and nature of anisotropy of the geologic materials and/or formations in the area?

What are the static water levels in the area? Consider water table and potentiometric surfaces. Where are the recharge areas for the aquifers? What is the general direction of ground water flow in each layer or formation?

What surface-water bodies exist in the area? How are they likely to interact with ground water?

Into what range are values of hydraulic conductivity (K) likely to fall? Consider aquifers as well as aquitards.

To what extent will fracturing influence the test? Is fracture permeability the primary source of water? How important is it?

What hydrologic boundaries are present? Consider both contributing boundaries (e.g., surface-water bodies, facies changes to higher-K formations, fracture permeability) and noncontributing boundaries (e.g., pinched-out formations, buried valley walls, facies changes to lower-K formations, or sealed fault zones).

What other wells are present that might influence or be affected by the test? Nonpumping wells should be considered as possible observation wells. Pumping wells should be considered as possible interferences, and their pumping rates and times should be monitored. Likewise, pumping the test well might cause a drop in water level in a private well; potential effects should be estimated, and the owners should be consulted.

Consider the Analytical Method

Consider the methods that might be used to analyze the data. Choice of analytical method determines the number and placement of wells, frequency of measurement, and pumping pattern. A variety of analytical methods are described in Walton (1987), Kruseman and de Ridder (1990), and Dawson and Istok (1991). Four methods are described here: the Theis, Jacob time-drawdown, distance-drawdown, and step-drawdown methods. Considerations about these and a few other tests are described in what follows.

Methods for Ideal Aquifers.   The Theis, Jacob time-drawdown, distance-drawdown, and step-drawdown methods apply when the formation being tested fits ideal assumptions. These assumptions are that the aquifer is isotropic and homogeneous, horizontal, fully confined, of constant thickness, and infinite in lateral extent; that it has no hydrologic boundaries and no interference from other pumping wells; that the well intake extends through the full saturated thickness of the aquifer; and that flow in the aquifer is laminar and horizontal.

When a pumping test is performed in nonideal conditions, the test's results do not conform to the ideal curves. The ways in which they deviate should guide the choice of analytical method. As a result, it helps to plot the data as for ideal conditions, and then to evaluate the nature of the deviation from the ideal.

Theis Method.   The Theis method involves plotting a curve of time versus drawdown data and matching the curve to type curve of the Well Function. Advantages of the Theis method are that the method is simple and well-known, and that deviations from the ideal are easily seen on the Theis plot. One disadvantage is that at least one observation (nonpumping) well is required if a value of S is to be calculated. Another is that the method involves curve matching, which invites interpretive error, although several computer software packages are available to match the curves numerically and obtain an optimal solution.

Jacob Time-Drawdown Method.   The Jacob time-drawdown method involves plotting data on a graph of time versus drawdown and fitting a straight line to the points. This method assumes the same ideal conditions as does the Theis method. In addition, it applies only to data for which the Jacob approximation is valid, that is, for which the parameter u, as defined in what follows, is less than or equal to about 0.01 (Lee and Fetter, 1994). This condition generally occurs when pumping times are long.

Advantages of the Jacob time-drawdown method are that the graphical technique is simpler than that of the Theis method, and that the effects of hydrologic boundaries may show clearly on the graph. In addition, the Jacob time-drawdown method may be used to interpret data from the pumping well, eliminating the need for an observation well. However, if no observation well is used, only transmissivity, and not storativity, can be determined. As with the Theis method, a disadvantage of the Jacob time-drawdown method is that it involves curve fitting, which invites interpretive error, although computer software for curve-matching may be used to obtain a statistically optimal solution.

Distance-Drawdown Method. The distance-drawdown method involves plotting drawdown versus distance from the pumping well. To use this method, at least three observation wells are needed, at specific intervals of distance from the pumping well.

An advantage of the distance-drawdown method is that only a few water-level measurements must be made. The major disadvantage is the need for several observation wells. In addition, this method involves curve fitting, which invites interpretive error, although as with the methods already described, some computer software packages may be used to minimize this error.

Step-Drawdown Test. The step-drawdown test involves pumping at several successively higher rates and observing the effects on drawdown. It is used specifically to determine well yield and optimum pumping rate for a particular well. Often, a step-drawdown test is conducted as a trial test to select an appropriate pumping rate for a subsequent pumping test. Calculating transmissivity and storativity from a step-drawdown test may be difficult or impossible. More information is included in the section titled "Determine the Pumping Rate," which follows.

Methods for NonIdeal Aquifers. When data from a pumping test are plotted using standard methods, and are found not to match idealized curves, it is an indication that one or more of the assumptions of idealized methods do not apply. This most commonly occurs because the aquifer is semiconfined or unconfined, the aquifer is not laterally extensive, the well does not fully penetrate the aquifer, or there is interference during the test by another pumping well.

Methods exist that offer ways to estimate the transmissivity and storativity of an aquifer even when ideal conditions are not met. Methods have been developed for semiconfined aquifers, with and without storage in the confining layer, as well as for unconfined aquifers. Some methods attempt to account for the effects of hydrologic boundaries or interference from other pumping wells. These methods should be considered when it becomes obvious that the data do not fit the type curve. Further information is available in Freeze and Cherry (1979), Walton (1987), Kruseman and de Ridder (1990), Dawson and Istok (1991), Fetter (1994), and Domenico and Schwartz (1998).

Determine What Is Needed To Do the Test

No matter what analytical method will be used, in order to conduct any pumping test, a pumping well must be available. For most analytical methods, and especially if storativity is to be estimated, there also must be at least one observation well. A means of measuring water level in all wells is needed. Although it might initially seem like a good plan to use the same water-level indicator for both the pumping and observation wells, in practice this may be impossible because of the physical distance between them and the rapid measurements needed early in the test. A means of disposing of the pumped water is needed; in some situations, regulations may specify that it be contained for later disposal by an approved method. A means of measuring discharge rate from the pumped well is needed. And, finally, a timepiece or stopwatch, and a notebook and/or recording forms are needed.

Design the Pumping Well

If no well exists in the study area, design and install one based on the considerations discussed in what follows. If an existing well is to be used as the pumping well, determine if it is suitable for the pumping test. In many cases, the construction details and location of an exisiting well make it inappropriate for the test, and another well must be designed and installed. If no information is available on the existing well, the well should not be used. Installing a new well provides the hydrogeologist with good subsurface geological data. In addition, it has the benefit of allowing the hydrogeologist to design a well that will provide good-quality pumping test data.

Diameter. Consider cost, the diameter of the pump, the discharge rate necessary to stress the aquifer and affect water levels in the observation wells, and what the well will be used for after the tests.

Depth and Placement of Intake. Design the well such that the screened or open portion is open only to the aquifer that is to be tested. Ideally, it should fully penetrate the aquifer.

Location. Frequently, choice of location is limited. But within the constraints of the site, try to choose a location that maximizes easy access, minimizes interference from other pumping wells, and optimizes data-analysis options.

Post-test Use of Well. If a well is to be abandoned after the test, it may be constructed of reusable materials. But if it is to be used for water supply or monitoring, the design may depend more on the final use than on its usefulness for the test. In any case, wells should be designed as described in Chapter 8, "Well Design and Installation."

Design the Observation Wells

If existing wells are to be used as observation wells, their construction details and locations must be reviewed. If new wells are to be installed for the purpose of the test, several factors must be considered, including well design, the number of wells, and locations.

Observation Well Design. Diameter and construction depend on what the well is to be used for after the test. If it is to be used only for the test, and will be abandoned thereafter, it may be constructed of reusable materials and may be of relatively small diameter. The diameter only need be big enough to accommodate a water-level indicator, so a 1- or 2-in. diameter might suffice.

Depth of the observation well(s) should be determined by geology: The intake portion of the well must be open to the same aquifer as is the pumping well. However, the intake need not fully penetrate the aquifer, assuming ideal conditions are met (horizontal, laminar flow). If conditions are not ideal, the observation wells should be open to the same portion of the aquifer as is the pumping well. Other design specifications should be considered as described in Chapter 8, "Well Design and Installation."

Number of Observation Wells. To characterize fully the hydraulic characteristics of a formation, several observation wells open to the aquifer are needed. In the case of a semiconfined aquifer, observation wells also should be installed in the overlying aquifer. However, cost is a major limiting factor in determining the number of observation wells. In some cases, the value of information gained from installing a well does not justify the expense.

No Observation Wells. A pumping test can be run with no observation wells. However, it will yield a limited amount of information. With no observation well, water levels must be measured in the pumping well, and turbulence within the well will affect quality of the data. Because of this, the recovery test becomes even more important. Neither storativity nor radius of pumping influence can be determined from a test with no observation wells. If the budget for the test is so limited that no observation wells can be installed, it may be more useful or cost-effective to abandon plans for a pumping test, install a few small-diameter wells, and use slug tests to estimate hydraulic conductivity.

One Observation Well. Use of one observation well may yield acceptable results in ideal conditions. However, if aquifer conditions are heterogeneous, a test with one observation well will be limited in its value, because conditions near the well may not represent conditions elsewhere in the aquifer. In addition, the distance-drawdown method cannot be used to interpret data from such a test, because three or more observation wells are needed to perform that analysis.

Two or More Observation Wells. Two or more wells should be used if conditions are laterally heterogenous and an assessment of the degree of heterogeneity is required. In this case, the wells ideally should be situated such that they are the same distance from the pumping well, but in positions where they will reveal the different conductivities (Fig. 5.6).

If conditions are laterally homogeneous, wells should be situated at different distances from the pumping well. For maximum benefit, their distances from the pumping well should be in different orders of magnitude (e.g., 10, 100, 1000 meters). The distance-drawdown method then will provide more precise and accurate information. A minimum of three wells is required for this analysis. Obviously, a limit exists beyond which a well will not provide useful information because it will not be affected by pumping. This "radius of influence" can be estimated before the test by using the equations for the distance-drawdown method.

Locations of Observation Wells. The wells must be close enough to the pumping well to register easily measured changes in water level within the first few hours of the test. (If the water level drops only a tiny amount throughout the entire duration of the test, then measurements are not likely to be precise enough to yield good results.) Observation wells must be appropriately placed to generate the curved portion on a Theis plot. The farther they are from the pumping well, the greater the portion of the aquifer that they test, but the longer in duration—and more expensive—the test must be. Of course, the observation wells must fall

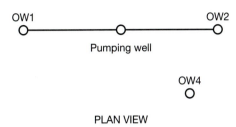

PLAN VIEW

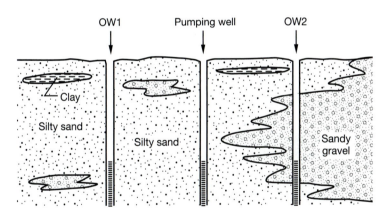

CROSS-SECTION

Fig. 5.6 In heterogeneous conditions, an assessment of the degree of heterogeneity may be made by using two or more observation wells, equidistant from the pumping well.

within the radius of influence of pumping. In addition, location of the observation well often is constrained by practical considerations such as property boundaries and drilling rig accessibility.

If multiple observation wells are used, they should conform to the spacing considerations described under the earlier "Number of Observation Wells."

With these competing demands, how can optimum distance of observation wells be determined? For some tests, it may be possible to estimate the optimum distance for a single-observation well. For this estimate to be most successful, the hydraulic conductivity of the aquifer must be already fairly well-established, and the pumping rate should be known. Optimum pumping rate should be established by performing a step-drawdown test, described later in this chapter.

If hydraulic conductivity and the pumping rate are known, to estimate the optimum distance to an observation well, follow these steps.

1. Multiply the hydraulic conductivity of the aquifer by the aquifer thickness (if the well intake fully penetrates the aquifer) or by the intake length (if it does not). This yields an estimate of transmissivity (T).

2. Estimate the order of magnitude of the storativity (S). Aquifer storativity ranges from about 10^{-7} to 10^{-1}. If the aquifer is perfectly confined, then as a preliminary estimate, use $S = 10^{-5}$. (This estimate will help during the design phase. But remember, one of the reasons for doing the pumping test is to determine the actual value of S, so the estimate is just a preliminary approximation, and very well may be proven to be inaccurate after the test.) If the aquifer is "leaky" (semiconfined), use $S = 10^{-3}$. If it is unconfined, use $S = 10^{-1}$ (or estimate a range of values for specific yield, S_y, from a table of common values such as Table 6.2, and use $S = S_y$).

3. Assume that drawdown ($h_0 - h$) in the observation wells should be at least 1 to 3 meters after about 12 hours of pumping. If it were any less than this, it would be difficult to measure the small changes in water level. This time period is the variable t in the equation in Step 5, which follows.

4. Now, use the Theis equation and the values estimated in Steps 1 to 3 to calculate values of the well function, $W(u)$, for the various pumping rates:

$$W(u) = \frac{4\pi(h_0 - h)T}{Q}$$

5. Use Appendix 3 to find corresponding values of u for the $W(u)$ values calculated in Step 4.

6. Use the following equation to calculate distances to the observation wells (r) for the values of u found in Step 5.

$$r^2 = \frac{4Tut}{S}$$

7. Now compare these r values with the field situation. What is practical, given the constraints of the site? Property lines, obstructions, buildings, and locations of existing wells all will influence the final decision.

Sample Calculation of the Optimum Distance to the Observation Well. A 24-hour-long pumping test is to be conducted in a well that is screened through the full thickness of a 20-m thick fully confined aquifer. The aquifer is composed of well-sorted medium sand, which from previous studies has been shown to have a hydraulic conductivity of 10^{-2} cm/sec. Water is to be pumped from the well by a submersible pump at the rate of 100 gpm. Because the aquifer is fully confined, a preliminary estimate of storativity is 10^{-5}. The results of the test, of course, will yield a more accurate estimation. A drawdown of 2 meters after 12 hours of pumping is desired. What is the optimum distance from the pumping well to the observation well?

To solve this problem, first, the value of transmissivity (T) is calculated, where b is the saturated thickness of the aquifer:

$$T = Kb = 10^{-2}\frac{\text{cm}}{\text{sec}} \times 20\text{m} \times 100\frac{\text{cm}}{\text{m}} = 20\frac{\text{cm}^2}{\text{sec}}$$

Then the value of the well function $W(u)$ is calculated:

$$W(u) = \frac{4\pi(h_0 - h)T}{Q}$$

$$= \frac{(4\pi)(2\text{ m})\left(\dfrac{20\text{ cm}^2}{\text{sec}}\right)\left(\dfrac{60\text{ sec}}{\text{min}}\right)\left(\dfrac{264\text{ gal}}{\text{m}^3}\right)\left(\dfrac{\text{m}^2}{100^2\text{ cm}^2}\right)}{100\,\dfrac{\text{gal}}{\text{min}}}$$

$$= 8$$

From Appendix 3, when $W(u) = 8.0$, the value of u is about 2×10^{-4}. Substituting this value into the equation for the radius,

$$r^2 = \frac{4Tut}{S}$$

$$= \frac{(4)\left(\dfrac{20\text{ cm}^2}{\text{sec}}\right)(2 \times 10^{-4})\left(12\text{ hr} \times \dfrac{3600\text{ sec}}{\text{hr}}\right)\left(\dfrac{1\text{ m}^2}{100^2\text{ cm}^2}\right)}{10^{-5}} = 6912\text{ m}^2$$

Therefore, $r = 83$ meters.

Ideally, then, the observation well should be located about 80 m from the pumping well. In practicality, however, the site boundaries, the lateral extent of the aquifer, and access for the drilling rig all must be considered before the location of the observation well can be finalized.

Determine the Length of the Test

For a confined aquifer, the typical duration of a test would include a pumping phase 12–24 hours long, and a recovery phase which runs as long as the pumping phase. For an unconfined aquifer, a typical test may run 3–7 days in the pumping phase and the same duration in the recovery phase. Longer durations might be needed, e.g., if the test is intended to reveal the effects of a distant hydrogeologic boundary. In any case, each phase of the test ideally should be continued until equilibrium is reached.

In planning the test, consider the data-interpretation method. Note that when drawdown values are plotted versus time on a logarithmic axis, points plotted after about 8 hours into the test are very close together (Fig. 5.7). Thus, if a test must be terminated early, it makes more sense to do so at the end of a workshift of 8 hours than at the end of two workshifts, or 16 hours. If terminated at the end of 16 hours, the final 8 hours of the test very likely will have yielded little additional useful information.

Consider the personnel who are to be onsite during the test. The greatest number of measurements are needed at the beginning of the pumping phase and again at the beginning of the recovery phase, so the greatest number of workers will be needed at these times. They will be needed to make simultaneous measurements at several different wells.

Finally, consider that longer tests give more information about hydrogeologic boundaries and secondary permeability than do shorter tests. Thus, the length of the test must take into account the objectives of the test.

Choose a Pump

The pump used for the pumping test must meet several criteria. Its intake must fit in the well, so diameter is critical. Consider what else will be downhole during the test, including the water-level indicator.

The equipment must provide the needed "lift." How far down is the water level? How far below that will the pump intake be positioned?

The pump also must deliver the required pumping rate (which is discussed in what follows). Pumps generally are designed with a specific purpose in mind, for example, sampling, purging, well development, or water supply. Their capacity depends on their design, the lift, the head of water above the intake, and the diameter, among other things. The pump's manufacturer can supply the needed information on pump capacity. Keep in mind that no vacuum pump will be able to lift water more than about 25 feet.

Ideally, the pump should deliver discharge at a steady rate. As a result, suction pumps, positive-displacement pumps, and submersible pumps may be suitable, but bladder pumps may not be. Bladder pumps operate by alternately filling, then discharging water from a bladder. The time between the fill and discharge cycles may be as long as several minutes. As a result, the discharge is not steady. Data from the recovery phase of the test will be more reliable than data from the pumping phase in such a situation. See Chapter 9 for more information on pumps.

The power source for the pump must be considered. Some pumps run off of a car or truck battery, but this is not likely to be a reliable source for a long pumping test. Others must be plugged into a household-voltage circuit. Is a power supply reliable and readily available? Will it remain so throughout the duration of the test? If not, a generator may be used; be sure to have plenty of fuel on hand to power the generator. Some pumps are gasoline- or diesel-fueled. Ensure that supplies are ample.

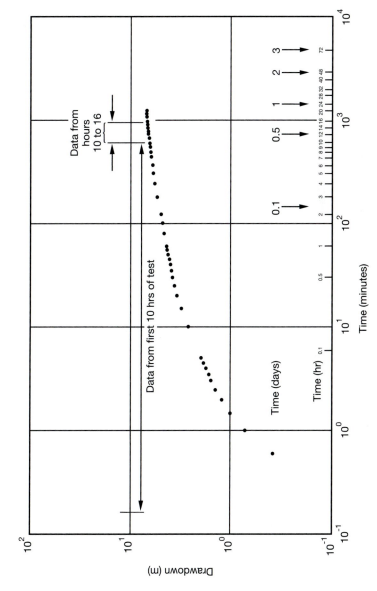

Fig. 5.7 When drawdown values are plotted versus time on a logarithmic axis, little information is gained between 8 and 16 hours after pumping began.

In some cases, and if a low level of sophistication and precision is acceptable, a pump may not be required at all. In these cases, the well must be able to be bailed by hand quickly enough to yield the required discharge. This is most practical if the wells are shallow and if the aquifer is of low conductivity. In deep wells, the time required to haul the bailer out of the well, empty it, and then lower it again is too long. If bailing is used, each time water-level measurements are made, they should be made at the same point in the bailing cycle. Bailing must be done at a constant rate throughout the test. It helps to have several field personnel performing this task, so that some may rest while another bails. Volume of water and time must be measured constantly throughout this type of test, as the bailer may collect different volumes of water each time. Obviously, bailing is not practical for tests more than a few hours long or for tests of very deep wells. Care must be taken during bailing tests that the cable for the water-level indicator is not disturbed as the bailer is raised and lowered. This type of test will yield very crude data, at best.

Determine the Pumping Rate

To choose the pumping rate, consider what the pump can handle as well as what rate will give the desired drawdown. As just indicated, the maximum capacity of the pump is determined by its design, size, the lift required, and the head of water above the pump's intake. Do not design the test such that the pump must run at its maximum capacity from the beginning of the test. The reason for this is simple: The yield of the pump will decline throughout the test, as the water level in the well drops. If the pump is already running at maximum capacity, its performance cannot be improved to increase the yield to its initial level.

How far below the water level will the pump intake be positioned? It must be far enough below the water level so that the pump won't run dry. In addition, if a pressure transducer is used, the pump intake should be separated from the transducer by a few feet to minimize turbulent interference.

What head of water will be above the pump intake? As the test goes on, this will decrease, which may decrease the yield.

Consider what the well is capable of yielding. Pumping at 200 gallons per minute (gpm) will immediately dry out a well that is only capable of producing 2 gpm. Estimate the transmissivity (T) of the aquifer; do this by estimating the hydraulic conductivity (K) and multiplying by the aquifer thickness. If the aquifer is much thicker than the portion penetrated by the well, then instead of thickness, use the length of the well intake. Also estimate storativity (S), as described in Step 2 under "Locations of Observation Wells." With the estimates of T and S, and with an estimate of the distance between pumping and observation wells, use the Theis equation to estimate the pumping rate that would yield an easily measured drawdown in the observation well at some time (t) after pumping begins.

Also consider how much water must be pumped out of the well before the Theis solution begins to be applicable. It is not enough to simply empty out the water that is standing in the well; new water must be drawn into the well and pumped out during the test. Calculate the volume of water standing in the well as volume = $\pi r^2 \times$ (height of water column), and then ensure that the pumping rate is sufficient to pump out all of this water and more during the first portion of the test.

Finally, it might be wise and ultimately cost-effective to conduct a pilot test to determine the optimum pumping rate. Do this by performing a step-drawdown test: guidelines for doing this are described under "Pretest," which follows.

Determine How to Measure Discharge

Measurements of discharge should be made at various time intervals throughout the test. At the beginning of the test, they should be made frequently; once flow stabilizes, they may be made less often. In the case of a bailing test, they must be made continuously. Therefore, a reliable method for measuring discharge must be employed.

The bucket-and-timer method works well if flow is relatively low (up to about 10 gpm, if using a 5-gallon bucket). A flow meter may be used by installing it in the discharge line; ensure that the expected discharge falls within the operating range of the meter. Orifice weirs may be used, but care should be taken that the discharging water is contained if necessary or routed away from the pumping well. For more information, see the section on measuring discharge from wells in Chapter 4, "Soil Water and Ground Water."

Disposing of Pumped Water

Water pumped out of the well during the test must not be permitted to recharge the pumped aquifer. This is of particular concern for unconfined aquifers. However, it is not necessarily an issue for certain confined aquifers, especially if there is no chance that water discharged onto the surface will change hydraulic head within the confined aquifer. Convey the discharged water away from the site through pipes or hoses. Be sure that the piping system is capable of carrying water at the proposed pumping rate. In some areas, it may be necessary to check beforehand to determine if any permits are required before releasing water into a stream or drainage way.

At some sites, particularly if the water is known to be contaminated, it may be necessary to contain the pumped water. If this is the case, depending on the pumping rate and duration of the test, it may be possible to contain the water in several 55-gallon drums. Alternatively, tank trucks may be used to haul the water away for appropriate disposal. Depending on test design, if tankers are used, it still may be necessary to temporarily contain the water in stock tanks or drums. In addition, consider having a backup truck on hand so that there is no delay in the test if one should fill up; consider the pumping rate and the truck's capacity to determine if this is necessary.

Depending on the water chemistry and the volume that is to be pumped, in some situations, it may be advantageous to obtain a permit to discharge the water to a local sewer or waterway. This might require obtaining a National Pollutant Discharge Elimination System (NPDES) permit. The appropriate environmental agency should be consulted for more information.

Determine How to Measure Water Levels

How often should measurements be taken? The early portion of the time-drawdown curve is critical in curve matching. As a result, it is important to make frequent measurements at the beginning of the pumping phase of the test, and then again at the beginning of the recovery phase. As the test proceeds, measurements may be made less frequently. Kruseman and de Ridder (1990) suggested the following schedule of measurements:

Time Since Start of Pumping (min)	Time Intervals of Measurements (min)
0–5	0.5
5–60	5
60–120	20
120–pump shutdown	60

In addition, consider that the analytical method influences the frequency of measurements needed. If the data will be analyzed by the distance-drawdown method, for example, only a few measurements are necessary, and these may be taken quite some time after pumping begins. For a step-drawdown test, a rapid series of measurements should be taken every time the pumping rate increases, but frequency may decline thereafter.

Use a water-level indicator to make the measurements. These are described in Chapter 4, "Soil Water and Ground Water." Considerations for choosing instruments appropriate for a pumping test are discussed here.

Measurements in the pumping well must be taken quickly, especially at the beginning of the test, because water level changes rapidly. Electric cable water-level indicators or pressure transducers are well suited to this task. However, a chalk line or steel tape may be difficult to use with much accuracy, because making one measurement typically takes more than a minute. In addition, turbulence in the well caused by the pump will degrade the quality of these measurements. Noise from the pump may render a popper useless.

Measurements in the observation well(s) involve different considerations. Can the same water-level indicator be used in the pumping well and observation well(s)? How close together are the wells? How easy is it to carry the instrument between them? Is it physically possible to make measurements in all wells with the same device, given that some measurements must be made simultaneously in all the wells? Some data loggers can accommodate several transducer inputs. If these will be used, is the transducer cable long enough to run between the wells and down the holes?

A checklist for pumping test planning appears in Box 5.1. For further reading on planning the field aspects of a pumping test, consult Walton (1987), Kruseman and de Ridder (1990), or Dawson and Istok (1991).

Performing the Test

Once the test and observation wells have been drilled, constructed, and developed, consider the record-keeping system. Determine who will keep the records during the test. Use standard forms (an example appears in Fig. 5.8) or set up tables in a

BOX 5.1

Checklist for Pumping Test Planning

Estimated T (range):
Estimated S (range):
Duration of test:
  Stage I (Pumping): _____
  Stage II (Recovery): _____
Pumping well:
  Elevation at surface
  Depth
  Screened or open interval: depth _____ to _____
  Aquifer(s) tapped (rock/sed type, or name)
  Location (draw on sketch map or site plan)
  Diameter
  Water-level indicator to be used
Number of observation wells:
Complete for each observation well:
  Observation Well No. _____
    Elevation at surface
    Depth
    Screened or open interval: depth _____to _____
    Aquifer tapped (rock/sed, or name)
    Location (draw on sketch map or site plan)
    Diameter
    Water-level indicator to be used
    Use (if existing well) and schedule of pumping
Other possible interferences (surface water, pumping wells, etc.):
Pump:
  Name, manufacturer, model no., type
  Capacity
  Diameter
  Power source
  Depth in well at which intake will be placed
Pumping rate:
Measurement of discharge:
  Method
  Supplies needed
Disposal of pumped water ■

field note book before the test. Several columns will be needed, with these headings as a minimum:

Pumping Well: Time, depth to water (or transducer reading), pumping rate, comments.

Observation Well: Time, depth to water (or transducer reading), comments.

Don't forget to include units of measurement in each column where appropriate.

Pumping Test Field Data Collection Form

Job _____ Field Personnel _____

Well Number _____ Date _____ Time _____

Well Location _____

Well Data: Casing Diameter_____ Total Depth of Well_____

Screen Diameter_____ Length of Screen _____

Depth of Screened Interval: from _____ to_____

Measurement Point: ____ Top of Casing (TOC) ____ Grade ____ Other(_____)

Water Level Measurement Devices:

For measuring depth to water_____

For measuring water levels during test _____

Depth to Water **Any trend/change in water**
Before test start _____ **level before test? Explain**_____

Purpose of Well (circle one): **Pumping well** **Observation well**

Pump Data (for pumping well only):

Pump type, size, Mfr., model #_____

Pump setting (depth)_____

Pumping rate measurement device _____

Storage/disposal of pumped water _____

Time	Elapsed Time	Water level (units:_____)	Pumping Rate (units:_____)	Comments

Fig. 5.8 Sample field data collection form for a pumping test.

The test should be conducted in several stages. Some hydrogeologists conduct a step-drawdown test before every pumping test. This pretest provides the information needed to determine the equipment, personnel, and time that will be required for the actual pumping test. This practice can save money and time and yield better results.

The pumping test itself consists of two stages: Stage 1, Pumping; and Stage 2, Recovery. The pumping stage involves pumping the well and recording the time and water level in the pumping and observation well(s). The recovery stage simply involves recording the water level in the pumping and observation well(s). This section gives the step by step method for performing the complete test.

Pretest: Step-Drawdown Test

A step-drawdown test (or "step test") is conducted for the purpose of determining the optimum pumping rate for a well. This type of test may be conducted before a standard pumping test, or it may be conducted in any production well when data on optimum pumping rates are needed.

A step-drawdown test is performed by pumping at several successively higher rates and observing the effects on drawdown. To do this, follow the field procedures for Stage 1 of a standard pumping test, as described in what follows. However, in a step-drawdown test, the pumping rate is varied stepwise through the test. This is done by beginning the test at a low pumping rate. Pumping at this rate continues for a period of 1–2 hours, or until drawdown has stabilized to some degree. Then, pumping rate is increased, and pumping at the new rate continues for the same length of time. Then, the pumping rate is increased again, ideally by the same factor as it was increased in the previous step. The test continues in this fashion for several "steps," or pumping rates. Walton (1987) suggested at least three steps, whereas Driscoll (1986) suggested five to eight steps.

To analyze the data, begin by plotting drawdown versus time on an arithmetic graph with, time on the horizontal axis and drawdown on the vertical axis; see Fig. 5.9 (a). In addition, plot discharge versus drawdown on an arithmetic graph, with drawdown on the horizontal axis and discharge on the vertical axis; see Fig. 5.9 (b). This second plot shows specific capacity, which is the well's yield per unit drawdown. On both graphs, connect the points with a smooth curve. Conduct the pumping test at a pumping rate chosen from the straight (early) portion of the specific-capacity curve. The point at which the specific-capacity curve flattens out is the maximum safe yield. It is not wise to operate the well at the maximum safe yield.

Stage 1: Pumping

Measure and record water levels in all wells.

Ensure that nearby wells will not be pumping during the test. Any wells in the expected zone of pumping influence of the test well should be shut down for the duration of the test.

Set up equipment.

Pump and power supply. Note that installing the pump raises the water level. Allow it to return to original before the test begins.

Water-level measuring devices. If using transducers, allow enough time after installing the probes in the wells for the water level to return to the original before the test begins.

Discharge-measuring system.

Discharge-disposal system.

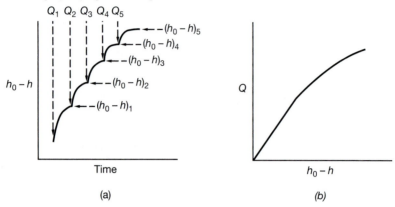

Fig. 5.9 Results of a step-drawdown test. (a) Drawdown vs. time. (b) Discharge vs. drawdown. This plot shows the specific capacity of the well.

Check with all field personnel to ensure that all duties, including record keeping, are clearly understood. Rehearse if necessary.

Synchronize watches. Determine which watch will be considered official time, and which will be backup.

Check and record time and water levels in all wells.

Start the pump and begin making measurements. The first several measurements must be made rapidly, so it is essential that field personnel know their jobs and be prepared to perform them immediately. In addition, the use of automatic recorders is strongly recommended for any wells near the pumping well, so there is no delay in taking measurements. Throughout Stage 1 (Pumping), monitor pump operation, discharge, and water levels. Try to anticipate problems before they occur.

Stage 2: Recovery

Near the end of Stage 1 (Pumping), prepare for Stage 2 (Recovery). Once again, measurements will have to be made rapidly at the beginning of the recovery period, so several workers should be on hand and well-rehearsed.

Before shutting down the pump, once again review duties.

If any of the observation wells are remote from the pumping well, and if a field worker is at that well and will begin making measurements there once the pump is shut off, ensure that the worker is aware of the timing of the shut off. Walkie-talkies may be useful for this, or hand or light signals.

Make and record a final discharge measurement and a final water-level measurement in all wells; these values are important because they will be used in analyzing the results. Then shut off the pump. Begin taking water-level measurements on the same schedule as the original measurements in Stage 1 (Pumping).

While recording the data, use the same column headings as during the pumping portion, except that no column is needed for discharge.

After it is stopped, the pump should be left in the well, as pulling it out will change water level in the well, and the time it takes to haul it out may interfere with the valuable early time measurements. In addition, pulling it out is likely to disturb the transducer, if one is being used. One problem with leaving the pump in the well is that water in the discharge line may drain back into the well. This should be avoided, because it changes the water level in the well. If necessary, a check valve may be installed near the bottom of the discharge line to prevent this from occurring. This must be done before the test begins.

Stop the test after the requisite time has elapsed.

Processing the Data

The field data must be processed to put them in a form that is useful for the analysis. Prepare to do this by constructing a table or computer spreadsheet with headings as shown in Fig. 5.10.

Convert the time data to "Elapsed Time" data. These values should give the elapsed time after pumping began, for Stage 1 (Pumping), and the elapsed time after pumping ended, for Stage 2 (Recovery) data. Use units such as seconds or minutes. Any time unit may be used, as long as it is used consistently.

Convert the water-level measurements to drawdown measurements. Use units such as feet or meters, and be consistent. This conversion is done slightly differently for pumping data than for recovery data, as described in what follows.

For Stage 1 (Pumping), if depth-to-water measurements (e.g., depth from the top of casing to the water level) were taken, find drawdown by subtracting depth to water *before* pumping from depth to water *during* pumping. If a transducer was used, calculate drawdown by subtracting height of water *during* pumping from height of water *before* pumping (Fig. 5.11). Some transducers or data loggers can be set to do this during the test, so that later conversion is not necessary, and this significantly simplifies the work.

For Stage 2 (Recovery) measurements, the quantity plotted is not drawdown, but recovery. Recovery is the distance that the water level rises after the pump is shut off. If depth-to-water measurements were taken, then calculate recovery as the depth to water *before* the pump was shut off minus the depth to water *after* the pump was shut off. If a transducer was used, then calculate recovery as the height of water above the transducer *after* the pump was shut off minus the height of water above the transducer *before* the pump was shut off (Fig. 5.12).

Analyzing the Data

Detailed discussions of data analysis may be found in several texts (e.g. Freeze and Cherry, 1979; Driscoll, 1986; Walton, 1987; Kruseman & deRidder, 1990; Dawson and Istok, 1991; and Fetter, 1994). As a first approach, however, make Theis and Jacob time-drawdown plots. If conditions are ideal, these plots will allow easy data analysis. If conditions deviate from the ideal, these plots will indicate the degree and nature of the deviation.

Analysis of Pumping Test Data

Depth to water immediately before pumping stopped: _____

Time	Elapsed time after pumping began	Depth to water during test (units:_____)	Drawdown (depth to water during test minus depth to water before pumping began) (units:_____)	Pumping rate (units:___)

Analysis of Recovery Test Data

Depth to water immediately before pumping stopped: _____
Pumping rate immediately before pumping stopped: _____

Time	Elapsed time after pumping stopped	Depth to water during test (units:_____)	Recovery (depth to water immediately before pumping stopped minus depth to water during test) (units:_____)

Fig. 5.10   Sample tables for analysis of pumping test and recovery test data.

Theis Method

A Theis plot may be constructed if at least one observation well was used. First, a type curve (or reverse-type curve) must be obtained. A reverse-type curve printed on acetate may be purchased from the National Ground Water Association. One may be photo-enlarged from Fig. 5.13; however, doing so will distort the axes somewhat. A computer-generated plot may be created by plotting values of $W(u)$ vs. $1/u$; however, printing the graph may distort the axes (depending on the software used). This problem sometimes may be solved by using the same software to plot both type curves and data curves. However, even so, it is important to check

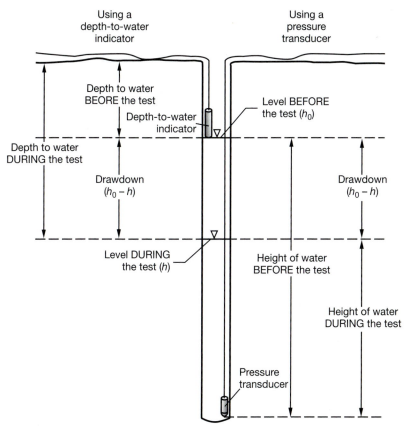

Fig. 5.11   Calculating drawdown ($h_0 - h$) during the pumping stage of a pumping test.

the plots to ensure that the axes match. It may be simplest to construct a hand-drawn curve. Although it is tedious, once this is done, it need never be done again (unless differently scaled graph paper is used).

To construct a hand-drawn curve, use a table of values of $W(u)$ and u, such as found in Appendix 3. Calculate $1/u$ for each value of u. Plot $W(u)$ vs. $1/u$ on 3×5–cycle log–log graph paper, with $W(u)$ on the y axis (three log cycles), and $1/u$ on the x axis (five log cycles). Several sheets may be cut and pasted together to cover the entire table of values. However, it is possible to fit the most critical portions of the curve on one sheet by plotting only those values for $u = 10^{-4}$ to 10^1 ($1/u = 10^{-1}$ to 10^4) and $W(u) = 10^{-2}$ to 10^1. Connect the points with a smooth curve, and darken it by going over it in black ink. Make a dot at the point $W(u) = 1$, $1/u = 1$ (this will be a convenient choice for a match point later on).

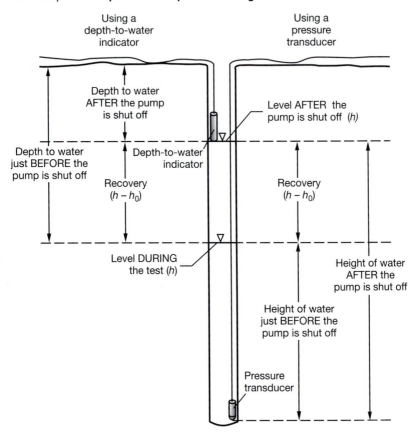

Fig. 5.12 Calculating recovery $(h - h_0)$ during the recovery stage of a pumping test.

Once the reverse-type curve is available, set it aside. Use a clean sheet of 3×5–cycle log–log paper to plot field data curves. If only one observation well was used, plot drawdown on the y axis (three cycles) vs. time on the x axis (five cycles) (Fig. 5.14). If more than one observation well was used, *and* conditions are laterally homogeneous, calculate time/(radius)^2 for all wells. Then plot the drawdown on the y axis vs. time/(radius)^2 on the x axis.

If the points do not fall along a fairly smooth curve with the same general shape as the reverse-type curve, the assumption of lateral homogeneity is likely to be false. However, given the natural variation in geologic data and in measurements of this nature, some deviation from assumed conditions may always be expected. This does not necessarily invalidate the test results.

If more than one observation well was used, and conditions may *not* be assumed to be laterally homogeneous, then plot drawdown on the y axis and either time or time/(radius)^2 on the x axis. It does not matter which quantity is chosen for

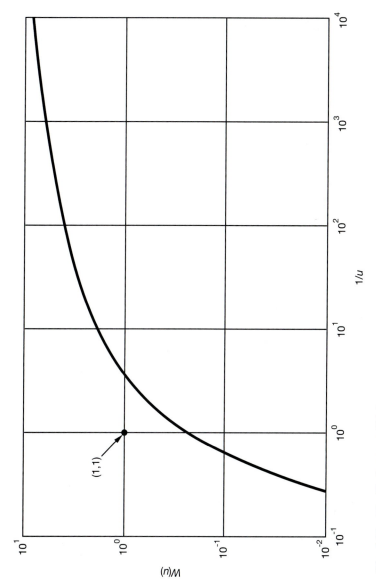

Fig. 5.13 The Theis nonequilibrium reverse-type curve.

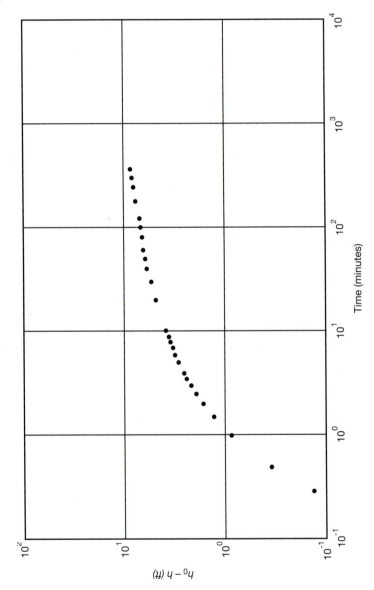

Fig. 5.14 Field data plot of drawdown (or recovery) vs. time since pumping began. If more than one observation well were used, plot drawdown (or recovery) versus time/(radius)^2.

the x axis, as long as it is clearly labeled, but using time/(radius)² and plotting the data from each well as a separate line may show the nature of the heterogeneities.

Plot a curve (or curves) for the pumping stage and a curve (or curves) for the recovery stage.

Once the field data curves are complete, overlay a field data curve on the reverse-type curve (Fig. 5.15). This may be done by using computer software, or it may be done by hand. If done by hand, the curves may be overlain on a light table or by taping the graphs to a sunny window. Keeping the two graphs' axes parallel, slide the field data curve up and down and right and left until the curves match each other. If the curves cannot be made to match, then conditions are non-ideal, and some specialized analytical method must be used.

If the curves can be made to match, hold them in matched position and select a match point. This can be any point on the graphs, whether on the line or not. It is convenient to choose the point $W(u) = 1$, $1/u = 1$, as this makes later calculations simpler. On the field data plot, make a dot at the match point, and note the values of $W(u)$ and $1/u$ at that point (write them directly on the graph for later reference). From the graphs, read the values of $W(u)$ and $1/u$ (already noted), and drawdown and time [or time/(radius)²]. Substitute these values into the Theis equations that follow. Be sure to include units of measurement!

$$T = \frac{QW(u)}{4\pi(h_0 - h)}$$

where T = transmissivity

Q = pumping rate during the test

$W(u)$ = value of $W(u)$ at the match point

$h_0 - h$ = value of $h_0 - h$ at the match point

Use the value of T calculated with this equation, and substitute it into the following equation to find S:

$$S = 4Tu(t/r^2)$$

where S = storativity

T = transmissivity value calculated in previous equation

u = reciprocal of the $1/u$ value at the match point

t = time value at the match point

r = distance from the pumping well to the observation well,

$t/r^2 = t/r^2$ value at the match point

Calculate T and S for the pumping data and for the recovery data, and for each observation well's curve.

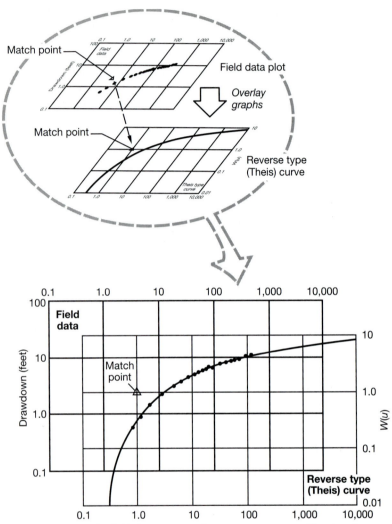

Fig. 5.15 Overlay the field data curve on the reverse-type curve. Keeping the axes parallel, match the two curves.

Jacob Time-Drawdown Method:

The Jacob time-drawdown method may be used with data from the pumping well or from observation wells. Plotting data from each observation well separately will show the nature of the heterogeneities. If conditions are laterally homogeneous, then a single line may be used for all observation well data.

Plot the field data on 3- or 5-cycle semilogarithmic paper. This may be done by hand or by using a computer spreadsheet. Plot drawdown on the y axis (arithmetic); ensure that the scale extends to zero drawdown. Some workers prefer to construct these plots such that the zero drawdown point is at the top, rather than the bottom, of the scale (Fig. 5.16). On the x axis (logarithmic), plot time if the data is from the pumping well or a single-observation well; plot time/(radius)^2 if the data are from more than one observation well. Ensure that the scale is constructed such that the plotted line crosses the line along which drawdown equals zero.

Draw a straight line through the plotted points. In drawing the line, ignore the early time data, which show a curved line. The assumptions of the Jacob method are not valid for the early values. If one straight line does not fit all the points (ignoring the early time data), break the line into segments of varying slopes. Most plots probably will have only one segment, but two or three segments are not unusual (Fig. 5.17).

The different segments indicate different values of T and S for materials at a distance from the well. They may also indicate a variation in storage characteristics with respect to time. The value of time at which one segment ends and the next begins indicates the time at which the storage characteristics changed *or* the time at which the cone of depression intersected a zone of the aquifer with a different T or S value. A different value of T or S should be calculated for each segment.

Determine the change in drawdown over one log cycle of time [or time/(radius)^2]. This is "delta drawdown": $\Delta (h_0 - h)$. Determine the value of time [or time/(radius)^2] at which the line (or line segment) crosses the zero drawdown line; this is t_0 ("time zero"), or t/r_0^2. Some computer programs can be made to calculate these values. Note that the early time data should be excluded from the calculation. Substitute the values into the Jacob time-drawdown equations, which follow. Units of measurement must be included carefully in the calculation!

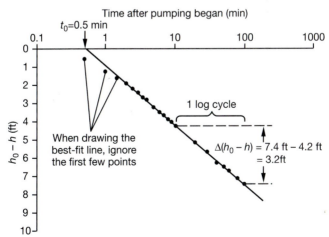

Fig. 5.16 Field data plot for the Jacob time-drawdown method, showing drawdown (or recovery) versus time. If more than one observation well were used, *and* if conditions were laterally homogeneous, plot drawdown (or recovery) versus time/(radius)^2.

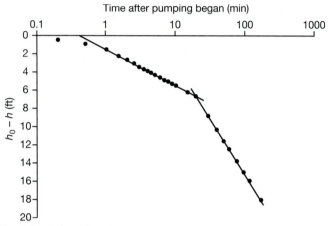

Fig. 5.17 A Jacob time-drawdown plot showing a change in conditions at some time after pumping began.

$$T = \frac{2.3\,Q}{4\pi\Delta(h_0 - h)}$$

where T = transmissivity

Q = pumping rate

$\Delta(h_0{-}h)$ = drawdown over one log cycle of time

$$S = \frac{2.25\,T\,t_0}{r^2}$$

where S = storativity

T = transmissivity, as calculated above

t_0 = the value of time at which the line crosses the zero drawdown line

r = the distance between pumping and observation wells.

Note that if no observation well was used, there is no value of r. Therefore, the Jacob time-drawdown method cannot be used to find storativity if no observation well was used.

Jacob Distance-Drawdown Method

This method employs drawdown measurements all taken at the same time, but from several different observation wells. The test must have used at least three observation wells; in general, more is better.

Plot the field data on 3- or 5-cycle semilogarithmic paper. Plot drawdown on the y axis (arithmetic scale); ensure that the scale extends to zero drawdown. Some workers prefer to construct these plots such that the zero drawdown point is at the top, rather than the bottom, of the scale (Fig. 5.18). On the x axis (logarithmic), plot distance from pumping to observation well. Ensure that the scale is constructed such that the plotted line crosses through the zero drawdown line.

Draw a straight line through the plotted points. Determine the change in drawdown over one log cycle of distance; this value is "delta drawdown": $\Delta(h_0-h)$. Determine the value of distance at which the line crosses the zero drawdown line; this is the "radius of influence of pumping": r_0. Substitute values into the Jacob distance-drawdown equation, which follows. Again, units of measurement must be carefully included in the calculation!

$$T = \frac{2.3Q}{2\pi\Delta(h_0 - h)}$$

where T = transmissivity

Q = pumping rate

$\Delta(h_0 - h)$ = change in drawdown over one log cycle of distance

Using the value of T calculated by this equation, find storativity by using the following equation:

$$S = \frac{2.25Tt}{r_o^2}$$

where S = storativity

T = transmissivity, as calculated above

t = the time at which measurements were made

r_0 = value of r (distance) at which the line crosses the zero drawdown axis (see Fig. 5.18); r_0 also is the radius of influence of pumping

Note that if the line breaks into two or three segments, the distance on the graph at which the change in slope occurs is the distance (in the aquifer) from the pumping well to the point at which the change in aquifer characteristics occurs. Calculate T and S for each segment.

Tracer Tests

A tracer test is a test of aquifer or aquitard characteristics using a chemical introduced into the ground water at one point and detected at another. In general, tracer tests are used to determine velocity of travel. But in some cases, particularly in karst aquifers, they may be used to determine the hydraulic interconnectedness of springs, streams, wells, surface water, and sinkholes.

For more detailed information on how to conduct a tracer test, consult Davis and colleagues (1985), Brassington (1988), and Domenico and Schwartz (1998).

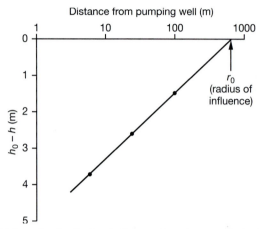

Fig. 5.18 Field data plot for the Jacob distance-drawdown method, showing drawdown (or recovery) vs. distance. The distance r_0 is the radius of influence of pumping.

Tracers come in various types. Chloride and bromide salts are relatively inexpensive and are chemically conservative (they do not interact readily with other chemicals or minerals). They are nontoxic in low concentrations, and their concentrations are easy to measure. Bromide is more conservative than chloride. Chemical dyes, for example, Rhodamine, Rhodamine WT, and others, are more expensive; some are nontoxic, whereas others are toxic. A spectrometer or fluorimeter must be used to detect one of these tracers. Radioactive isotopes may also be used as tracers, whether they were deliberately or accidentally introduced into an aquifer.

Many factors must be considered in designing a tracer test. The following questions should be answered before undertaking a tracer test.

What is the purpose of the test? It must be defined clearly before beginning the test, or the test is likely to yield no useful information.

Is a tracer test the best way to accomplish the goals? Perhaps the information can be collected in some other way. What would nondetection of the tracer indicate? Keep in mind that nondetection does not necessarily mean that the tracer is not present.

What is the geology of the test site? What are all of the possible connections among springs, streams, and so on? Where else might the tracer go? Is it possible that too much dilution will take place, rendering the detection method ineffective? Might there be other sources of the tracer that would obscure the meaning of the readings? Is there a possibility that the tracer might never be detected at the outflow point?

Which tracer should be used? Can a "natural" tracer be used, that is, can water chemistry be used to trace the flow paths, without introducing another agent? How much will the tracer cost? How easily can it be detected? Might it be taken up by plants along the flow path? How toxic is it? How rapidly will it degrade? What chemical interactions with the aquifer or sediments might take place (adsorption, precipitation, dissolution), and how might they alter tracer concentrations? Will the tracer flow at the same rate as the water?

How will you know when to take readings, and at what location? Dilution may render even a colored dye tracer invisible; how will you know when it appears at the outflow point? How frequently will readings be taken? Can the readings be taken in the field, or will samples have to be taken to a laboratory? If a lab, how rapidly will results be in, and how will you know when you have taken enough samples?

If a tracer is to be used to determine hydraulic conductivity, the time of travel will need to be estimated. Distance along the flow path must be determined; this is not necessarily the same as map distance, but can be measured if an accurate flow net is available. The rate of travel is the distance traveled divided by the time of travel. Will the first appearance of the tracer indicate the time of travel, or will the appearance of the peak concentration be used?

Once velocity of travel is determined, hydraulic conductivity can be found. Hydraulic conductivity is the rate of travel multiplied by the hydraulic gradient and divided by the effective porosity.

Describing Sediments and Rocks

Chapter Overview

To learn about subsurface water, we must examine sediments and rocks and their texture, composition, and structure. Is a certain formation an aquitard or aquifer? Will water flow through it along preferential flow paths, or uniformly? Will the composition of a rock or sediment affect ground water chemistry? Answering these questions requires the hydrogeologist to obtain a thorough description of the rock and sediment characteristics.

To learn about flow paths, chemistry, or contamination of ground water, we must assess rock and sediment composition, layering, fracturing, structure, and hydraulic conductivity. All of the skills we possess as geologists come into play in these investigations. This chapter reviews sediment and rock description.

Describing Sediments and Soils

A geologist's skill at constructing a sediment boring log increases dramatically with experience. The cautious beginner, however, can improve her or his initial efforts by keeping a checklist of items to be recorded in the field notebook or on a preprinted form. Many companies require that logs be recorded on a standard form (Fig. 6.1). These field logs may be taken back to the office, and after the required laboratory test results are available, may be reproduced in a more polished form for inclusion into a final report (Fig. 6.2). Well construction or abandonment procedures may be recorded on separate forms. Examples of each are given in Figs. 6.3a, 6.3b, and 6.4.

At the bare minimum, the following must be noted whenever a log is produced:

Date

Job name/number and company

Names of geologists, drillers, and other personnel present

Location of the site

Borehole number and location (sketch maps may help)

Surface elevation

Weather conditions

Equipment used

Sampling device/type

Drilling fluids used

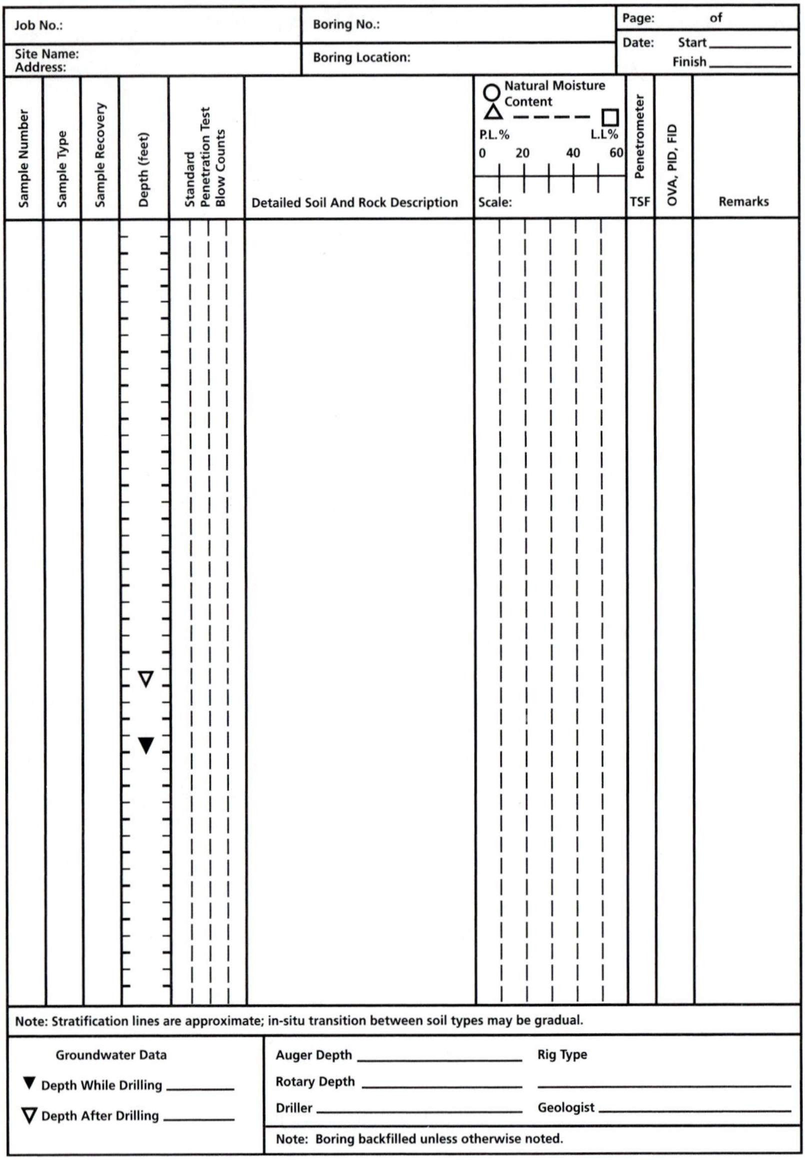

Fig. 6.1   A sample form for logging a borehole.

Job No.: 90349	Boring No.: B-10	Page: 1 of 1
Site Name: Modera Inc. Address: 6442 Elm, Elmhurst, IL	Boring Location: 17'N & 15'E of SW Property Corner	Date: Start 8-13-97 Finish 8-13-97

Sample Number	Sample Type	Sample Recovery	Depth (feet)	Standard Penetration Test Blow Counts	Detailed Soil And Rock Description	Natural Moisture Content — P.L.% 0 20 40 60 L.L.% — Scale:	Penetrometer TSF	OVA, PID, FID	Remarks
			1'		12" Sand & Gravel Fill			ppm 29.8	
	S	24"		1 1 1 1	brown-gray silty CLAY-soft orange mottling trace sand & gravel slight petroleum odor		2.5		
S-010	S	24"	3'						
			5'	1 2 1 1				32.9	
	S	24"	7'	1 1 1 2	olive-gray silty CLAY-soft trace sand		2.5	26.2	
	S	0"	9'	3 6 6 7	NO RECOVERY				
	S	18"	11'	2 2 2 2	gray-brown silty CLAY-soft trace sand		4.0	28.2	
	S	12"	13'	7 9 10 9	gray CLAY-very stiff			12.7	
	S	12"	15'	▽ 7 8 9 9				2.7	
	S	24"	17'	▼ 8 9 10 11	6" gray CLAY-moist 18" light brown SAND saturated			1.4	
	S	18"	19'	11 11 10 11	gray CLAY-very stiff		4.0	.8	
			21'		EOB — 19 FEET				
			23'						

Note: Stratification lines are approximate; in-situ transition between soil types may be gradual.

Groundwater Data	Auger Depth _____	Rig Type
▼ Depth While Drilling 16'	Rotary Depth 19'	Hollow stem auger
▽ Depth After Drilling 14'	Driller C&G	Geologist Rick Polad
	Note: Boring backfilled unless otherwise noted.	

Fig. 6.2 A completed boring log, based on the field boring log, typed and including the results of laboratory tests. Such a log may be included in a final report to a client.

Well Completion Report

Incident No.: _____ Well No.: _____

Site Name: _____ Date Drilled Start: _____

Drilling Contractor: _____ Date Completed: _____

Driller: _____ Geologist: _____

Drilling Method: _____ Drilling Fluids (type): _____

Annular Space Details **Elevations – .01 ft.**

Type of Surface Seal: _____

Type of Annular Sealant: _____

Type of Bentonite Seal (Granular, Pellet): _____

_____

Type of Sand Pack: _____

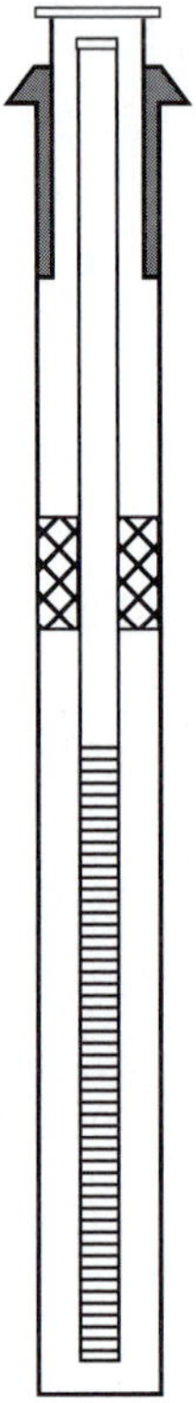

_____ Top of Protective Casing
_____ Top of Riser Pipe
_____ Ground Surface
_____ Top of Annular Sealant
_____ Casing Stickup

Well Construction Materials

	Stainless Steel Specify Type	PVC Specify Type	Other Specify Type
Riser coupling joint			
Riser pipe above w.t.			
Riser pipe below w.t.			
Screen			
Coupling joint screen to riser			
Protective casing			

_____ Top of Seal
_____ Total Seal Interval
_____ Top of Sand
_____ Top of Screen

Measurements

Riser pipe length	
Screen length	
Screen slot size	
Protective casing length	
Depth to water	
Elevation of water	
Free Product thickness	
Gallons removed (develop)	
Gallons removed (purge)	
Other	

_____ Total Screen Interval

_____ Bottom of Screen
_____ Bottom of Borehole

Completed by: _____

Fig. 6.3a A sample well construction form.

Bit size and type

Signature of logger

Other pertinent information

As the drilling and sampling proceed, the following should be recorded:

Depth of lithology described

Well Completion Report

Incident No.: 94079
Site Name: Modera, Inc.
Drilling Contractor: C&G Drilling
Driller: Mark Bellwood
Drilling Method: Hollow Stem Auger

Well No.: MW-4
Date Drilled Start: 6-13-96
Date Completed: 6-13-96
Geologist: Rick Polad
Drilling Fluids (type):

Annular Space Details

Type of Surface Seal: Cement/Bentonite
Type of Annular Sealant: Bentonite
Type of Bentonite Seal (Granular, Pellet): Pellet

Type of Sand Pack: #4 Filter Pack

Elevations – .01 ft.

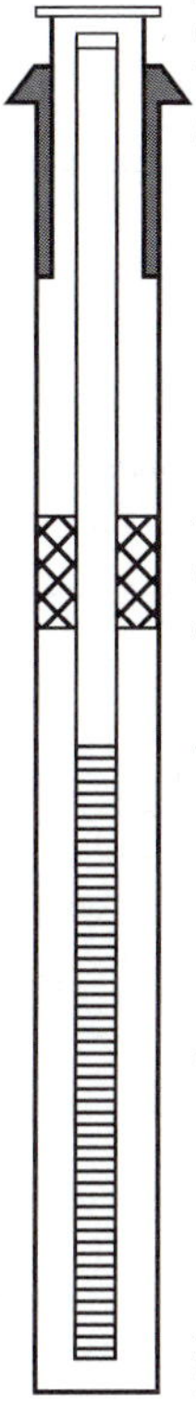

99.00' Top of Protective Casing
98.67' Top of Riser Pipe
99.00' Ground Surface
98.31' Top of Annular Sealant
Flush Casing Stickup

Well Construction Materials

	Stainless Steel Specify Type	PVC Specify Type	Other Specify Type
Riser coupling joint		threaded	
Riser pipe above w.t.		Sch. 40	
Riser pipe below w.t.		Sch. 40	
Screen		Sch. 40	
Coupling joint screen to riser		threaded	
Protective casing			Steel

98.31' Top of Seal
12.31' Total Seal Interval
86.00' Top of Sand

84.00' Top of Screen

Measurements

Riser pipe length	14.67'
Screen length	5.00'
Screen slot size	0.010'
Protective casing length	1.0'
Depth to water	9.83'
Elevation of water	88.84'
Free Product thickness	None
Gallons removed (develop)	8
Gallons removed (purge)	8
Other	

Completed by: *Rick Polad*

5.00' Total Screen Interval

79.00' Bottom of Screen
78.50' Bottom of Borehole

Fig. 6.3b   A sample well construction form (completed).

Description of lithology (see what follows)

Number of blow counts (if the Standard Penetration Test, [SPT] is performed)

Presence of water in the hole (saturation of cuttings as well as water level at various times during and after drilling operations; more information on this is provided elsewhere in this chapter)

Depth interval and length of any samples taken

Water Well Abandonment Form

Date of Well Abandonment _____

Owner_____

Well Location _____
                 (Description and Address)

                _____
                 (City, County, State)

                _____
                 (Legal Description, Township, Range, Section)

Year Drilled _____

Drilling Permit Number and Date_____

Well Construction: Diameter_____ Total Depth_____
                Screened or Open Interval _____ to _____
Sealing:

    Sealing material_____ , installed from _____to_____depth

    Sealing material_____ , installed from _____to_____depth

    Sealing material_____ , installed from _____to_____depth

    Sealing material_____ , installed from _____to_____depth

Casing removal: Depth to which upper portion of casing was removed_____

Licensed water well driller or other person approved to perform well sealing

    _____   _____
    Name   License Number

    _____
    Address

Fig. 6.4 A sample well abandoment form.

Field time is often a limiting factor in determining the degree of detail included in a field log. The hydrogeologist is in charge of the drilling operations, and is the one who sets the pace of work. However, experienced drillers and backhoe operators may work efficiently and quickly. If this is the case, an inexperienced hydrogeologist may be the only factor slowing the process of a project, and this can lead to the hydrogeologist feeling pressured to rush the job.

If you find yourself in this situation, remember that you are in charge and that the quality of the entire project is likely to rest on the quality of your logs, so do not allow yourself to be rushed into doing sloppy work. However, also remember that advance practice and preparation and review of procedures will help minimize problems.

During split spoon sampling (described in Chapter 7, "Sediment and Rock Sampling and Drilling Methods"), the first few samples may be taken in rapid succession. Because they come from shallow depths, the driller has less work to do in adding and removing sections of drill stem. As a result, it is particularly easy to have work back up at the very beginning of a borehole.

Major Descriptors of Sediments and Soils

Soil and sediment samples may be produced by augering, digging, coring, or other methods. These methods are described in Chapter 7. However, no matter what the sampling method, certain sediment and soil characteristics are particularly important to record on the boring log or field record. These characteristics are color, grain size, sorting, and water content. This section describes how to assess these characteristics; the next section reviews some additional factors.

Color

Color of a sediment or soil depends on its chemical and mineralogic composition, oxidation conditions, and organic content. The presence of certain contaminants also may affect color. Sediments or soils are commonly described as being brown, dark brown, tan, black, orange, red, gray, blue, yellow, or white. In some cases, color may be noted quickly and without further investigation. However, because one person's "red" may be another's "orange," sometimes it is helpful to have a more precise means of specifying color. If a more precise description is needed, it may be appropriate to use a color chart such as the Munsell Soil Color Charts (Munsell Color, 1994). These charts can be purchased from geotechnical, forestry, or wetlands science equipment suppliers. The Munsell Soil Color Charts essentially are paint chips of various colors for comparison to the sample, allowing identification of colors numerically by their hue and chroma.

Because color of a sample may change dramatically as it dries, any time a color is noted, it also should be noted whether the sample is moist or dry. Likewise, light conditions should be noted, because some materials appear different under direct sunlight than under cloud cover or other light conditions..

Color may be mottled, that is, the sample may have a matrix that is one color, but mottles or clumps that are another. Some samples, especially those with high clay content and/or high iron content, may be gleyed. A gleyed soil is one with blue-gray mottles or concretions within the soil. Gleying is due to reduction and oxidation of iron and manganese. It occurs when the soil is alternately saturated and unsaturated, and indicates periodic saturation, often caused by a seasonally fluctuating water table. The brown or orange colors in mottles are due to oxidized iron, indicating unsaturated conditions; the blue-gray color of gley is due to reduced iron, indicating saturated conditions. The Munsell Soil Color Charts contain a separate chart for identification of gleyed soils.

In some cases, color can be used to give a first approximation of the seasonal position of the water table. The change from orange to blue-gray is particularly pronounced in iron-bearing sediments and clays. Because the color change is not immediate, it cannot be used as a precise indicator of water table location at any given time. If a perched water table or extensive capillary fringe exists, sediments may be saturated even though they are above water table. A leaky pipe or drainageway that emits water that percolates to the water table below can cause the same conditions to occur.

Grain Size and Grain Sorting

Sediments fall into four general groups depending on their grain size: gravel, sand, silt, and clay. Some sediments fall in between these designations, for example, silty clay, silty sand, sandy gravel, and so on. Grain size and grain sorting play a very important role in determining a sediment's hydraulic conductivity, porosity, and other hydraulic characteristics, so they should be noted carefully.

Suggestions for differentiating between various grain sizes are given in this section. As hydrogeologists gain experience, they learn to make rapid assessments of grain size and sorting. However, for the inexperienced hydrogeologist, it is worthwhile to practice describing sediments before going out in the field on the first assignment. Box 6.1 suggests one way to practice estimating grain sizes.

The first step in identifying sediment grain size is to determine the grain size of the largest fraction of grains. Use the following rules: Gravel is sediment with a

---

BOX 6.1

Practicing Grain Size Estimation

To practice determining grain size and sorting, collect a number of samples from geologically varied areas. Some colleges, companies, and agencies maintain a "library" of soils for exactly this purpose. Another possibility is to visit several field areas and auger several different samples, to obtain a variety of materials. Using the instructions on field identification of sediments given later in this chapter, classify each sample according to its grain size and sorting. Then, check your work by performing the following simple test on each sample.

Remove gravel from the sample, and place some of the remaining sample in a glass jar or test tube. The size of the jar is not important, but it *is* important that the sample fill approximately one-third of the jar. Use a grease pencil or china marker to mark the height of the sample on the side of the jar.

Next, add water to the jar, filling it nearly full. Cap the jar and shake it, or vigorously stir the mix to make sure that the sediment thoroughly disaggregates and goes into suspension. Then suddenly stop the shaking or stirring, allowing the particles to settle, and start timing. Twenty seconds after stopping the shaking, mark the height of the layer of sediment that has settled out at the bottom of the jar. Allow the sample to settle for another few minutes, and mark the top of the settled material.

The 20-second mark shows the proportion of sand in the sample; the difference between this mark and the next shows the proportion of silt. Clay will remain in suspension for hours or days, so estimate its proportion by measuring the difference between the silt mark and the original mark showing sample volume. Now compare your results with your original assessments. With practice, you will be able to correctly identify sediment grain size and sorting without using the shake test. ∎

diameter of greater than about 2 mm. Sand grains have a diameter smaller than 2 mm, but the grain edges are still visible. For a person with normal vision, it actually is possible to make the eye trace the edges of sand grains. Silt grains may sparkle in the sun, but they are too small for the unaided eye to pick one out and trace its edges. Differentiating between silt and clay may be difficult for the novice. Box 6.2 gives some suggestions on how to make the distinction.

Once the grain size of the greatest fraction of the sediment has been identified, look at the remaining fraction. Be aware that because our attention is naturally drawn to the largest of objects in a collection, we are likely to overestimate their importance (Fig. 6.5).

Water Content

Water content of the sediment should be noted in any sediment or soil description. The following are some possibilities [see also ASTM Standard D2488, on sediment description]:

Dry: No water (no moistness) apparent; sample does not change color or texture after air drying

Moist: Moistness apparent; sample changes color or texture after air-drying

Wet: Free water apparent or film of water apparent

Note that a wet sample does not necessarily mean that the sediment is wet *in situ*. Instead, it may be wet from water or other drilling fluids that were added during the drilling process, or from water that entered the hole from a saturated layer that lies above the sampling zone.

Other Descriptors of Sediments and Soils

Degree of *rounding* or angularity of sediment grains may be noted, where the grains are large enough to see the rounding.

BOX 6.2

How to Tell Clay and Silt Apart

Differentiating between clay and silt takes a bit of practice. Rubbed between the fingers, silt feels gritty. Dry lumps of silt are easy to smash, because they have no cohesion. When wet lumps or pats of silt are jolted or tapped, water rises quickly to the surface, producing a wet sheen. This can be tested by forming a pat of wet sediment in the palm of the hand and striking the side of that hand with the empty hand several times.

In contrast to silt, when clay is rubbed between the fingers, it feels slippery or smooth, especially when wet. Clay particles are far too small for the unaided eye to pick out individual grains. Clay smears and stains stick to the fingers when wet. On an auger, wet clay makes smooth ribbons. Wet clay also can be rolled into thin threads. Dry lumps are hard to smash. When a pat of wet clay is jolted or tapped, no wet sheen appears, as it does on silt. ■

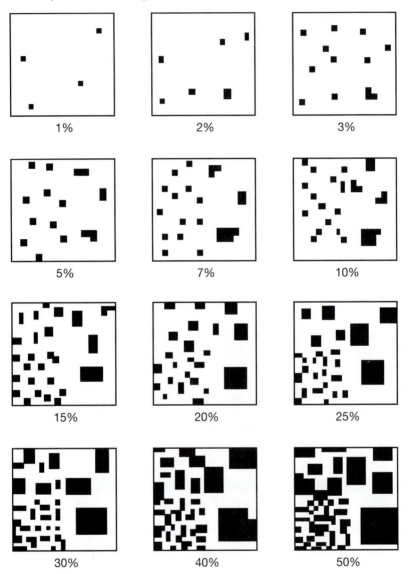

Fig. 6.5 A chart for estimating what percentage of a sample is composed of a particular material. Note that each quarter of each square in the diagram has the same percentage of black area. (*Source:* USEPA, 1991.)

The presence of *organic matter* should be noted. This may include plant roots or undecomposed plant and animal material, especially but not exclusively at the top of a soil profile. It may be peat, which is a fibrous mass of plant material. In certain glaciated areas, multiple soil layers may have developed during successive interglacial episodes. In places where earthmoving equipment has been operating, such as in surface mines and quarries, a well-developed soil profile may have been covered with other sediment, resulting in more than one layer with abundant organic matter.

Odor of the sediment may be of importance. Natural odors might be found in organic-rich soils; other sediments may have their own distinctive odors. Contaminated sediments may have the odor of the contaminant; gasoline, solvents, pesticides, ammonia, or other fertilizers all have a distinctive odor. Never vigorously inhale the odor of any sediment, because even natural materials may have harmful fumes.

Cementation of the sediment may be worth noting, because it may have a strong effect on hydraulic conductivity or strength of the mineral. *Mineralogy* of the sediment and/or the cement might be easily determined. Common cements are silica, calcite, clay, and iron oxide.

Bedding and sedimentary structures should be noted where observed. Bedding thickness may range from thin laminations to very thick beds of more than a meter thickness. Examples of sedimentary structures are cross-bedding, ripple marks, mud cracks, burrows or bioturbation, and slumping structures, among others.

Plasticity may be important, especially if the project involves engineering foundations or other structures. Plasticity of sediment or soil is related to its ability to hold a shape without crumbling when wetted. A highly plastic clay, for example, can be molded into almost any shape, whereas a handful of low-plasticity silt will crumble when worked. More information on plasticity is given under "Unified Soil Classification System," which follows.

Unconfined compressive strength is a measure of the sediment's stiffness or resistance to crushing and molding. It may be estimated with a pocket penetrometer or, less precisely, by a "rule-of-thumb" estimate. A pocket penetrometer is a small hand-held instrument that is pressed into a sediment sample (Fig. 6.6). A calibrated spring inside the penetrometer compresses; when the nose of the penetrometer penetrates the sediment a distance of 1/4 inch, the instrument is removed, and the unconfined compressive strength is read from a scale. The rule of thumb for describing consistency of fine-grained soils is given in Table 6.1.

It may be useful to classify the sediment according to one or more *classification* schemes. Because it is used commonly in hydrogeologic, engineering, geotechnical, and environmental work, the Unified Soil Classification System (USCS) is described here. However, other classification systems, such as the American Association of State Highway and Transportation Officials (AASHTO) classification, might be used as well.

Unified Soil Classification System (USCS)

The Unified Soil Classification System (USCS) is used to classify sediment with a two-letter code that describes its grain size and one of two engineering properties.

The two-letter code should be assigned based on the results of laboratory tests (specifically, grain size analysis and Atterberg limits). For more information on these tests, see any basic text on geotechnical or soil engineering, such as Holtz and Kovacs (1981), Spangler and Handy (1982), West (1995), and Rahn (1996).

Fig. 6.6 A pocket penetrometer.

TABLE 6.1		
Estimation of Compressive Strength of Fine-Grained Soils		
Description	**Consistency**	**Approximate compressive strength (kg/cm^2)**
+++		0.1
Soil extrudes from between fingers when squeezed	Very soft	
+++		0.5
Soil is molded by strong finger pressure	Medium	
+++		1
Soil is readily indented by thumb, but penetrated with great effort	Stiff	
+++		2
Soil is readily indented by thumbnail	Very stiff	
+++		4
Soil is indented by thumbnail with difficulty	Hard	
+++		10

Source: Dennen and Moore, 1986.

Although the soil classification cannot be confirmed without laboratory testing, with practice, the hydrogeologist can learn to perform a field classification that may be useful as a shorthand way of delineating hydrostratigraphic units in the field. This section describes this field classification. For further information, see the ASTM Standard on Description and Identification of Soils (Standard D 2488).

Engineers tend to describe sediment grain size distribution in terms of grading, whereas geologists usually describe it in terms of sorting. The USCS is primarily an engineering classification, so it uses the term *grading*. Sediment that is poorly graded is well-sorted, and sediment that is well-graded is poorly sorted.

Finding the First Letter of the Code

For most soils, the dominant grain size gives the first letter of the code. The following are the major grain size classes:

G = gravel
S = sand
M = silt
C = clay

Two special types of soil merit their own first letters:

O = organic material made almost entirely of decaying plants (musty in odor, nonfibrous, often black in color). If the sediment is organic, determine its second letter by following the steps for silt and clay.

Pt = peat (musty in odor, with a spongy or fibrous texture). For peat, the entire code is Pt, and no secondary characteristic is determined.

Finding the Second Letter of the Code

The second letter of the code is determined on the basis of further observations and follow-up tests. The various options for the second letter depend on the grain size of the dominant fraction of the sediments. Once this is known, the grain size or the plasticity of the nondominant fraction must be investigated. The following sections describe how to choose the second letter in the case of dominantly coarse-grained material and in the case of dominantly fine-grained material.

Dominantly Coarse-Grained Material.   When the sediment or soil is mainly composed of coarse-grained material (gravel or sand), the size of the nondominant portion must be determined. If it, too, is coarse-grained, then the grading of the finer portion determines the second letter. However, if the nondominant portion is fine-grained (silt or clay), then the size of the finer portion determines the second letter. These two possibilities are examined more fully in what follows.

To determine the size of the nondominant fraction, remove any gravel and saturate the sediment. If the sediment does not stain the hands, and if weak casts (e.g., thumbprints) cannot be formed because the material crumbles, it probably contains little fine-grained material. Beach sand is an example; it usually is poorly graded, without an appreciable amount of fine-grained material.

Both the Dominant and Nondominant Fraction Are Coarse-Grained If the sediment contains more than 88% coarse-grained material, then the second letter of the code is chosen based on whether the sediment is poorly graded or well-graded. To determine this, spread some of the sample out on a flat surface and observe the grain sizes. If the sediment contains all one size of grains (or two sizes with no intermediate sizes), it is poorly graded, and the second letter of the code is "P." If the sediment contains grains of a variety of sizes, it is well-graded, and the second letter of the code is "W." For example, "GP" indicates poorly graded gravel, and "SW" indicates well-graded sand.

The Dominant Fraction is Coarse-Grained, But the Nondominant Fraction Is Fine-Grained. If the gravelly or sandy sediment contains a significant amount (>12%) fine-grained material, then the second letter of the code tells if the fine-grained material is mainly silt (M) or mainly clay (C). To determine this, perform two tests.

In the first test, wet the sample and mold it between the fingers; if necessary, add enough water so that the sediment does not stick to the hands. Roll the sediment into a thread, and then take hold of both ends of the thread and gently try to pull it apart. If it stretches instead of breaking, the fine-grained material probably is mostly clay, and the second letter of the code is "C." If the sample thread is easy or moderately easy to pull apart, the fine-grained material is mostly silt, and the second letter of the code is "M." More evidence in support of this preliminary assessment may be gained by performing the next test.

In the second test, roll the sample into a thread about 4 cm long. Keep rolling, removing excess sample as needed, until the 4-cm thread just crumbles as you pick up one end of it. At this point, determine the diameter of the thread. Because clay is more cohesive than silt, if the sediment contains a great deal of clay, it may be possible to roll a very thin thread without it crumbling. If the crumbly thread is more than 4 mm in diameter, the fine-grained fraction is mainly silt, and the second letter of the code is "M." If the crumbly thread is less than 4 mm in diameter, the fine-grained material is mostly clay, and the second letter of the code is "C." For example, "GM" is gravel with a significant amount of silt, and "SC" is sand with a significant amount of clay.

Remember that laboratory tests are needed to make a definitive assessment of the classification.

Dominantly Fine-Grained Material. When the dominant material is fine-grained (silt and clay), the second letter of the code is based on the plasticity of the sediment. Plasticity can be determined definitively only by performing laboratory tests of the material's Atterberg limits. However, a field classification may be made as a preliminary assessment. This field assessment is modified later, after the laboratory test results are available. To make the field assessment of plasticity, perform three tests on the sample.

In the first test, wet the sample and mold it between the fingers; if necessary, add enough water so that the sediment does not stick to the hands. Roll the sediment into a thread, and then hold the thread by both ends and try to pull it apart. If the thread does not pull apart, but instead stretches, the fine-grained material has a high plasticity, and the second letter of the code is "H." If the sample thread is easy or moderately easy to pull apart, the fine-grained material has a low plasticity, and the second letter of the code is "L."

In the second test, wet the sediment and mold it with the fingers. Form a smooth pat of sediment in the palm of the hand. Strike the hand with the sample several times against the other hand, watching the top surface of the sample for a sheen indicating that water has risen to the top of the sample. Then, squeeze the sample and note if the water disappears quickly or not. If water appears immediately and disappears quickly, the sediment is probably silt of low plasticity (ML). If water appears slowly and disappears slowly or not at all, the sediment is probably clay of low plasticity (CL) or silt of high plasticity (MH). If there is no visible change either when shaking or squeezing, the sediment is probably clay of high plasticity (CH).

In the third test, use a clod of dry sediment no smaller than 2 cm in its smallest dimension. Crush the clod between the fingers. If it is easy to crush the clod, the sediment is probably silt of low plasticity (ML). If it is moderately difficult to crush, the sediment is probably clay of low plasticity (CL) or silt of high plasticity (MH). If it is difficult to impossible to break, the sediment is probably clay of high plasticity (CH).

Table 6.2 summarizes sediment classification using the USCS.

Describing Rocks

A hydrogeologist may need to describe rocks in an outcrop or as drill cuttings or core. This text cannot substitute for performing laboratory work, studying petrology texts, or gaining field experience in rock identification. However, some characteristics that might be noted in the field log are described here. Compton's book *Geology in the Field* (1985) is an excellent reference for further reading.

Describing Rocks in Outcrops

The first step in rock identification is to classify the rock as igneous, sedimentary, or metamorphic. In order to do this, obtain a fresh surface of the rock by digging away loose debris and chiseling or hammering away the weathered surface. Look at the rock's texture, layering, relationships with other rocks in the outcrop, and crystallinity. Refer to and consider previous work in the area and geologic maps of the area. The basic principles of rock description are not unique to hydrogeology and are discussed in other sources. See Compton (1985) for a comprehensive discussion of characteristics that might be described. Some of those characteristics are listed below.

Igneous Rocks

Rock type
Crystal size

TABLE 6.2

Summary: Field Classification of Sediment Using the USCS

Step 1: Find the first letter of code

Determine grain size of dominant grains.

Gravel (greater than 2-mm diameter)
→ GW, GP, GC, or GM

Sand (smaller than 2-mm diameter, but grain boundaries are visible to the unaided eye)
→ SW, SP, SC, or SM

Silt (grain boundaries are not visible to the unaided eye, but material feels gritty; dry lumps are easy to smash; water rises quickly to the top of wet pats of material when jolted, causing a wet sheen)
→ ML or MH

Clay (feels slippery when wet; smears, stains, and sticks to the fingers when wet; makes smooth ribbons on an auger, can be rolled into thin threads; dry lumps are hard to smash; water does not rise to the top of wet pats of material when jolted, resulting in no wet sheen)
→ CL or CH

Organic (made almost entirely of decaying plants, but not spongy or fibrous)
→ OL or OH

Peat (spongy or fibrous texture)
→ Pt (no second letter is necessary)

Step 2: Find the second letter of the code

Is the first letter G or S?

Yes, the first letter is G or S.
→ Sample is gravel or sand. Now, perform this test:

Remove gravel

Saturate sample

Does sample stain hands? Can weak casts be formed?

No → not much fine-grained material
Are grains poorly graded (well-sorted)?
Yes → GP or SP
No → GW or SW

Yes → significant fine-grained material(>12%)
Determine grain size of fine-grained material:
Test one (Plasticity):
Wet sample; mold it between fingers (add enough water so sample does not stick to hands)
Roll sample into a thread
Try to pull thread apart:
hard to pull apart → GC or SC
medium to pull apart → GM or SM
easy to pull apart → GM or SM

Test Two (Crumbling Thread):
Roll sample into thread until 4-cm-long section just crumbles when you try to pick up thread by one end
What is the thickness of thread?
>4mm → GM or SM
<4mm → GC or SC

(continued)

TABLE 6.2   (*continued*)

No, the first letter is not G or S
  → Sample is silt, clay, or organic soil, so first letter is M, C, or O. (Peat is an
  exception readily recognized by its fibrous texture; if the sediment is peat,
  the code is Pt, and the following tests need not be performed.)
To determine the second letter, perform these tests:
Determine plasticity of materials
  Test one (Plasticity):
  Wet sample; mold it between fingers (add enough water so sediment does
  not stick to hands)
  Roll sediment into a thread
  Try to pull thread apart
  hard to pull apart: MH, CH, or OH
  medium to pull apart: ML, CL, or OL
  easy to pull apart: ML, CL, or OL
  Test Two (Crumbling Thread):
  Roll sample into thread until 4-cm-long section
  just crumbles when you try to pick up thread by one end
  What is thread's thickness?
  >4mm → ML, CL, or OL
  <4mm → MH, CH, or OH
  Test Three (Dry Crushing Strength):
  Use a dry clod of sediment
  How easy is it to crush the clod between the fingers?
  easy → ML
  moderately difficult → CL or MH
  difficult or impossible → CH

Color

Minerals

Degree of weathering

Occurrence and degree of fracturing

Attitude of structural features (joints, faults, and other similar features)

Sedimentary Rocks

Clastic

Rock name

Grain size

Color

Minerals

Degree of weathering and solutioning

Cementation

Sedimentary structures and bedding

Fossils

Occurrence and degree of fracturing

Attitude of structural features (joints, faults, bedding planes, and other similar features)

Chemical

Rock name

Minerals

Sedimentary structures and bedding

Crystal size

Degree of weathering and solutioning

Cementation

Fossils

Occurrence and degree of fracturing

Attitude of structural features (joints, faults, bedding planes, and other similar features)

Metamorphic Rocks

Rock name

Minerals

Foliation (nature and orientation)

Recrystallization

Degree of weathering

Occurrence and degree of fracturing

Attitude of structural features (joints, faults, foliation, and other similar features)

An understanding of structural features may be particularly important in delineating flow paths and in determining controls on subsurface water movement. This is particularly true when fracture flow is significant within the hydrogeologic system. In addition, for engineering studies, information on the characteristics of the discontinuities in the rock may have a direct bearing on the stability of a rock mass. As a result, careful mapping of the attitudes of rock units and the orientations and character of bedding planes, fractures, joints, foliation planes, and similar structural features within an outcrop may be the most critical aspects of the field study.

Strike and dip measurements, attitude and orientation of rock and soil contacts, descriptions of brecciated zones, and measurements and descriptions of offset sections all may provide important clues to the nature and geometry of a flow system. Compton (1985) gave detailed instructions on how to gather data for mapping geologic structures, and the reader should consult this source for more information.

Discontinuities within rock masses, such as joints, faults, foliation planes, shear zones, and bedding planes, provide pathways for water flow and exert strong control over the strength of the rock mass. When mapping discontinuities, the field hydrogeologist should take particular note of their orientation, spacing, length, surface roughness, weathering or alteration of the surface, degree of separation of the two sides of the discontinuities, and presence or absence of water. Johnson and DeGraff (1988) reviewed important aspects of discontinuities, with an emphasis on how they affect rock mass strength.

Describing Rocks in Chips, Cuttings, or Bulk Samples

Describing rock samples in the form of chips, cuttings, or bulk samples is more difficult than describing rocks in an outcrop, because the size of the sample is small. To the extent possible, these rocks should be described following the same general guidelines as for outcrops. For the most part, it will be difficult to describe sedimentary structures or relationships between different rock formations. Do the best you can, and resist the temptation to make interpretations that are not supported by the evidence at hand.

Describing Rock Cores

Rock coring may be performed where it is necessary to have a core for laboratory testing, or where a continuous record of the rock is needed. Cores should be numbered and labeled as they are retrieved. They should be stored in core boxes, sturdy boxes built for this purpose. Be sure that when boxing or examining cores, they are oriented and labeled such that the "up" direction is known. Mark this direction on the cores and their boxes.

In general, rock cores should be described in the same way as and using the same descriptors as for rock outcrops. Cores can be more difficult to describe because the smooth surface of the core does not show natural fracture faces. However, stratification often is clearly visible in cores, and structures, bedding, and fossils may be identifiable. Davis and colleagues (1991) gave a thorough review of rock core geologic description.

A few other characteristics of rock cores also may be described. These characteristics, which follow, are particularly important in engineering, geotechnical, or environmental work.

Rock Quality Designation (RQD)

Rock Quality Designation (RQD) is a rough measure of the degree of fracturing of a length of rock core. To find the RQD, measure all pieces of core 4 inches long or longer. Sum up and record the total length of all those pieces put together. Then, divide this by the total length of the core "run". Core run is the length of the core barrel that was drilled into the ground.

Percent Recovery

Percent recovery refers to the fact that as core drilling proceeds, some sections of core may not be retrieved. This may occur for one of several reasons. In some cases, especially in karst regions, it is possible that no rock was there to be retrieved. In other cases, the rock is so friable that it crumbles during the coring, and as the core barrel is pulled from the hole, the pieces drop out, resulting in an empty barrel.

To find Percent Recovery, first, find the total length of the core that is recovered. Divide this by the total length of the core run. The core run is the length of the core barrel that was drilled into the ground.

Values of Rock and Sediment Characteristics

Tables 6.3 and 6.4 give typical ranges of values for various characteristics of rock and sediment. These values are simply summaries of reported ranges and should not be assumed to represent a specific sample or geologic unit.

TABLE 6.3

Ranges of Values of Porosity and Effective Porosity or Specific Yield[a]

Sediment/Rock	Porosity (%)	Effective Porosity or Specific Yield (%)
Clay	40–60	0–5
Silt	35–50	3–19
Sand		
Fine or silty	20–50	10–28
Coarse or well-sorted	20–50	20–35
Gravel	25–40	13–26
Shale		
Intact	1–10	0.5–5
Fractured/weathered	30–50	
Sandstone	5–35	0.5–10
Limestone and dolomite		
Nonkarst	0.1–25	0.1–5
Reef or karst	5–50	5–40
Chalk	5–45	0.05–0.5
Anhydrite	0.5–5	0.05–0.5
Salt	0.1–0.5	0.1
Basalt	1–50	
Unfractured igneous and metamorphic rocks	0.01–1	0.0005
Fractured igneous and metamorphic rocks	1–10	0.00005–0.01

Sources: Johnson, 1967; Norton and Knapp, 1977; Freeze and Cherry, 1979; Maidment, 1993; Shen and Julien, 1993;.Smith and Wheatcraft, 1993; Fetter, 1994; Domenico and Schwartz, 1998.

[a]Although effective porosity and specific yield are not synonymous, for practical purposes, they may be estimated to be approximately equal in value.

TABLE 6.4
Ranges of Values of Hydraulic Conductivity and Permeability

Sediment/Rock	Hydraulic Conductivity (cm/sec)	Intrinsic Permeability (cm^2)
Clay	10^{-9} to 10^{-6}	10^{-14} to 10^{-11}
Silt	10^{-7} to 10^{-3}	10^{-12} to 10^{-8}
Sand		
Fine or silty	10^{-5} to 10^{-3}	10^{-10} to 10^{-8}
Course or well-sorted	10^{-3} to 10^{-1}	10^{-8} to 10^{-6}
Gravel	10^{-1} to 10^{+2}	10^{-6} to 10^{-3}
Till		
Dense/unfractured	10^{-9} to 10^{-5}	10^{-14} to 10^{-9}
Fractured	10^{-7} to 10^{-3}	10^{-12} to 10^{-8}
Shale		
Intact	10^{-11} to 10^{-7}	10^{-16} to 10^{-12}
Fractured/weathered	10^{-7} to 10^{-4}	10^{-12} to 10^{-9}
Sandstone		
Tightly cemented	10^{-8} to 10^{-5}	10^{-13} to 10^{-10}
Loosely cemented	10^{-6} to 10^{-3}	10^{-11} to 10^{-8}
Limestone and dolomite		
Non-karst	10^{-7} to 10^{-3}	10^{-12} to 10^{-8}
Reef or karst	10^{-4} to 10^{+4}	10^{-9} to 10^{-1}
Chalk	10^{-6} to 10^{-3}	10^{-11} to 10^{-8}
Anhydrite	10^{-10} to 10^{-9}	10^{-15} to 10^{-14}
Salt	10^{-12} to 10^{-5}	10^{-17} to 10^{-9}
Basalt		
Unfractured	10^{-9} to 10^{-6}	10^{-14} to 10^{-11}
Fractured/vesicular	10^{-4} to 10^{+3}	10^{-9} to 10^{-2}
Unfractured igneous and metamorphic rocks	10^{-12} to 10^{-8}	10^{-17} to 10^{-13}
Fractured igneous and metamorphic rocks	10^{-8} to 10^{-4}	10^{-13} to 10^{-9}

Sources: Freeze and Cherry, 1979; Maidment, 1993; Smith and Wheatcraft, 1993; Fetter, 1994; Domenico and Schwartz, 1998.

Sediment and Rock Sampling and Drilling Methods

Chapter Overview

Hydrogeologists who wish to identify aquifers, aquitards, and flow directions must investigate the physical characteristics of the sediment and rock. Geological reports, maps, and cross-sections may provide information that can be applied on a regional scale. However, many projects require site-specific information. In these cases, hydrogeologic features may be mapped and examined by sampling from outcrops, trenches, pits, mine adits, and quarries. However, sometimes the only way this information can be gained is by drilling boreholes at the site.

Sampling and drilling programs are important in several types of hydrogeological projects. For example, characterizing subsurface lithology often requires the drilling of soil or rock borings and extraction of samples. Some soil tests are best performed during drilling operations. And most water wells in the United States are installed in boreholes that are drilled. As a result, hydrogeologists need a working knowledge of methods for sampling and drilling sediment and rock.

This chapter investigates the aspects of a sampling program that must be considered at the outset of a hydrogeologic project in order to achieve the project goals. It introduces the components of sampling plans and commonly used sampling protocols. Finally, it describes typical sampling tools, drilling methods, and soil-testing procedures performed during drilling operations. More information is available in numerous other sources, including Campbell and Lehr (1973), Driscoll (1986), Clark (1988), Aller and colleagues (1989), Roscoe Moss Company (1990), Nielsen (1991), Maidment (1993), and Australian Drilling Industry Training Committee (1997).

Considerations For Selection of Sampling Methods

Many different sampling and drilling tools and methods are available, and are described later in this chapter. To set the stage for selection of the appropriate tools and methods, a number of other factors must be considered: the budget and schedule of the project, purpose of the borings, geologic and hydrogeologic conditions, sampling requirements, site access, disposal of cuttings and fluids, and health and safety considerations. Those considerations are presented in this section.

Budget and Schedule

The budget and work schedule will help determine the type and quantity of samples to be taken. In addition, these factors will help determine the drilling and sampling methods to be used. Drilling and sampling methods vary in cost. Some

199

methods require more time than others, which in turn raises the costs. In addition, geologic conditions will control the suitability of various methods for the necessary tasks. A drilling contractor who is experienced in a given area and with the appropriate drilling methods may be able to provide estimates for different situations.

Some portions of a job may be bid on a time and materials basis. In this case, the drillers are paid for the materials they use and according to how many days or hours they work. For these items, the costs of unexpected delays in the drilling process and unforseen geologic conditions are passed on by the drillers to their clients.

Other portions of a job, or a whole job, may be bid on a lump-sum basis. For these portions of the job, the cost is the same no matter what conditions are encountered nor how many borings are to be drilled. For example, "mobilization" is one such item. Because it costs the drillers just as much to move their rig to the site for one hole as for 100 holes, this item is often charged as a lump sum. In lump-sum contracts, the cost of unseen conditions is borne by the driller. Alternatively, if the job goes smoothly, the driller realizes the profit.

To make bidding realistic and fair, the drillers should have as much information as possible on the anticipated geologic conditions that might affect their progress, as well as the specifics of sampling, well construction, and health and safety requirements. Driscoll (1986) presented a thorough examination of the construction of water well drilling contracts, and Aller and colleagues (1989) provided a useful summary of considerations for contracts and specifications of monitoring well drilling and installation.

Purpose of the Boring

A borehole may be drilled for any number of reasons. Some of these are rock coring, soil sampling, water sampling, vapor sampling, soil venting, air sparging, monitoring well installation, extraction well installation, or piezometer installation. In some cases, the holes may be temporary; in other cases, permanent wells will be installed in the holes. Each purpose will dictate its own set of requirements. For example, a boring that is drilled for the purpose of retrieving rock core samples clearly cannot be made by a method that produces only cuttings, not cores.

Geologic and Hydrogeologic Conditions

Along with the objectives of the field program, geologic and hydrogeologic conditions will determine what drilling method is used, what type of bit is used, how rapidly the drilling will progress, and what special materials will be required at the site. For example, it is more difficult to drill in heaving sands (sands below the water table, which tend to push up into the drill stem during drilling) than in well-compacted clays. In rock formations, auger drilling is impossible, so the driller will need to be prepared for rock drilling. Drilling conditions are a function of rock or sediment type as well as hydrogeologic conditions. Procedures will vary depending whether the drilling is to be performed above or below the water table, in artesian or nonartesian conditions, or in changing or varied lithologies.

To help the drilling go most efficiently, the total depth of the hole should be estimated, as well as the type and thickness of sediments, depth to rock, and location of the water table. This information should be obtained before talking with the drilling contractor. Before making a bid or mobilizing the equipment, the driller will want to know the type of sampling and testing to be done, what equipment will be needed, and the conditions that will likely be encountered.

Sampling Requirements

Determine what sampling method will be used. To do this, consider what the samples will be used for. What tests will be conducted? Will samples need to be packaged and shipped, or will onsite tests suffice? Will samples need to be taken from several depths? Will continuous sampling be required? Will sample "cores" be needed? Will cores need to be oriented, that is, retrieved in such a way that their original geographic orientation is known?

Site Access

Some borings may be made in easily accessible locations. But at some locations, access may be difficult. The following conditions may cause problems for drilling rigs: sites inside buildings (some drillers have small rigs with low masts that can work indoors with no problem), overhead power or phone lines that might restrict raising the mast, viaducts or overhead pipelines that are too low to permit the rig to pass, dense brush or trees that might block access for a truck-mounted rig; marshy conditions, unstable soils, steep slopes, locations under water (some drillers can drill from an anchored barge or platform). Often, these problems are not insurmountable. But they may require special equipment or procedures.

Before drilling, the site should be checked for underground utilities such as gas, electric, phone, cable television, sewer, or water lines. The local utility companies or a local hotline that checks for buried lines usually can perform this check within a few days. Hitting utilities with the drilling rig or backhoe may cause anything from a minor inconvenience for the local residents to a fatal accident. And whether it results in an accident or not, hitting utilities inevitably will slow down the job.

Disposal of Cuttings and Fluids

Some methods produce no cuttings or fluids at all; others produce so much material that the drillers will have to add another member to the crew just to keep up with clearing the cuttings away. Will cuttings have to be drummed and disposed of by a special waste handler? Will they be returned to the hole from which they came? What will be done with the excess cuttings?

Health and Safety

The drilling and sampling should be planned and carried out in such a way that workers are not exposed to hazards or put at risk. This may require special precautions; personal protective equipment (PPE); or special procedures before, during, and after the drilling and sampling. Safety considerations and commonsense rules should be attended to. At outcrops, extreme caution should be used near walls of sediment or newly blasted quarry walls. Collapse of these walls is common and can be deadly. Pits and trenches are dangerous; no one should enter any excavation deeper than waist deep. Loose or unstable rocks or sediments should not be climbed. Material falling from above may be a hazard; working directly downslope or upslope of another person should be avoided.

In underground mines, walls and ceilings should not be climbed or hammered. Mine gases may be dangerous; safety equipment and monitors should be carried, and safety regulations followed. Workers at mines and quarries should learn the signals that indicate that a blast or charge is about to be set off. Workers should know the locations of emergency shelters, first aid kits, and communications stations. At any outcrop or in any mine, vehicular traffic should be given wide clearance, as the vehicles may have poor visibility and may require long stopping distances.

Sampling Plans and Protocols

Sampling involves extracting a portion of a rock or sediment formation in order to examine it or perform some test or analysis. No samples should be taken unless the reason for doing so is well-understood, because drilling and sampling are expensive, and because the reason for sampling determines how, where, and when the sample is taken. Choice of sampling plan, protocol, method, and container all depend on what is being sampled and the reason for sampling.

Why might a sample be taken? Sampling is done for the purpose of gaining site-specific information about the geology or subsurface conditions so that the hydrogeological conditions can be characterized. Sometimes, samples are taken to delineate the extent of soil contamination. In some cases, a simple field description of the material is needed (a discussion of field methods for describing such parameters as color, grain size, moisture content, and other factors appears in Chapter 6). In these situations, the sample may be disposed of immediately after it is described in the field. But in other cases, further analysis, either onsite or in a laboratory, may be necessary. When further tests are required, samples must be retrieved and packaged for the subsequent analysis.

Laboratory analyses include physical, chemical, and biological testing. Analyses of physical properties include tests of such characteristics as permeability, grain size distribution, porosity, specific gravity, organic content, shear strength, unconfined compressive strength, moisture content, Atterberg limits, or mineralogy, among others. Chemical analyses might investigate composition of organic, inorganic, or radioactive components of gases, fluids, or solids. Biological analyses might examine bacteria, viruses, or various other organisms in ground water or soil. Whatever the tests required, the reason for sampling should be known before the sampling plan is developed.

Sampling Plan

The sampling plan should be designed around the specific goals of sampling. At a minimum, the plan should include type, location, depth, and number of samples, along with sampling method and protocol, whether disturbed or undisturbed samples are required, sample identification system, sample containers to be used, sample preservation, and timing requirements. The remainder of this section describes these factors.

The plan also should describe how, when, and where the samples will be examined. It is most common for an initial field description to be made at the site where the sample is taken. However, in some cases, such as the logging of rock cores, the description may be done at some other designated location. For some analyses, samples must be packaged and transported to a laboratory. No matter where the description is to be performed, the hydrogeologist should be present in the field when the sample is taken.

Type of Samples

Samples may be composite or representative. A *composite sample* is a mix of materials from several depths within a borehole. These are taken to make a general characterization of an interval or area. For example, composite samples may be useful at the beginning of a remedial investigation. In these situations, the purpose of composite sampling is to answer the question of whether contamination exists at the site. At this stage, the precise amount of a contaminant at a given location or depth is less important than answering the question of whether the contaminant is present at all. Composite samples are typically retrieved in the form of drill cuttings, rather than cores.

In contrast to a composite sample, a *representative or grab sample* is taken from a specific depth and for the purpose of representing conditions at that depth and location. For example, representative samples might be used to determine the extent of soil contamination, determine whether a particular geologic formation exists at a given location and depth, or to characterize the hydrogeologic and engineering properties of the geologic materials at a specific point.

Location, Depth, and Number of Samples

As with most sampling considerations, the choices of location and depth of sampling depend on the purpose for which the information will be used. How much will sampling and analysis cost, and how much is budgeted? When the sampling budget is limited, as it virtually always is, it is important to extract the maximum amount of information from the minimum number of samples. Selecting sampling points judiciously should help minimize the number of samples needed to fulfill the goals of the sampling plan.

Is the purpose of the sampling project to determine the characteristics of each new formation encountered during a drilling project? If so, and if each formation is fairly consistent in composition, it might be enough to take one sample from each formation. Or is it, perhaps, more important to get a point-by-point record of subsurface conditions? In such a situation, it might be more useful to take samples continuously, or at constant intervals, for example, every foot, or every 5 feet. Answering these questions will help define the sampling plan.

In some cases, the minimum number of samples may be dictated by environmental or other regulations. For example, regulations on monitoring ground water in the area of a landfill require a certain number of wells that produce water samples representing background or upgradient conditions, and a certain number of wells that produce samples representing downgradient conditions.

Sampling Method and Protocol

Choice of sampling method must be governed first and foremost by safety considerations. For example, climbing up rock walls or down into trenches should be avoided; safety glasses should be worn if rocks are to be hammered; protective clothing should be worn if the risk warrants it. Many other precautions might be necessary, and for this reason, anyone who is to perform sampling should have completed a training course in health and safety and should follow a health and safety plan. See Chapter 2 for more information on health and safety plans.

Sampling tools and methods should be specified in the sampling plan. Important features of the various tools and methods available are described in more detail later in this chapter.

Sampling protocol should be established ahead of time and stipulated in the sampling plan. Sampling protocol depends on the purpose of the sampling program. For any of the sampling protocols that follow, be sure to make a record in the field notebook of each sample taken, giving borehole number, location, depth, and any other pertinent information.

Sampling Procedure for Volatiles

If the sample is to be analyzed for *volatile contaminants* (e.g., gasoline or light oils), it is important to minimize contact of the samples with the air. When the material sampled is rock, the sample should be placed quickly into the sample container, and the container capped without delay. Once the container is capped, it should not be opened again, and should be disturbed as little as possible. Some sampling plans call for the use of an organic vapor analyzer (OVA). If this is the case, the OVA should be used at this point to check the level of gases in the soil. The reading should be recorded in a field notebook. The sample container should be labeled and placed in a cooler at 4°C. Then, the next sample may be taken. When all samples have been taken, the cooler should be sealed with tamper-evident tape, the chain-of-custody forms completed, and the cooler shipped or transported to the laboratory.

For soils and sediments, the procedure is basically the same. The USEPA (1991) gave the following suggestions for sampling soil for volatiles. Tube samplers should be used whenever possible, rather than augers. Tube samplers are pushed or jacked into the soil to extract a core of soil. They cause less disturbance to the structure of the soil than do augers, and therefore cause less disturbance to the gases held in the pore spaces.

The first sample should be taken, keeping it as intact and undisturbed as possible. It should be placed immediately into a vial or widemouth glass bottle. The container should be quickly filled to the top and capped. Once the sample container is closed, it should not be opened again. If an organic vapor analyzer is to be used to measure amounts of gases in the borehole, it should be used at this point, with the readings recorded in the field notebook.

The container should be labeled as described in the section on sample identification (in this chapter). The sample may be put in a plastic bag if necessary, and then placed in a 4°C cooler. Then, the next sample may be taken. When all samples have been collected and placed in the cooler, the cooler should be sealed with tamper-evident tape, the chain-of-custody forms completed, and the samples shipped or transported to the laboratory.

Sampling Procedure for Semivolatiles and Metals

If a sample is to be analyzed for anything other than volatiles, such as *semivolatiles* or *metals,* then the handling of samples may be different. In this case, there is minimal concern that the components of interest will escape into the air. Instead, the major concern is obtaining a sample that provides information on the actual conditions in the subsurface.

Rock samples should be extracted, placed in the appropriate containers, sealed, labeled, and sent to the lab. Samples of soil and sediments may be treated basically the same way. However, when the situation calls for it, a bulk sample may be mixed thoroughly while still in the field. By mixing the sample, it is possible to obtain fairly consistent and uniform subsamples, each of which may be used for a different analysis (e.g., semivolatiles, metals, duplicates, and so on).

The USEPA (1991) provided the following guidelines for sampling and mixing soils for analysis of semivolatiles and metals. First, the sample (or samples, if a composite sample is needed) is collected. Sampling tools and containers should be made of a material that will not compromise the analytical results. That is, the materials should not contain the constituent that is to be analyzed in the sediment.

Next, if the sampling plan calls for it, the samples are composited by mixing them together in a bucket or other container. If the samples are not to be composited, they should be kept intact as they were taken. The gravel is removed by screening the material in the field through a clean stainless steel or Teflon®-lined screen with a 2-mm (#10) mesh. The screen material should be chosen on the basis of which analyses will be performed on the sample.

Once the gravel has been removed, the sample is mixed in a mixing container, using a tool such as a spatula. Then, the sample is placed in the middle of a sheet of plastic or canvas, and then is mixed by lifting the corners of the sheet and rolling the sample about. When the sample is completely mixed, it is spread out on the sheet and divided into quarters or "splits." These sections then are quartered or split into smaller and smaller subsections, until the necessary size sample is obtained. By using a clean scoop, spoon, or trowel, the sample is scooped into a sample container.

Finally, the container is labeled and placed in a cooler at 4°C. All sampling tools should be decontaminated between samples. When all samples have been collected, the cooler should be sealed with tamper-evident tape, the chain-of-custody forms completed, and the cooler shipped or transported to the laboratory.

Sample Identification and Labeling System

Each sample should be labeled with the date, time, boring number (if appropriate), and depth interval sampled (if appropriate), as well as with the initials of the sampler. A permanent marking method should be used, but caution is advised when using a marker that contains volatile solvents (or other volatile chemicals) if the sample is to be analyzed for volatiles.

Choosing a sample identification method takes some forethought. It might appear that the simplest method would be to number the samples consecutively as they are taken. However, this method is unsatisfactory for many reasons. With this method, one needs a key to determine which sample came from what location and depth; if the key is lost, the samples are worthless. In addition, the sample identification provides no information in itself.

To be more useful, the sample identification system should be informative and self-explanatory. For example, the samples may be numbered with the borehole number, then a hyphen, and then the depth of the sample. Sample labels also should give the job name or number, date, time, and initials of the sampler. Entries should be made in the field notebook to keep a record of samples taken.

Sample Containers

Choice of sample container is dictated by the reason for taking the sample. An analytical lab can provide guidance on the appropriate type of container for each test.

In choosing a container, consideration should be given to how much sample is required to perform the test, as well as chemical reactivity of the container. Common sense should be used; for example, a metal container should not be used if the plan is to analyze metals. As another example, a water- and vaportight container should be used if the plan calls for an analysis of moisture content. An alternative in this situation is to seal an undisturbed sample with paraffin.

Food-grade plastic bags are inexpensive and easily available, but must be sealed tightly to prevent mixing of samples. If moisture content is to be determined, the bag must be water- and vaportight. However, because moisture content can change even in a sealed bag, this type of analysis must be performed very soon after sampling.

Cloth sampling bags may be particularly appropriate for rocks and dry sediments. Widemouth glass sampling jars or vials may be watertight, and therefore more useful for moist samples. Glass is less likely than plastic to cause inadvertent contamination of the samples by off-gassing of vapors. Glass jars are available in sizes that conveniently hold cores or split-spoon samples.

Samples are preserved by isolating them from air, water, and other agents, and placing them in a 4°C cooler. In general, rock and sediment samples need no preservation, but in some cases, they may need to be stabilized by impregnating them or adding some chemical stabilizing agent.

Timing

Because conditions such as temperature and weather can change rapidly, and because these might affect the chemistry of the sample, it may be important to take samples at a particular time or with a certain rapidity. This should be considered in the sampling plan.

Documentation

Chain-of-custody forms show exactly who had possession of the samples at what point in time, and to whom possession was relinquished. These forms help the hydrogeologist keep track of what has happened to the samples. They are particularly important in situations where contamination is known or suspected, because they provide some measure of accountability if sample tampering is alleged. A sample of a chain-of-custody form is given in Fig. 7.1(a) and a completed chain-of-custody form is shown in Fig. 7.1(b).

Shipping

Samples that will undergo chemical analysis, those for which moisture content will be determined, and those containing contaminants should be shipped to the lab on the day they are taken. Some parameters are unstable and may change with time, so these analyses must be performed immediately. Instructions on holding times for various parameters can be provided by the analytical lab. Holding times are the time periods within which samples must be analyzed in order to achieve valid and reproducible results. Some samples, such as rock cores, may be held onsite in core boxes for logging.

Sampling Tools and Drilling Methods

A wide variety of hand tools and powered machines are available as sampling tools. Choice of sampling tool depends on the type of sample to be taken; the accessibility of the soil, sediment, or rock to be sampled; the type of material to be sampled; and the tests to be performed. Will the sample be of sediment or of rock? Are undisturbed samples required? Is a core sample needed or can the sample be taken from drill or auger cuttings? Will the sample come from the surface or subsurface? If subsurface, from what depth? Cost and availability of the sampling tool must be considered, as well as site access for the tool.

This section reviews sampling tools and methods of drilling, with particular attention to the aspects that are most pertinent to the work of the hydrogeologist. Driscoll (1986) gave a comprehensive treatment of drilling methods and specialized techniques, and Aller and colleagues (1989) provided a thorough review of considerations particular to monitoring wells.

Common Hand Tools

Common hand tools such as trowels, spades, shovels, and scoops may be used for sampling soils or rocks or for digging holes for wells.

Soils may be sampled by hand trowel or scoop. If contamination due to metal tools might be a problem, plastic scoops should be used. When using shovels, spades, or picks, use proper techniques in lifting. Lift with the leg muscles instead of the back, and do not twist while holding a loaded shovel.

Rocks may be sampled with a geologist's hammer (rock hammer). For highly indurated rocks, use a "crack hammer" and rock chisel. (A crack hammer is a short-handled heavy sledge with hardened faces.) A sledge and wedge might also

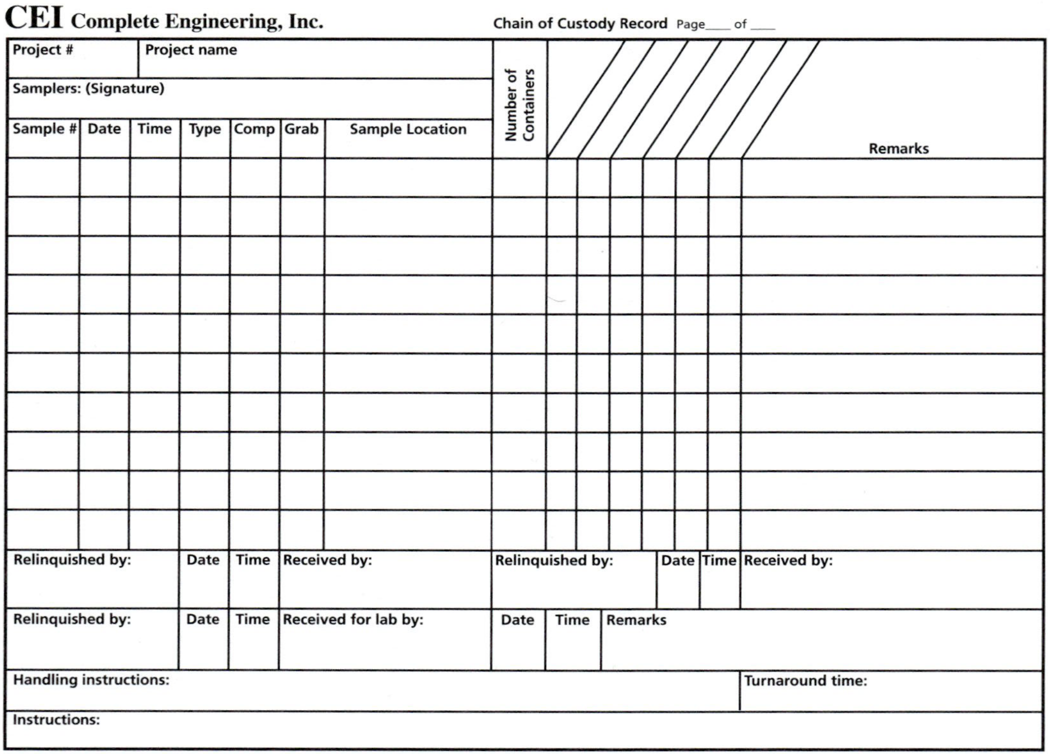

CEI Complete Engineering, Inc.

Chain of Custody Record Page ___ of ___

Project #	Project name								
Samplers: (Signature)									

Sample #	Date	Time	Type	Comp	Grab	Sample Location	Number of Containers	Remarks

Relinquished by:	Date	Time	Received by:	Relinquished by:	Date	Time	Received by:
Relinquished by:	Date	Time	Received for lab by:	Date	Time	Remarks	

Handling instructions:

Instructions:

Turnaround time:

(a)

Fig. 7.1(a) A sample chain-of-custody form.

CEI Complete Engineering, Inc. Chain of Custody Record Page 1 of 1

Project #	Project name						Number of Containers	BETX	PNAs	TCP Lead	Flashpoint	paint Filter	pH	Remarks
97512	Modera, Inc.													
Samplers: (Signature) Rick Polad														
Sample #	Date	Time	Type	Comp	Grab	Sample Location								
S-001	8-17 95	10 AM	Soil		X	South wall 6' BGS	2 / 40 oz.	X	X					
S-002	8-17 95	10 AM	Soil		X	North wall 6' BGS	2 / 40 oz.	X	X					
S-003	8-17 95	10 AM	Soil		X	West wall 6' BGS	2 / 40 oz.	X	X					
S-004	8-17 95	11 AM	Soil		X	East wall 6' BGS	2 / 40 oz.	X	X					
S-005	8-17 95	10 AM	Soil		X	South floor 6' BGS	2 / 40 oz.	X	X					
S-006	8-17 95	10 AM	Soil		X	Noth floor 10' BGS	2 / 40 oz.	X	X					
S-007	8-17 95	4 PM	Soil	X		Soil Stockpile	2 / 40 oz.			X	X	X	X	

Relinquished by:	Date	Time	Received by:	Relinquished by:	Date	Time	Received by:
Rick Polad	8-18 95	11 AM	Mike Bell				

Relinquished by:	Date	Time	Received for lab by:	Date	Time	Remarks
Mike Bell	8-18 95	2 PM	Mary Brennan			

Handling instructions:	Turnaround time:
Keep cool and sealed	Standard 5 day

Instructions:

Fig. 7.1(b) A completed chain-of-custody form.

be used. Geologists working with shales or thinly bedded materials sometimes find it convenient to use mason's or roofer's hammers; these have a broad flat blade that can be used to split layers apart. When using any kind of hammer on any kind of rock, wear eye protection and insist that other people do likewise or stay well out of the range of flying chips.

Hand Augers

A hand auger is a device that may be used to sample unconsolidated materials, generally at shallow depths, and generally above the water table. Hand augers consist of a bit, shaft, and handle (Fig. 7.2). The bit is twisted into the sediment and withdrawn to yield a sample.

Various bit types may be used, depending on soil type. The bucket-type auger bit is most useful in moist sands and loose sediments; the screw type, or continuous flight auger bit, is most useful in clays. Bits usually connect to the shaft by means of a threaded connection, bayonet mount, or other sturdy connection. Extensions may be fitted to the auger for greater depth. If the connections are threaded, two pipe wrenches should be taken into the field to use in adding and removing extensions. A maximum depth of about 30 feet may be reached in some situations, but the practical limit may be less, or, only rarely, more. Handles may be hardwood or metal, welded, threaded, or otherwise connected to the shaft. The longer the handle, the more torque may be applied to the bit and shaft—and the stronger the handle must be. Handles with threaded connections are weakest in the area of the threads.

Above the water table, hand augers generally work well, but hand augering is by no means easy work. Dry clay or till can be rock-hard, as can caliche or hardpan. Gravel, whether a layer or a single piece, can stop auger progress entirely.

Below the water table, augers work well in silt and clay, although suction may be a problem in saturated clay. However, hand augers work poorly in sand and

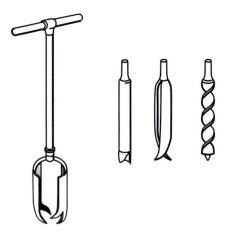

Fig. 7.2 Hand augers.

gravel below the water table. Under saturated conditions, sand is likely to collapse into the hole each time the auger is withdrawn, and as a result, sometimes no progress can be made with the auger. In some cases, casing may be driven ahead of the auger, and the sand augered out from within the casing. Below the water table, gravel is very difficult to auger. As with sand, this is because of the collapse of the hole during augering. But in addition, the large grain size of gravel means it is not easily carried up the flights or into the bucket of an auger.

Although hand augers are simple devices, using them incorrectly can cause injury, damage to the auger, or a lost bit. Following a few simple rules will help avoid these problems. To auger a hole, the auger should be turned in a clockwise direction. The bit should be lifted out to empty the bucket or clean the flights. A short distance should be augered at a time, not more than about 6 inches. After augering this distance, the auger should be retrieved and cleared of sediment. When lifting the auger out of the hole, the leg muscles should be used, not the back muscles.

If the auger does get stuck, it should be twisted and pulled at the same time. If the auger uses threaded connections, the auger should never be turned counter-clockwise, even if it is stuck. Turned counterclockwise, the bit may unscrew from the shaft and be lost down the hole. Instead, it should be turned clockwise, at the same time, pulling to free it.

If the auger is still stuck, two or three people should be enlisted to help twist and pull at the same time. In some soils, it may help to dig away a few feet of soil to help reduce frictional drag on the shaft. If need be, a jack such as a car or truck jack can be fitted under the auger handle to provide additional upward force. Whenever an auger is to be used, a shovel should be packed with the field equipment, in case a lost bit needs to be dug out.

Once the auger has been twisted into the soil so that the handle is only a foot or two from the ground, an extender may be added to bring the handle up to chest or waist level again.

To sample sediment from a particular depth, the hole should be cleared of sediment collapsing from the walls. When this sediment is cleared, augering continues into the undisturbed sediment, and the sample may be taken.

When examining an augered sample, it should be kept in mind that cuttings may have fallen into the hole from above. As a result, what appears on the threads or in the bucket may not have been cut from the apparent depth of the auger. This problem cannot be avoided altogether, but one method for minimizing this problem is to twist the auger into the soil a few inches, and then pull it directly up without turning it. The top inch or two of material retrieved should be ignored, as it may consist of nothing more than cuttings from above. The rest of the material stuck to the threads or in the bucket may then be examined.

Small Motorized Augers

Small motorized augers consist of a small engine (similar to a lawnmower engine) that turns an auger bit (Fig. 7.3). Some augers are not mounted on wheels, and some are mounted on a hand truck, dolly, or small trailer. Most have extensions

Fig. 7.3 A small motorized auger.

that allow the shaft or bit to be lengthened. In ideal conditions, these augers may drill to a depth of about 30 feet. Larger power augers are mounted on drilling rigs; these are discussed later in this chapter.

Soil Sampling Probes (Direct Push)

Soil sampling probes may range from simple hand tools to truck-mounted, hydraulically operated models. The basic concept is the same for all of these samplers: The tool is pressed into the sediment, filling the tube, and then the tool is withdrawn. The procedure results in a "core" of sediment that may be preserved for lab tests. Split-spoon samplers fit this description of a direct-push sampler, but are described in a separate section because they have unique characteristics and applications.

Direct-push hand tools, in particular, often have an opening lengthwise down the sampling tube, and as a result, layers may be observed easily. Direct-push tools that are hydraulically driven have a much greater depth capability than do hand tools. Examples of these are the Hydropunch® and Geoprobe®, which are truck- or van-mounted. Like the hand tools, these tools press a bit into the sediment. The bit may be fitted with a device for soil sampling or it may have a vapor sampler. These devices also may be used to take water samples at discrete depths.

Some direct-push samplers may be loaded with a thin-walled sampling tube such as a Shelby tube, which retains the core and may be capped at both ends to preserve moisture (Shelby tubes are described in detail later in this chapter). One of the primary benefits of taking a direct-push sample, however, is that because it retrieves a relatively undisturbed sample, it allows the observation of soil structure and layering.

Hydraulically operated direct-push devices have the advantage of easy mobility and easy access to some sites that are unaccessible to larger drilling rigs. However, direct-push samplers may have difficulty in retrieving samples in sand, especially sand below the water table, as the sample is likely to run out of the tube as the tool is withdrawn. In addition, they are limited to use in unconsolidated materials.

Split-Spoon Sampling

A split-spoon sampler is a tool for extracting a somewhat disturbed, core-shaped sample of sediment (Fig. 7.4). When a split-spoon sampler is used, the sample can be examined in the field, or portions can be saved for later analysis. Because the sample retrieval method involves some disturbance of the sediment, split-spoon samples are not suitable for representative strength tests such as triaxial tests.

Split-spoon samplers are used both as samplers and as devices to help assess, *in situ,* the unconfined compressive strength of sediment. In this assessment, the samplers are used to perform the Standard Penetration Test (SPT), which is described in this chapter. Because split spoons are commonly used in environmental and geotechnical work, and because they may be unfamiliar tools for most inexperienced hydrogeologists, their use is described here in more detail.

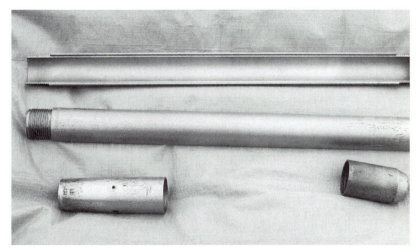

Fig. 7.4　A split-spoon sampler.

A split-spoon sampler essentially is a length of pipe split into two pieces, lengthwise. The ends of the sampler are threaded, so that when the two pieces are placed together to form a cylinder, a short length of pipe can be screwed on to each end, holding the two halves of the cylinder together. One of these short lengths has a tapered bit and holds a disposable "core catcher." At the opposite end of the sampler is another short length of pipe that simply holds the two split sides of the sampler together and connects the whole sampler to a section of drill stem. This end also may have a check valve.

The bit end of the sampler is advanced into the sediment through the use of a specialized hammer, which is operated by a drilling rig (for more information, see the section on the Standard Penetration Test later in this chapter). One way this is commonly done is by inserting the sampler through the stem of a hollow stem auger. The hollow stem auger, described in more detail later in this chapter, remains in the borehole during the sampling operation. The sampler is lowered all the way to the bottom of the hollow stem, and then is driven into the sediment at the bottom of the hole. The result is that the sampler is driven into soil that has as yet not been disturbed by the auger.

As the driller drives the sampler into the sediment, a "core" of sediment is pushed up into the sampler. When the sampler is retrieved, the sample remains in it. The core catcher keeps the sediment from falling out as the sampler is removed.

It theoretically is possible to fill the entire sampler with a core of sediment. In practice, that may not be achieved. The first few inches of material (on the "up" side of the sampler) may be sediment from higher up in the hole that has fallen to the bottom of the hole, and then is forced into the sampler as it is driven deeper. If the sediment is loose or sandy, especially if it is saturated, the sample may fall out of the sampler as it is raised out of the hole. If the sampler was placed on a cobble at the bottom of the hole before it was driven, the cobble may block the opening of the sampler so that no sample is retrieved. Or it may be that consolidated rock has been reached; a split-spoon sampler will not penetrate intact rock.

Once the driller retrieves the split-spoon sampler, the geologist opens the sampler, examines the sample, and records the observations in a log. To make this examination simpler, when space is available, a table or makeshift platform may be set up so that the sampler can be placed on it while being inspected. It is helpful to have the drillers always place the sampler on the table in the same orientation, that is, with the top of the sampler consistently to the same side of the table. The geologist always should check the orientation before starting the examination. The geologist then examines the sample, describes it, performs any required field test, and if needed, preserves part or all of it for later analysis.

The geologist should examine the sample in these steps:

First, the threaded portions that hold the sampler together are unscrewed, and the two sides of the sampler are pulled apart. The "core" should be inside, resting in one of the sides. If no material is found inside the sampler, one of several things may have happened: sand may have run out of the sampler as it was being pulled to the surface; the sampler may have reached bedrock, which it is incapable of sampling; or the sampler may have caught a piece of gravel that became wedged in the bit, keeping other sediment from moving into the sampler. Sometimes, there is no apparent reason for failure to retrieve a sample.

Next, the geologist should check again to see which end of the sample is "up." A knife or spatula then is used to remove any "slough," or cuttings, that fell into the hole from above the sampling depth. This may appear as an inch or more of loose material at the top end of the sample, although there may be no slough at all. Slough should be discarded because it does not represent conditions at the sampling depth.

A gross description should be recorded next, especially noting any changes in lithology, variations in color or texture, breaks, or discontinuities. Next, the length of the sample should be measured. Percent retrieval may be measured and recorded, if required, by determining what length of sample was retrieved and dividing it by the distance the sampler was driven into the sediment.

Next, the sample is cut in half lengthwise. A sharp, sturdy knife or spatula may be used for this. Making several successive passes with the knife, instead of making only one deep cut, results in less disturbance to the soil structure. The sample should be cut only halfway through its thickness; then the two halves of the sample should be pried apart the rest of the way. This will show a fresh face of the sample, not smeared by the knife blade.

After the sample is pried apart, layering or structures in the sample should be noted. Caution should be used in this examination, as some structures may have been disturbed or destroyed during the sampling process. Likewise, new apparent "structures" may have been created, because the sample may have been distorted as the sampler was advanced into the sediment.

Observations of color, moisture content, grain size, and other sediment characteristics should be recorded. More information on how to make these observations is given in Chapter 6.

If samples are to be kept for later examination or lab analysis, the portions to be kept should be selected, placed in labeled containers, and treated as described under "Sampling Plans and Protocols" earlier in this chapter.

Finally, the unused cuttings should be disposed of, and the split spoon should be cleaned and decontaminated. Depending on the agreement between the geologist and the drilling crew, this may be done by the geologist or by the driller's helper.

Thin-Walled Sampling Tubes (Shelby Tubes)

A thin-walled sampling tube is used to retrieve a "core" of sediment or soil and contain it until further analysis can be made (Fig. 7.5). This is a particularly useful device if precise moisture content information is needed, or if intact samples are needed for other lab analysis. These tubes are commonly called Shelby tubes.

Thin-walled sampling tubes have some disadvantages. When a Shelby tube is used, the sample cannot be examined in the field, because the sampler is opaque, so it is impossible to see the sample or subsurface characteristics. In addition, because the tube is thin-walled, if it is driven into gravel or hardpan, it may bend or collapse, so that no sample is retrieved.

Fig. 7.5 Thin-walled sampling tubes (Shelby tubes). Tubes are capped or sealed with paraffin after sampling. The tube at right is still connected to the drive fitting. At left is a specialized tube designed to fit inside a split spoon sampler.

One fairly convenient way to sample with a thin-walled sampling tube is to use it during hollow stem augering operations. In this situation, the auger remains in place in the borehole. A thin-walled tube is lowered all the way to the bottom of the hollow stem. At this point, it is pressed into the sediment that is as yet undisturbed by the auger. Pressing the sampler into the sediment is preferable to driving it, because pressing it in is less likely to disturb the sediment structures or damage the tube. As the sampler advances, sediment fills the tube, which is then retrieved.

If all goes well, when the sampler is withdrawn, it will contain a core of sediment. The sampler then is capped at both ends, and the caps secured with tape, or it may be sealed with paraffin. The caps and sealing prevent moisture from escaping from the tube, so that moisture content may be measured more accurately.

The tube should be labeled carefully with all pertinent information, including a marking on the tube that indicates which end of the sampler is up. The samples should be kept at 4°C by packing them in a cooler with ice, and then should be transported or shipped to the lab.

Well Points (Sand Points)

Well points, also called sand points or drive points, are essentially a length of sturdy well screen with one threaded end and one tapered or pointed end (Fig. 7.6). Strictly speaking, installing a well point might not involve using a drilling method, nor can samples be taken with this method. However, well points are described here because in some situations, they are installed for use as production wells or wells used to monitor water table level. When appropriate conditions and project goals apply, well points may provide a simple, low-cost alternative to installing a well using a drilling rig.

Fig. 7.6 A well point (also called a sand point or a drive point).

In granular sediments, well points may be jetted into place. This procedure does not involve a drilling rig. Instead of drilling, water is pumped down into a jetting point, and emerges at a velocity sufficient to wash sand away from the well point. This allows the well point to be advanced farther into the sediment. This method is most practical at fairly shallow depths (up to about 25 feet).

Well points also may be driven in, in which case the term "drive point" may be used. A drive point is literally driven into the ground by hammering it. Fence post drivers or sledge hammers may be used for this purpose. Hammering should not take place directly on the threaded end of the drive point, as this would damage the threads. Instead, a "drive cap" may be placed over the end. A drive cap is a short, heavy-gauge steel pipe with one end closed. The drive cap may be threaded and screwed onto the end of the drive point, or it may be large enough to be placed over the point without screwing it in. If it is threaded, two pipe wrenches should be used to tighten the joint frequently while driving it. Otherwise, the joint will loosen, increasing the risk of damaging the threads. Some drive points have a drive plate welded to the screened portion, allowing the installer to place a section of pipe on this plate and drive it into the ground, carrying the well point with it. Because the plate takes the force of the driving, this eliminates the risk of damaging the threads.

When the top of the point has been driven nearly to ground level, the drive cap is removed, a section of well casing is threaded on, and driving continues until the desired depth is reached.

Drive points work best in sandy sediments. Some problems are associated with drive points. If the point encounters gravel during driving, it may bend, rendering the well useless. It is difficult to retrieve a damaged or abandoned well point without using a drilling rig. The depth limit in ideal conditions is about 75 feet; in practice, it may be much less than this. Diameter of the point is limited to a few inches. Keeping the well vertical and plumb may be difficult, especially if the sediments contain gravel or boulders, which can deflect the point. No filter pack can be constructed about the screen, and no sediment samples can be retrieved in this type of well installation.

Advantages of this method of well installation are that it is relatively inexpensive and that materials are easily available (some ordinary hardware stores carry well points and drivers). In addition, points can be driven by one or two people, and they can be driven in spots where drilling rig access might be impossible.

Power Shovels and Other Excavators

Power shovels, backhoes, and other excavating equipment may be used to dig trenches or pits. They can expose layering or other soil structures. If samples are to be retrieved from the walls of these excavations, they should be retrieved by the operator, who will use the machine to take the sample and bring it up to the surface. No one should climb into the pit, because newly excavated sediment walls are unstable and could collapse.

Jetting

Jetting is a method of advancing a borehole and/or installing a shallow well or well point. In this method, water (or air) is pumped through a hose into a jet point. The water flows out of the jet point at high velocity, jetting away sediment. As the jetting proceeds, the screen and casing is advanced down the hole.

This method requires a pump, power source, and water source at the site, as well as a hose and jet point. Jetting is slow compared to some other methods of advancing a borehole. It introduces water into the formation, which makes it of limited use in monitoring well installation, because introducing water can alter the chemistry of the formation water. In ideal conditions, the depth of a jetted well is limited to about 200 feet, for wells of up to a 4-inch diameter (Driscoll, 1986). In practice, the limit may be considerably less.

An advantage of the jetting method is that the capital cost is low; a jetting rig costs far less than a rotary or cable tool rig.

Cable Tool Drilling

In cable tool drilling, a heavy chisel is hung at the end of a cable. The chisel is dropped repeatedly, chopping up the formation (Fig. 7.7). The chisel is turned slightly before each drop, so that the hole it makes is circular. At regular intervals, cuttings are retrieved from the hole by means of a bailer. Casing may be advanced into the hole ahead of the actual drilling.

Information about the formation that is being drilled comes from two sources: the cutting speed and the bailed cuttings themselves. However, experienced cable tool drillers may be able to tell the characteristics of the formations being drilled by noticing the sound or vibrations of the drilling rig and the speed at which the cutting progresses.

This type of drilling is slower than rotary drilling. However, it provides good subsurface information, as the bailer removes the cuttings essentially in place

Fig. 7.7 Cable tool drilling.
(Courtesy of Evans Well Drilling.)

(though pulverized). Holes drilled with cable tool rigs are 6 inches or more in diameter. Cutting speed varies with variations in formation characteristics.

The cuttings that a cable tool rig produces are chopped-up bits of rock and loosened sediment. These cuttings may be collected in sample containers and should be carefully labeled. Examining the cuttings is made easier with a hand lens or binocular microscope.

Auger Drilling
Augers can be used only in soils and unconsolidated materials. This section refers to large power augers that are driven by drilling rig engines (Fig. 7.8). Smaller, portable augers, with lawnmower-sized engines, were described earlier in this chapter.

Solid Stem Augering (SSA)
Solid stem augering (SSA) uses an engine to turn a continuous flight, or screw-type, auger into unconsolidated material (Fig. 7.9). The sediment travels up the flights as the auger advances. Samples may be retrieved off the auger flights. The amount of engine torque and the type of sediment determine the maximum depth and diameter of the hole.

Fig. 7.8 Auger drilling.

Advantages of solid stem augering are that it is simple, rapid, and no drilling fluids are needed. Disadvantages are that it works only in unconsolidated materials (not rock), that it can be difficult to advance a hole in heaving sands (coarse material below the water table), and that it is very difficult to retrieve a sample from a specific depth, as soil tends to mix with other soil as it travels up the auger flights. To try to avoid this mixing, the driller should be instructed to drill to just above the zone of interest and to clean the hole of any loose cuttings. Then the driller should insert the augers into the hole, drill just far enough to auger into the zone of interest, and then lift the augers out of the hole without turning them. The sample should be on the augers at the bottom. This will not work if the sediment is loose.

Hollow Stem Augering (HSA)

Hollow stem augering (HSA) is a drilling method that uses an engine to turn a continuous-flight auger into unconsolidated sediments. It differs from solid stem augering in one important respect: In hollow stem augering, the center stem of the auger is hollow, allowing several important functions to be performed downhole (Fig. 7.10).

With the hollow stem augers in place in the borehole, holding the hole open, soil testing can be performed undisturbed. Samples can be retrieved, and wells can be constructed. A sampler may be driven into the intact sediments ahead of the augers, inside the augers, so that undisturbed material may be retrieved. These capabilities make hollow stem augering particularly versatile and effective in environmental drilling programs.

Like SSA rigs, HSA rigs with engines having a great deal of torque are well-suited for augering. The auger flights generate significant drag or friction as they move through the sediment, necessitating high torque. This factor, along with soil

Fig. 7.9 Solid stem auger.

type, determines the depth and diameter capabilities of a given rig. Rigs with less torque but more lift are better suited to rotary drilling. Using HSA can be particularly difficult in drilling through zones of gravel, as the gravel may bend or otherwise damage the flights of the auger.

Hollow stem augering works this way: A plug is put in the bottom of the hollow auger bit, and augering proceeds. The plug keeps sediment from moving up into the hollow stem. When the desired depth is reached, the plug is removed. Two kinds of plugs may be used: One is a metal plug on the end of a string of drill rods; this type is removed by hauling the string of drill rods, with the plug still attached, out of the hole, section by section.

The second type of plug is a disposable "knockout plug"; this type is removed by pounding a sampler or other device into it, knocking it out, and allowing sampling to proceed without hauling the drill rods out of the hole. Using knockout plugs obviously allows drilling to proceed faster. The trade-off is that the plugs themselves are expensive and can be used only once, because they are destroyed in use. In situations where sand heaves up into the auger stem as soon as the plug is removed, knockout plugs are the only practical solution.

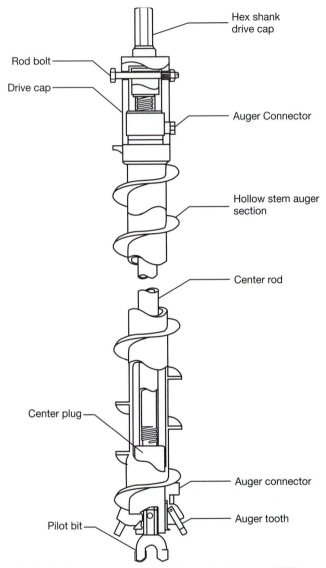

Fig. 7.10 Hollow stem auger (after Aller and colleagues, 1989).

Once the plug has been removed from the bottom of the hollow stem auger bit, soil testing and sampling can proceed. When the Standard Penetration Test (SPT) is performed, testing and sampling occur at the same time. The SPT is described later in this chapter.

Rotary Drilling

In rotary drilling, a bit is rotated in the borehole, producing samples that are lifted to the surface for collection and observation (Fig. 7.11). The geologist gathers the samples, examines them, often using a hand lens or microscope, and keeps a log of formation characteristics as drilling progresses.

Depending on the type of bit used, rotary drilling may produce samples in the form of cuttings or cores. Rotary drilling generally uses one of two kinds of bits: tricone roller bits (Fig. 7.12) or coring bits. Tricone roller bits are used when intact cores are not required. Coring bits are used when the project requires that intact cores be extracted from the formation of interest for some particular or specialized purpose. Coring is slower and more expensive than drilling with a tricone roller bit, and so coring is done only when there is some particular need for it. In most projects, the necessary information can be gained by using the less expensive method. Rotary drilling may also include percussion, in which the bit is hammered into the rock while rotating.

In rotary drilling, fluids such as water, mud, air, or foam are circulated through the drill stem and the borehole. The fluids carry cuttings to the surface, cool the bit, and in many cases, help hold the hole open and prevent it from collapsing. In general, the more prone the formation is to collapsing during drilling, the more viscous the drilling fluids must be.

Fig. 7.11 Rotary drilling. (Courtesy of Northeastern Illinois University.)

Water is a common drilling fluid because it is cheap, easy to use, and easily available. For deep holes, drilling fluids made chiefly of clay may be needed, because density and viscosity characteristics of clay improve the drilling process and help stabilize the borehole.

Various additives such as polymers and surfactants may be mixed with the fluids to impart particular characteristics appropriate to the specific situation. One type of common additive is a natural polymeric colloid such as guar gum, xanthan gum, or similar natural polymers that increase the viscosity of the drilling fluids. These materials are biodegradable. As a result, they will break down in time, or after the addition of other chemicals.

Any drilling fluids used must be removed from the well and surrounding material before a chemically representative sample may be extracted. As a result, well development is particularly important when drilling fluids and additives have been used. In environmental work, such as the drilling of monitoring wells, water, clays, and polymers are not ideal drilling fluids because they affect the chemical composition of formation water. In situations where this is a concern, the use of compressed air should be considered.

Fig. 7.12 Tricone roller drilling bits. (Courtesy of Ackerman International Corp.)

Drilling for Cuttings

Cuttings are produced when a tricone roller bit is used in rotary drilling. Tricone roller bits are made of three cones that are studded with hardened teeth and that rotate together to grind up the formation, producing cuttings. The cuttings then are pushed to the surface by a drilling fluid.

In conventional rotary drilling, drilling fluids are pumped down the drill stem and then return to the surface, carrying the cuttings, through the annulus. In reverse rotary drilling, the fluids are pumped down in the annulus and return to the surface, with the cuttings, through the drill stem. Reverse rotary drilling is especially useful for large-diameter holes. It is not commonly used in environmental work (Davis and colleagues, 1991).

A geologist who is making a log of cuttings produced on a rotary rig retrieves a container full of cuttings, washes the drilling mud away (if mud is being used), examines the lithology, and records the observations. It is critical to note that there is a lag between the time of actual cutting of the rock and the time that the cuttings appear at the surface. This lag is due to the time it takes the cuttings to travel to the surface. Therefore, it is more pronounced in deep holes. Also in deep holes, cuttings being carried to the surface may segregate on the basis of density, so the cuttings coming out at the top of the hole may not accurately represent the composition of the formation. Because of this, a wellsite geologist's log from rotary drilling may not be absolutely reliable representations of the subsurface lithologies. By geophysically logging the hole after drilling has finished, better control over depth and formation data may be gained.

Drilling for Rock Cores

In some projects, samples of cuttings will not satisfy the goals of the drilling and sampling program. For these projects, cores may be more appropriate. For example, in some projects, intact cores of rock are needed so that detailed descriptions of lithology may be made. As another example, the goal of some coring programs is to learn the nature, frequency, and orientation of the discontinuities within the rock. These factors might be important when the aim of a sampling program is to determine the stability of a rock mass, or to estimate the hydraulic properties of fractures within a rock formation on which a dam and reservoir are to be sited.

When a core sample of a formation is needed, a coring bit is used to drill the borehole. Coring bits are essentially cylinders of hard metal with diamonds embedded at the end. The bit is rotated and cuts the formation, forming a cylindrical core that is pushed up inside the bit and core barrel as they advance.

After it is cut with the bit, a core is hauled to the surface by pulling the bit and core barrel up on a cable. The core is then extruded from the core barrel, and the bit and core barrel are lowered down into the hole again to take the next core sample. Because this must be done for each section of the core sample, core drilling is a slow, and therefore expensive, process. In contrast, drilling with a tricone roller bit produces cuttings continuously, so drilling need not cease in order to collect a sample.

In some cases, it may be desirable to extract a core in such a way that its geographic orientation is known. For example, it might be important to a project that the strike and dip of bedding planes or discontinuities are known. Oriented cores may be obtained by using special coring equipment that scribes the orientation on the core as drilling proceeds. However, this method is often unsuccessful in highly fractured rock. In addition, sections of core obtained this way are shorter than those obtained in conventional coring. Because the bit and core barrel must be removed from and reinstalled in the hole more frequently, drilling with this method is slower and therefore more expensive than conventional coring.

Once the core has been carried to the surface, the geologist examines the core and then stores it for later analysis. If moisture content is of interest, the core may be sealed with paraffin or inserted into a plastic sleeve. Measurements such as Rock Quality Designation (RQD) or Percent Recovery may be taken. These measurements are described in Chapter 6, "Describing Sediments and Rocks."

Cores should be stored in core boxes, which are wooden boxes with divisions constructed in them to cradle the cores safely. The cores should be placed in boxes with the "up" end always at the same end of the box; this direction should be marked on the boxes and on the core, using permanent markers. The depth at which the core sections were taken should be clearly given, as well. Other identifying marks may be drawn on the core before sealing. See ASTM Standard D2113 for more information.

Soil Testing and Sampling During Drilling (The Standard Penetration Test)

In some projects, it may be desirable to learn something about the unconfined compressive strength of the sediment while the drilling operations proceed. At the same time, it may be desirable to obtain a sample of the sediment. The Standard Penetration Test (SPT) is a test that meets both of these goals. It is commonly used in geotechnical and environmental projects (see ASTM Standard D1586).

To perform the SPT, a split-spoon sampler is attached to a length of drill stem and lowered down the hole. More sections of drill stem are added as needed, until the sampler is at the bottom of the hole, resting on the sediment surface. Then, the sampler is driven into the sediment. Sediment fills the sampler as it advances.

The sampler may be driven in one of two ways: On many rigs, the sampler is literally hammered by a 140-pound weight that is lifted a distance of 30 inches and then dropped. This type of hammer is a cylindrical weight that fits loosely within a tube. The tube is slipped over the drill stem, and the weight is lifted within the tube and then dropped.

On many rigs, especially older rigs, lifting of the weight is done by means of a rope that passes over a pulley, comes down to shoulder height, and then is wrapped several times loosely on a shiny spool. This spool, called a cathead (pronounced "cat head") spins rapidly under the power of the rig's engine. The driller pulls the end of the rope, thereby tightening the rope against the cathead. When it has tightened enough, friction between the cathead and rope causes the rope to rotate along with the cathead. This pulls on the rope, lifting the hammer. When it has lifted 30 inches, the driller lets the rope go slack again, and the rope is no longer able to rotate along with the cathead. The result is that the hammer drops. This sequence is repeated over and over, at the rate of about one hammer drop per second.

The rapidly spinning cathead and rope are among the most dangerous features of a drilling rig. Should the driller get caught in the rope, or should it tangle and not go slack when expected, serious injury or death could result. Some rigs are equipped with an accessory that lessens the danger: a hydraulic hammer, which produces the same amount of force per blow as the other type, but with less risk to the driller.

Hydraulic hammers have another benefit: They generally perform consistently each time. Inexperienced drillers using the cathead and rope method may not be aware that it is important that the SPT is performed the same way every single time. As a result, they may not recognize how important it is that the hammer be lifted and dropped precisely 30 inches for each blow. The hydrogeologist is responsible for overseeing the test and ensuring that it is performed properly.

The amount of force per blow is an important factor, for it is one of the standardized aspects of the Standard Penetration Test. The SPT tests how many blows of the hammer it takes to advance the sampler a total of 2 feet, in 6-inch intervals. It is the job of the geologist to count and record the blows as the test proceeds. Often a preprinted borehole log form will have blank spaces for the blow counts, so the geologist can simply fill in the appropriate blanks.

Before drilling begins, the geologist should discuss the test with the drillers. The geologist is responsible for ensuring that the test is performed properly and that the hammer blows are counted carefully during the SPT. Many drillers are experienced at counting the blows, but it is the geologist who must sign the log indicating that the counts are accurate. To count the blows accurately, the geologist must communicate clearly with the drillers and pay careful attention to their activities, because a missed test cannot be repeated. Once they have placed the split-spoon sampler down the hole, the drillers make chalk marks on the drill rods. These marks are at 6-inch intervals, for a total of 2 feet. Then, the drillers pound the sampler in, and the chalk marks disappear down the hole, one by one. A count is made of the number of blows of the hammer it takes to make each chalk mark disappear, that is, how many blows it takes to advance the sampler 6 inches.

Blow counts may be used to determine the "N count," which is simply the sum of the blow counts for the third and fourth 6-inch intervals. The first 6-inch blow count readings are not included in the N count, because as the sampler advances through the first several inches, it may simply be advancing through slough. The N count roughly correlates with unconfined compressive strength of a clay sediment, and relative density of a sand sediment (Table 7.1). It takes more blows to advance the sampler through stiff or dense soils, and fewer to advance it through soft or loose soils.

If 40 or 50 blows are struck and the sampler still does not advance a full 6 inches, the notation "Refusal" is made in the boring log, and the test stops. Refusal may indicate that bedrock has been reached. Refusal may also indicate that the sampler is pressed against a large cobble or boulder, especially in gravel, till, or in the weathered zone near the bedrock surface.

TABLE 7.1

Correlation Between Sediment Characteristics and N Count (from the Standard Penetration Test)

Soil Type	Penetration Resistance (N blows/ft)	State	Approximate Unconfined Compressive Strength	
			psf	kN/m²
Sand	0–4	Very loose		
	4–10	Loose		
	10–30	Medium		
	30–50	Dense		
	50	Very dense		
Clay	2	Very soft	500	25
	2–4	Soft	500–1000	25–50
	4–8	Medium	1000–2000	50–100
	8–15	Stiff	2000–4000	100–200
	15–30	Very stiff	4000–8000	200–400
	30	Hard	8000	400

Source: West (1995).

Depths at which the SPT is to be performed are determined in advance of the drilling. The more frequent the testing is performed, the more information is gathered. However, the more frequent the testing, the more slowly drilling proceeds, and the more expensive it is, as well. More testing also produces more samples to be examined, which also may slow down the work. The maximum amount of information is obtained by testing continuously as the hole advances. However, this is rarely necessary, and the expense of continuous testing is rarely warranted.

However, if required, sampling may be performed continuously. This might be necessary when very precise information about soil conditions at depth is required, or in heterogeneous formations where conditions change rapidly with depth. Sensitive engineering designs, such as foundations for nuclear power plants, may require continuous testing. However, in environmental work, it is not unusual to test and sample at intervals of 5 feet or more. Alternatively, depending on the purpose of the investigation, it might be enough to sample only when a new formation is encountered; or it might be appropriate to sample at widely spaced intervals for part of a boring, but at closely spaced intervals throughout a zone of particular interest or importance.

A typical pattern might be this: Test and sample the uppermost soil layer, then drill to a 5-foot depth, then test and sample again, then drill to a 10-foot depth, and test and sample again, and so on. Additional samples may be taken when a change in lithology is encountered. (It may be possible to recognize these changes by differences in the rate of advancement of the drill bit, losses in drilling fluids, or similar factors.) More or fewer sampling intervals may be specified. When specific data are needed on only one zone of particular interest, it may be cost-effective to test and sample infrequently until that zone is reached, and then to test and sample at more frequent intervals. Whatever the plan, the geologist should discuss it clearly with the drillers before drilling begins.

Problems Encountered During Drilling

During a drilling program, a wide variety of geologic conditions may be encountered. Some of these conditions make drilling particularly difficult, and it is not unusual for problems to develop as the boring progresses. Working together, the hydrogeologist and the driller can evaluate the causes of the problems and develop potential solutions. An experienced driller can be invaluable in these situations. For this reason, bid documents should be written in such a way as to ensure that drilling contractors who lack experience with the expected hydrogeologic conditions will be excluded from bidding.

The hydrogeologist should be familiar with the problems that might be encountered during drilling, because it will facilitate communication and troubleshooting. Some of these problems are described here, but this is only a partial list. More information on drilling practices and problems is available in Driscoll (1986), Roscoe Moss Company (1990), and Australian Drilling Industry Training Committee (1997).

Loss of Circulation

"Loss of circulation" is a condition in which a drilling fluid is pumped into the borehole during normal drilling operations, and then exits from the borehole into the formation and does not return to the surface. The driller might notice a sudden disappearance of the fluid from the borehole, or might simply notice that less fluid is returning than was pumped down the hole. Loss of circulation may occur when the borehole penetrates a cavity or crevice, extensively fractured or faulted zone, or other subsurface opening. These zones have clear hydrogeologic significance as potential preferred pathways for flow. As such, they may be critical in predicting or determining contaminant flow directions.

One way to deal with loss of circulation is to add an agent that will have the effect of plugging the borehole. In this case, grout may be added to the hole to form a plug, and then the plug is drilled through. Some polymeric gels may be added to stop loss of circulation. The gel materials may be added to the borehole in ungelled form; another agent is then added to cause gelling. When the zone of lost circulation is a zone of large caverns or cavities, or when there is a concern about adding the gel materials to a potentially contaminated aquifer, the gels might be inappropriate solutions. Because of this, the hydrogeologist should discuss with the drillers the options for solving the problem.

Another solution to the problem of lost circulation is to drill a large-diameter hole through the zone of lost circulation, install a large-diameter casing, and then continue drilling with a smaller-diameter bit past the problem area. A cable tool rig may have the capability to drill through the problem zone, whereas a rotary rig may not. When more information about the zone is needed, it may be possible to investigate the nature of the cavity by using downhole cameras.

Borehole Collapse and "Blow In"

In unconsolidated sediments, the borehole may collapse when the drilling tools are removed from the hole. To avoid this, drillers may drive casing into the hole to keep it open, and then continue drilling inside the casing. In some cases, the casing is driven ahead of the drill bit, so the bit essentially works to rout out the casing.

The depth to which casing may be driven is limited by the friction of the sediments on the walls of the casing. When this limit is reached, it may be possible to insert a smaller-diameter casing inside the larger casing, and to continue drilling, with a smaller bit, inside the smaller casing.

Another potential solution to the collapse problem is to use heavy drilling muds, which support the walls of the borehole during drilling. This may not be desirable in the drilling of monitoring wells, because the fluids may interfere with the ability to obtain a representative sample of the formation waters.

"Blow in" can occur during drilling of "heaving" or "running" sands. These are sands that are below the water table or that occur in a confined aquifer. Saturated fine sands, in particular, are unstable. When blow in occurs, the unstable sand collapses into the borehole, so that as soon as the drilling tools are removed, the hole fills in with sediment. Even while drilling is in progress, if the sand is under artesian pressures, it may be forced up through the bit and into the drill stem. This freezes drilling progress and makes sampling very difficult.

Heavy drilling muds may be useful to counteract the pressures that are forcing the sands into the hole. Also, hollow stem augers may be useful for drilling in situations like this, because the auger holds the hole open, and the knockout plug keeps the sand from being forced up through the bottom of the hollow stem. When a sample is needed, it can be retrieved by knocking out the plug and quickly taking the sample through the hollow stem.

Stuck Drilling Tools

Drilling tools may get stuck when unconsolidated sediments collapse around the tools. Stuck tools may occur when drilling without fluids, such as in augering or shallow rotary drilling. To prevent this, a drill rig with sufficient torque to turn the augers or rotary bit, even in unconsolidated materials, should be used, and the depth limitations of the rig should be known and respected. Tools also might become stuck during rotary drilling if the drilling fluids stop circulating, whether because the fluid pump stops working or because of loss of circulation to a creviced or cavernous zone. To prevent this, it is essential to keep the fluids circulating. This may involve such things as keeping a careful watch on the fuel level for the fluid pump, or researching the area before drilling to determine whether loss of circulation is likely to occur. A driller with experience in a particular area should be able to provide this information, as might other experienced hydrogeologists.

Because of the possibility of tools being stuck in unconsolidated sediments, drilling tools should never be left downhole for any significant time without circulating fluids or driving casing to support the hole. The driller should "trip out," or remove the tools, if there is any possibility of collapse. Again, experienced drillers will strive to avoid these problems.

Drilling Through Boulders

Boulders can present difficult drilling problems. Boulders can deflect a string of drilling tools, causing the hole to progress out of plumb. If drilling proceeds past a boulder that projects from the side of the borehole, eventually, the boulder may be worked free, and may fall on the bit, trapping it down the hole. In rotary drilling, if the bit is drilled into a boulder, the boulder may lock around the bit and spin freely as the bit spins, but not allow any further downward progress. Boulders may be encountered in glacial tills, alluvial materials, conglomeratic deposits, beach deposits, or in a weathered bedrock zone.

To deal with the problem of boulders, it may be possible to destroy the boulder during drilling. It also may be possible to cement it in place and then drill and case through it. In some cases, it may be necessary to blast the boulder apart, but this should only be done if there is assurance that damage to the borehole will not result. It may be most cost-effective to simply abandon the hole and start over again.

Fishing

"Fishing" refers to the process of attempting to retrieve "fish," or items that have fallen into the borehole. It is easy to lose small objects like bolts, wrenches, pens, or keys by accidently dropping them into the hole. Bailers and water-level indicators may be lost if their cables break. At the other end of the spectrum is the loss of strings of drilling tools that break during drilling, so that part of the string of tools can be pulled from the hole, but the other part falls to the bottom.

Drillers generally carry fishing tools that are appropriate to their equipment. These may be specialized tools. For example, one type of fishing tool is a pipe with tapered threads that is lowered into the hole and slowly fed into a lost piece of drill stem, rotated so that the threads engage the lost piece, and then pulled back to the surface. Other fishing tools may have hooks, spears, powerful magnets, or devices that are particular to a certain kind of fish. In some cases, it may be possible to drill through the fish. This destroys the lost item, but may be the only way to continue progress on the borehole.

Even after the well is completed, fishing may be necessary. The field hydrogeologist might lose an item down the well when sampling or monitoring water level. To be prepared for this situation, the hydrogeologist should carry a magnet (packaged carefully to keep it away from computing equipment or diskettes), strong fishing line, and a few stout fishing hooks in the toolbox.

Working with Drillers

Good drillers are skilled professionals who are an essential part of most ground water field investigations. They may be the key component in making a project go smoothly, or they may present the thorniest issues and problems in the project. Geologists and drillers depend on each other to make a project go well. Following these suggestions will help smooth the way.

Learn as much as possible about drilling methods and machinery. This will make for better communication with the drillers and will help improve understanding of their suggestions and duties.

Communicate expectations, needs, and goals clearly, and to the member of the drilling crew who has decision-making responsibility. Often, drillers have suggestions and information that will help make the job more efficient or will help get a better sample.

When needed, ask drillers for advice. Then, take responsibility for your own decision making.

Respect drillers as professionals, and ensure that your own attitude remains professional.

If you think you see drillers making a mistake, talk with them early on, before it is too late to remedy the situation.

Consider the drillers and yourself as part of the same team, working toward the same goal.

At the same time, remember that the responsibility for the collection of accurate hydrogeologic data belongs to the hydrogeologist, not the drillers. It is the duty of the hydrogeologist to ensure that the appropriate samples are taken from the appropriate depths. The hydrogeologist is in charge of the operation.

Well Design and Installation

Chapter Overview

Measurements of water level and chemical analyses from well waters provide the raw data on which much hydrogeologic work is based. The methods for drilling boreholes were discussed in Chapter 7. However, a well is much more than a simple borehole. Wells are constructed of materials such as well casing, well screen, filter pack, and annular seals, all of which are carefully placed within boreholes or pits.

Wells should be designed and constructed in such a way that they provide the necessary samples, data, or water supply for which they are intended. Many aspects of well construction should be considered before the well is built. This chapter examines basic procedures for designing, installing, developing, and abandoning wells. More information is available in numerous other sources, including Campbell and Lehr (1973), Driscoll (1986), Clark (1988), Aller and colleagues (1989), Roscoe Moss Company (1990), Nielsen (1991), and Maidment (1993).

Well Design

Many decisions and choices must be made in designing a well. The design stage is critical. No matter how carefully constructed, if a well is not designed properly, it may be essentially useless. In designing a well, it is important to consider the purpose of the well or well network, the site characteristics, and the well location. Once these considerations are thoroughly explored, the specifications of the well may be determined.

Purpose of the Well or Well Network

The purpose of a well is a key factor in determining the specifics of its design. An individual well may be installed for several reasons. An *extraction well*, also called a production or supply well, provides water for domestic, industrial, or agricultural uses. A *recovery well* recovers contaminants or contaminated water from the ground water. A *monitoring well* is installed for the purpose of obtaining chemically representative water samples, and in some situations can give useful information about water level. A *piezometer* is essentially a well with a short intake, installed for the purpose of measuring hydraulic head, pressure head, or pore water pressure. Finally, an *injection well* is installed to provide a conduit for the injection of water into a formation.

Like individual wells, networks of two or more wells may be designed for various purposes. Some networks are designed for the purpose of providing a water

233

supply, and others may be networks of monitoring wells used to delineate a contaminant plume. Some are designed to capture contaminants. Some networks are installed simply to determine flow conditions throughout an area.

Site Characteristics

Site characteristics must be evaluated carefully before a well can be designed. The following characteristics should be investigated. When it is possible to do so, precise data on these characteristics should be gathered. But even when precise information is not available, the characteristics should be investigated and if applicable, approximations should be made.

> *Geology:* Lithology of sediments and rocks, elevation and thicknesses of formations, lateral extent of formations, and structure of formations
>
> *Hydrogeology:* Hydraulic conductivity (vertical and horizontal), saturation conditions, water levels, effective porosity and/or specific yield, storativity, flow directions, existing wells (locations, depths, pumping rates)
>
> *Topography:* Topographic features that might affect hydrogeology, accessibility for drilling rigs, when applicable
>
> *Human-Made Features:* Buildings, basements, foundations, excavations, septic systems, sewers, buried utility lines, pumping and injection wells, drainfields, agricultural activities, roadways, rail lines, and other similar features
>
> *Surface-Water Hydrology:* Surface-water features (perennial and ephemeral), climatic conditions such as precipitation and evaporation, seasonal changes in water levels
>
> *Potential (or Confirmed) Contaminant Sources:* Locations, volume of release, nature of release (continuous or slug), chemical and physical characteristics of contaminants

Well Location

Choosing the location for a well depends on the purpose of the well, among other factors.

Extraction Wells

Extraction wells are used to extract water from an aquifer. Extraction wells may be installed for several different purposes. For example, extraction wells may provide a water supply for domestic, industrial, or agricultural use. They also may be used for dewatering, such as dewatering construction sites during the building of a foundation.

In the case of a water supply well, the location depends on several factors. Convenience and access for the drilling rig and well servicing equipment should be considered, as should the requirements for water distribution lines and storage tanks. Ideally, the location should maximize the well's performance and production, while minimizing the costs of installing and operating the well. In estimating costs, both initial and long-term costs should be considered. Initial costs include drilling and well installation, as well as installation of distribution lines, storage tanks, treatment facilities, and pump or pumping stations. Long-term costs include maintenance, repairs, and the cost of pumping and treating the water.

The location of water supply wells may be restricted by property lines or other legal considerations. Some communities, counties, and states have ordinances specifying a minimum allowable distance between a well and a property line. If a septic system exists on the property, county or state regulations will require that the well be located a certain distance from the septic system. This "setback distance" varies from state to state; the local health department should be able to provide the pertinent regulations.

A water supply well or well field should be located where there is a ready supply of water. Finding a location with sufficient yield requires careful study of the site geology. The hydrogeologist should scrutinize geologic reports and well logs from other wells in the area to determine the type, geometry, and variability of subsurface materials. If no logs are available, test borings may be made. Fetter (1994) gave a thorough review of hydrogeologic conditions in various types of geologic materials.

Large areas of the United States are covered with sediments or sedimentary rocks, many of which hold and transmit water in appreciable amounts. If aquifers of sand, gravel, fractured rock, sandstone, porous limestone, or other permeable materials are available, finding a suitable location for an extraction well may be a fairly simple task. In these aquifers, precise location may not be critical in determining well yield. However, in regions of karst aquifers, or regions with low permeability materials, precise location may be critical, since wells spaced only a few feet apart from each other might intercept totally different flow systems or may produce dramatically different amounts of water.

In regions without sedimentary aquifers, where igneous or metamorphic rocks form the bedrock, conditions may vary markedly. Some igneous rocks, particularly volcanic rocks, may have sufficient permeability to support an extraction well. Igneous or metamorphic rocks that are extensively fractured may have developed enough secondary permeability to provide sufficient yield. In regions of igneous or metamorphic rocks that do not have extensive fracturing, the precise location of the well may have a significant effect on its yield.

Even in regions with poor aquifers, it might be possible to design a well with satisfactory yield. Extending the intake (screened or open interval) such that it intersects several permeable zones might be sufficient. Zones of weathering at the top of the bedrock might be investigated, as they may have higher permeability than underlying rock, and might provide sufficient flow. Another possible alternative is to site the well such that its intake intersects a fracture or set of fractures. Using structural geology maps may be helpful in selecting these locations and estimating the appropriate depths. The identification of fracture traces in the area also might be helpful.

A *fracture trace* is the surface expression of a fracture that is thinly covered with sediment, possibly appearing as a zone of different colored vegetation, a shallow "ditch" or slight dip in ground surface, or a zone of soil that is wetter than the surrounding soil. Fracture traces often are linear, and might be aligned with local or regional structural features such as joint sets or faults. The variation in ground characteristics caused by fracture traces may be subtle, making them difficult to identify in the field. Fracture traces tend to be most easily discerned on aerial photos, and aerial photographic or photogeological interpretation provides an excellent means of targeting suitable locations for wells, particularly in rock aquifers.

Buried pipelines or utilities sometimes may be confused with fracture traces. To avoid this confusion, it is essential to ground-truth all photointerpretation work, and to correlate it with local geologic data. Fracture traces commonly run in sets, trending in the same general direction; regional geologic information may describe these joint sets. Buried pipelines may be identified by markers, flags, or access ports (e.g., sewer hole covers) along the pipeline. Utility companies may be consulted to identify their buried lines. Where fractures or fracture traces intersect the ground surface, spring lines or seeps may indicate their presence.

Extraction wells used for the purpose of dewatering are designed such that their cones of depression, which may overlap, draw ground water levels down to the necessary depth so that excavation, construction, or similar activities can proceed in the dry materials. Dewatering systems may consist of deep wells of a series of shallow well points, often arranged in lines or circles and connected to a common header. Locations of these wells are determined by engineering design. Watson and Burnett (1995) and Rahn (1996) provided more information and worked examples on dewatering. Freeze and Cherry (1979) provided further discussion as well.

Recovery Wells

Recovery wells are used to recover contaminants or contaminated ground water. For example, recovery wells might be used to pump out gasoline, a solvent, or another product that has leaked into the subsurface. Recovery wells also might be part of a pump-and-treat system, in which a network of wells is used to pump contaminated ground water out of the subsurface and route it to a treatment facility.

Many of the considerations described under the heading of "Extraction Wells" also apply to recovery wells. However, the design of a recovery well differs in two important ways from the design of a water supply well. The first difference depends on the characteristics of the formation that the well taps. A water supply well is designed so that the well taps a productive aquifer or productive zone within an aquifer. However, for a recovery well, the choice of what interval to tap is very limited. A recovery well must be designed so that it taps the formation that holds the material to be recovered. In some cases, this might be a productive aquifer. But in other cases, it may be a formation of moderate to low productivity. This might necessitate the use of larger diameters, longer well intakes, or other design alterations.

The second difference between an ordinary extraction well and a recovery well is that in the design of an ordinary extraction well, the fluid of interest is water. In the design of a recovery well, the fluid of interest is a contaminant or a mixture of several contaminants, any of which might have fluid properties very different from those of water. In these situations, the hydrogeologic design data must be considered along with data on contaminant sources, contaminant chemical and physical properties, and contaminant transport mechanisms. Design of these systems is complex, and is beyond the scope of this text. For an overview, see National Research Council (1994).

Monitoring Wells

A monitoring well is a well installed for the purpose of obtaining a chemically representative sample of formation water. This section describes considerations for siting upgradient monitoring wells, downgradient monitoring wells, and piezometers. Nielsen (1991) and Aller and colleagues (1989) are excellent sources of further information on this topic.

Upgradient Wells. Upgradient wells are wells placed in locations that allow them to intercept water flowing into a site so that its chemical characteristics might be determined. By comparing the chemistry of water flowing into the area to that of water flowing out of the area, a determination can be made about whether water chemistry is being affected at the site. The concept seems simple, but in practice, it may be difficult to determine where an upgradient well should be placed. Several factors should be considered, as discussed below. Sara (1991) provided a thorough and detailed discussion of the placement of upgradient and downgradient wells.

Direction of flow must be known if the upgradient direction is to be determined. A flow net or water-level contour map may be used to determine direction of flow under a given set of conditions. Direction of flow may be different in different formations, so water levels used to construct the map must be representative of the formation of interest. Direction of flow may change or fluctuate over time, depending on conditions such as season of the year, surface-water conditions (e.g., flooding or changes in the levels of nearby streams, lakes, or oceans), local pumping or injection activities, and the effects of vegetation (Fig. 8.1).

In some situations, it may not be possible to site an "upgradient" well. This may occur when the site overlies a recharge zone. In these zones, water recharges the aquifer by moving downward into the formation from the vadose zone, instead of migrating toward it from another, offsite direction. When this is the case, instead of an upgradient well, a "background" well may be used to characterize water in the area. Depending on the project and on the size of the recharge mound, a background well might be situated just offsite or at the site boundaries, or it might be situated some distance away from the site, at a point where ground water is not affected by the recharge mound.

Vertical gradients and flow paths must be considered in siting an upgradient well. Flow nets should be drawn in cross-section (as well as plan, or map, view) in order to characterize vertical flow.

Other factors that must be considered in siting upgradient wells include the effects of fracturing, layering, perched water, solution cavities, and karstification. Geologic structure should be considered as well. Finally, the depth of the intake (screened or open) portion of the well must be designed so that it fits the hydrogeological characteristics of the aquifer. More information on this is included in the section on designing screen depth, later in this chapter.

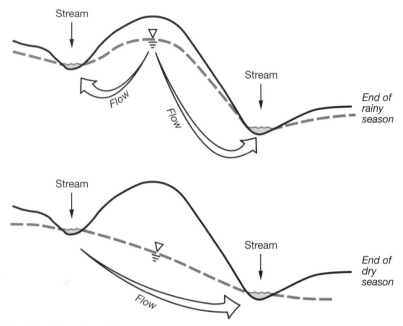

Fig. 8.1 Direction of flow may change over time. In this example, the change is due to seasonal factors.

Downgradient Wells. Downgradient wells are placed at locations that allow them to produce samples characteristic of water leaving a site. Depending on local, state, and federal regulations, and on the conditions and complexity of site hydrogeology, several downgradient wells may be required.

A downgradient well might be installed for the purpose of intercepting a potential plume of contaminant that might issue from a potential contaminant source at the site. In addition, a downgradient well might be installed before construction begins at a site, in order to monitor ground water quality.

To detect potential migration of a contaminant plume from the site, downgradient wells should be placed between the potential contaminant source and the property boundary, as well as downgradient from the source. To determine the appropriate locations for these wells, flow directions at the site must be known.

As described in the earlier section on upgradient wells, many factors influence the directions in which ground water might flow. In addition to those factors, the potential directions of subsurface contaminant movement also should be considered in order to place downgradient wells in the most useful locations. Contaminant movement flow directions will depend on the physical and chemical properties of the contaminants, as well as the hydrogeology.

Key characteristics of a contaminant that determine where and how it might move are the miscibility, density, solubility, volatility, and sorption characteristics of the contaminant. The miscibility of the contaminant is the extent to which it mixes with water. If a contaminant is miscible with water, and if the formation is moderately to highly permeable, the bulk of the contaminant will move by advection, along with groundwater flow. But if the contaminant is miscible and the formation has a low permeability, diffusion may be the dominant transport mechanism. This might be the case, for example, in a tightly compacted landfill liner. These factors all should be considered in siting downgradient wells.

If the contaminant is immiscible with water, its behavior will depend to some extent on its density. Light nonaqueous phase liquids (LNAPLs) tend to "float" on top of the saturated zone, so the downgradient well should be designed to intercept these floating contaminants. Dense nonaqueous phase liquids (DNAPLs) sink to the bottom of an aquifer and collect in pools or migrate downhill, even if this is not in the same direction as the flow of water (Fig. 8.2). In these situations, downgradient wells should be designed with their well intakes at the bottom of the aquifer, so that they successfully detect DNAPLs.

Depending on its particular chemical characteristics, a contaminant also might partition into gaseous, dissolved, or sorbed phases. Volatility of a chemical will determine the extent to which it will partition into a gaseous phase and migrate into soil vapor. Solubility of a chemical will determine to what extent it will dissolve into the ground water; if it dissolves, it may migrate with the ground water.

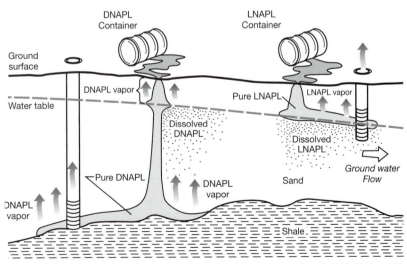

Fig. 8.2   LNAPLs tend to float on top of the saturated zone, whereas DNAPLs tend to sink and collect in pools or migrate downdip or downhill.

Sorption characteristics of a chemical determine the extent to which it will adsorb to or desorb from the formation materials. The situation is more complex if the contaminant is composed of more than one type of chemical, when reduction-oxidation reactions are significant, or when biological activity alters chemical composition. These should be considered before siting the downgradient well. More information on contaminant transport is given in Freeze and Cherry (1979), Fetter (1993), National Research Council (1994), and Domenico and Schwartz (1998).

In addition to chemical considerations, a number of hydrogeological factors might complicate any simple model of contaminant transport, and thus might affect the decision on downgradient well location. Seasonal or vegetation-induced changes in flow direction, flooding, nearby pumping or injection operations, kars-tification, and fracturing all may affect flow direction. Likewise, vertical gradients, structural controls on flow, and layering of sediments of various permeabilities also must be considered. Other potential conduits for flow, such as buried utilities or pockets of fill material, should be investigated as well.

Piezometers

Piezometers, which are essentially small-diameter wells with very short intakes or screens, are used to determine the hydraulic head at a specific point in an aquifer. This information is used to determine ground water flow directions, evaluate stability conditions (as at dam sites or for projects under construction), and to gather data needed to make flow nets.

Determining the optimal location of a piezometer depends on what information is needed and what it is needed for. When information about hydraulic head at a particular point is required, the piezometer should be installed with its intake set at that point. In some cases, particularly when the aquifer of interest is unconfined, information on surface-water levels may be combined with hydraulic head information from piezometers to produce a flow net. In some cases, wells also may provide limited data.

Nested piezometers may be used to measure vertical gradients. Nested piezometers are multiple piezometers installed side by side, with their intakes set at different depths (Fig. 8.3). The piezometers may be installed in separate boreholes or they may all be placed in the same borehole. In this case, each intake is open to a different level within the borehole, and low-permeability seals separate them from each other.

Injection Wells

Injection wells are used to inject fluids into the subsurface. Some injection wells are used for the disposal of waste fluids, for example, in pump-and-treat systems. In a pump-and-treat system, water is extracted from a contaminated aquifer, is treated onsite, and then is reinjected into the formation. The objective of such a system is to use the aquifer to assist in the remediation process. This reduces the cost of remediation by eliminating the need to transport, treat, or dispose of large volumes of water offsite. Injection wells are a key component of these systems.

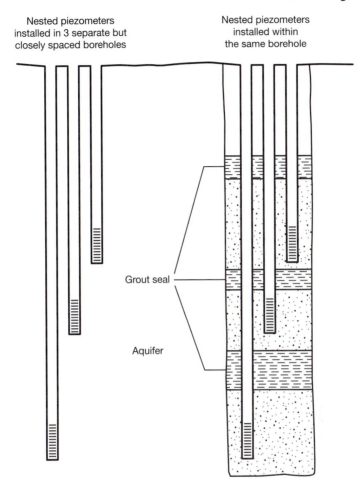

Nested piezometers installed in 3 separate but closely spaced boreholes

Nested piezometers installed within the same borehole

Grout seal

Aquifer

Fig. 8.3 Nested piezometers.

Another example of the use of injection wells for waste disposal is that of oil-field brine injection. In some states, oilfield brines are commonly disposed of by injecting them into deep formations that hold nonpotable water. A key considera-tion in the design of these systems is the chemical compatibility of the injected fluid with the formation water and mineralogic constituents.

Yet another type of injection well is designed to change the hydraulic gradients in an area so that the direction of ground water flow is changed or even reversed. For example, if water is flowing from a contaminated area toward a water supply well, a network of injection wells may be installed between the contaminant source and the water supply well. Injecting water into these wells at sufficient pressures forces the contaminated water to flow back in the direction of its source. This protects the water supply from the contaminant plume.

Designing Well Specifications

When the purpose of the well, the site characteristics, and the location of the well are known, the details of the well construction may be specified. This should be done before going into the field, so that the appropriate materials and tools may be obtained in advance.

The following specifications should be made for each well (Fig. 8.4): borehole diameter; depth and length of well intake (screened or open interval); diameter of the well intake; filter pack type, grain size, and volume of material needed; well screen slot size and type; well screen material; well casing material; well

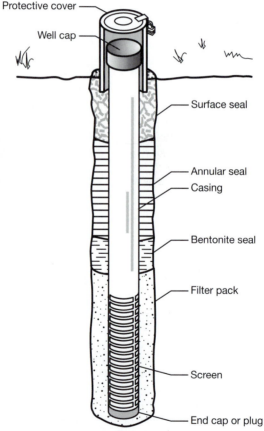

Protective cover

Well cap

Surface seal

Annular seal

Casing

Bentonite seal

Filter pack

Screen

End cap or plug

Fig. 8.4 Components of a well.

casing diameter; type of annular seal(s); types of surface seals; and, if appropriate, pump type and depth. This section describes considerations for making these specifications.

Borehole, Casing, and Well Intake Diameters

The borehole must be wide enough to allow a well to be built inside of it. Therefore, the casing and well intake diameters must be determined before the borehole diameter is determined. The casing must have a large enough inner diameter to admit devices such as pumps, bailers, and water-level measuring instruments to the required depth. In deep holes, or in holes in which the collapse of sediment from a shallow formation makes drilling into deeper bedrock difficult, casing may be "telescoped" or "stepped down", with a larger diameter at the top and a smaller diameter at the bottom.

Common casing diameters used in the United States are as follows: piezometers, 1–2 inches; monitoring wells, 2, 4, 6, or 8 inches; domestic supply or recovery wells, 4, 8, 12, or 16 inches; and public water supply wells, 8 inches to several feet. Hand-dug wells usually are several feet in diameter. In the case of nested piezometers, several piezometers, each having a diameter of 1–2 inches, are installed to different depths within a single borehole. Thus, the borehole must be large enough to accommodate all of the piezometers at once.

The well intake may be an open borehole or it may be a well screen. If a screen is to be used, screen diameter must be specified. (Not all wells require screens; see the next section on "Well Intake.") Screen diameter may be the same as casing diameter, but it need not always be so. The screen must have a large enough diameter to admit the pump or bailer, if one is to be used at the depth of the screen. The screen must have openings with enough area to permit the needed flow, but not so much that the entrance velocity of the water exceeds 0.1 ft/sec. (For a discussion of why this value is the recommended maximum, see Driscoll, 1986.) Entrance velocity (v) is the expected pumping rate (Q) divided by the total area (A) of the openings in the screen, or $v = Q/A$.

Large-diameter screens are expensive and admit only a small fraction more water than small-diameter screens. In monitoring wells, 2-, 4-, 6-, and 8-inch-diameter screens are common; in piezometers, 1- and 2-inch-diameter screens are common.

To determine optimum casing and screen diameter, consider the size of the pump or bailer that will be used in the well. Also, consider the pumping rate and transmissivity of the formation. Many household well systems include a storage tank or pressure tank in which water is stored during off-peak use hours so that it is available during peak use hours. This makes lower pumping rates possible.

Finally, in determining casing and well intake diameters, the costs of well materials, installation, and operation must be considered. Costs of well materials and drilling increase rapidly as diameter increases. A large-diameter well might provide a steadier source of water, but the costs of drilling and materials might be prohibitive. If the same supply can be provided in some other way, specifying a large diameter is not warranted. In general, the well design should specify the smallest diameter that will provide the supply required for present and future needs.

Once the casing and intake diameter are chosen, the borehole diameter may be specified. Borehole diameter must be larger than screen and casing diameter, to allow room to construct the well inside the hole. In sediments that collapse when the drill bit is withdrawn, the well may have to be constructed inside a hollow stem auger or larger-diameter casing, which is then removed to allow the sediment to collapse around the well. Will an artificial filter pack be used? If so, the borehole must be 4 inches or more larger than the largest casing or screen diameter, to accommodate the filter pack.

If the borehole is to be drilled using a hollow stem auger, it should be kept in mind that the auger cuts a hole several inches wider than the hollow stem itself. The well is built inside the hollow stem. Thus, although the auger diameter may be large, the well built inside the auger will have a significantly smaller diameter. For example, to build a 2-inch-diameter well, an auger of at least 6 ¾-inches diameter must be used. To build a 4-inch-diameter well, an 8 ½-inch-diameter auger must be used.

Well Intake

The well intake is the segment of the well through which water flows from the formation. The intake should be designed to optimize flow into the well from the formation of interest, at the depth of interest. To accomplish this, the purpose of the well should be considered when specifying the design, depth, and length of the intake.

A well intake is made up of either a well screen and filter pack or an open borehole. A well screen is essentially a slotted or perforated pipe that allows water to enter the pipe, but prevents the hole from collapsing. Types and materials are described in the section on "Casing and Well Screen Material" later in this chapter. A filter pack is granular material, usually sand, that is installed so that it surrounds the screen. A filter pack has the dual effect of enhancing the flow of water into the well and, at the same time, filtering out fine material that might otherwise enter the well and clog it or the pump. Together, the screen and filter pack facilitate the flow of water into the well. Filter packs are described in more detail in the section titled "Filter Packs" later in this chapter.

Not all well intakes require the use of a screen or filter pack. In consolidated rock that is relatively unfractured, stable, and sufficiently permeable, no screen or filter pack is needed. In these wells, the borehole is left open along a given length so that water may flow freely into the hole. However, if the geological formation is unconsolidated, a filter pack must be used in addition to a screen. If the formation material is consolidated, but is highly fractured or otherwise unstable, pieces of it may fall into the borehole either immediately or over the course of several years. This will close off the hole, fill it with sediment, or damage the pump or discharge lines. In these cases, a screen and filter pack must be used. Even if the formation is consolidated and stable, a filter pack might be desirable for its effects in enhancing the flow of water into the well.

Depth and Length of Well Intake.   The depth and length of the well intake should be designed such that the well taps the formation of interest at the depth of interest. Well intakes for extraction wells and monitoring wells are described in this section.

Extraction Wells.   If the well is a supply well, it should be designed in such a way as to maximize yield while minimizing costs. The most productive portion or portions of the aquifer should be used as the well intake segment(s). These segments either should be left open and uncased or they should be screened. If an aquifer contains some productive and some nonproductive segments, only the productive portions should be screened. Because screen is expensive, extending the screen though long sections of nonproductive aquifer material may not be cost-effective.

For extraction wells, the well intake should not extend above the water level during pumping. This might cause air to enter the pump and lines, contributing to excessive pump wear and overheating. As a minimum, to determine depth of the well intake, the amount of drawdown should be determined that would result from pumping at the maximum rate at which the well is likely to be pumped. The top of the intake then should be set a few feet below that point. Alternatively, the drawdown that would result from pumping for 6 months at the expected rate should be determined, that amount of drawdown should be doubled, and the top of the intake should be set a few feet below that point. The effects expected from the pumping of nearby wells should be included in the calculations.

To optimize well intake length, consideration should be given to the current and future pumping needs, as well as the hydraulic characteristics of the aquifer. Driscoll (1986) recommended the following for installing screens in various types of aquifers:

1. *Unconfined:* If thickness is less than 150 feet, screen the bottom one-third to one-half of the aquifer. If the aquifer is particularly thick and deep, up to 80% may be screened.

   If the formation has several zones of varying productivity, only the highly productive zones should be screened, and casing should be used in the unproductive or aquitard zones. In order to keep from drawing aquitard material into the well, the screens should end 1–2 feet from the aquitard. If the aquifer is nonhomogeneous and several short screens are used, the total length should equal about one-third of aquifer thickness.

2. *Confined*: About 80–90% of total aquifer thickness should be screened, or if the aquifer is non-homogeneous, 80-90% of the portions with high permeability should be screened. If possible, the well should be designed so that the water level does not drop below the top of the aquifer during pumping. If it does, the aquifer will act as if it is unconfined, and the pores may be drained (Driscoll, 1986). These recommendations apply most simply to layered sedimentary aquifers and highly fractured aquifers. In karst aquifers or those in which water is produced from widely spaced fractures, the productive portions of the aquifers should be screened, and the remainder should be cased.

Monitoring Wells. If the well is a monitoring well, the intake should be set at a depth where it is most likely to intersect contaminants. Contaminant density and miscibility should be considered, as well as aquifer characteristics. It should be kept in mind that LNAPLs are likely to float at the top of the saturated zone, whereas DNAPLs are likely to sink and travel downhill or collect in pools at the base of an aquifer. Miscible contaminants are likely to move with the flow of ground water. Are there fractures or other preferred pathways for flow? Depending on the size of pore spaces in the matrix, contaminants might be found in the fractured areas before they appear in other locations. Vertical as well as horizontal gradients should be considered in determining flow paths. Drawing flow nets in cross-section as well as in plan view will help define the potential pathways of flow.

The lengths of monitoring well screens may be specified by federal, state, or local regulations. Whatever the length of a screen, it should satisfy the goal of monitoring the zone of interest. In addition, the dimensions and characteristics of the contaminant plume must be considered.

If the contaminants are miscible with water, caution should be used to avoid specifying a screen that is too long. A screen that is too long will admit water along its whole length (Fig. 8.5), potentially diluting contaminants (Fetter, 1994). If samples are to be drawn from a particular interval that is narrow, the screen should extend through that interval and not beyond. In some cases, inflatable packers may be used to isolate a zone of interest so that a sample may be withdrawn from that specific zone.

In some situations, monitoring wells should be designed with a screen that extends above the water table. If the contaminant of interest is an LNAPL, the screen must intersect and extend above the top of the saturated zone. This way, when the water level rises, the screen will still allow LNAPL to enter the well (Fig. 8.6). Likewise, the screen must extend some distance below the water table, so that when water table drops, the screen still will intercept the floating contaminants.

Filter Packs. Filter packs may be natural or artificial. Both of these types of filter packs have the same purpose: They are installed to enhance the productivity of the well by allowing water to flow into the well easily, while filtering out clay and silt so that these fine sediments do not flow into the well or clog the screen.

Whether natural or artificial, the filter pack should be designed so that it extends to at least 1 foot above the well screen. Extending the filter pack above the screen keeps formation materials from being drawn into the screen. Fetter (1994) recommended installing it to 2–3 feet above the top of the screen to allow for settling, and Nielsen and Schalla (1991) recommended extending it to 3–5 feet for monitoring wells. If the filter pack is too long, however, water from above the aquifer could travel down the filter pack and into the screen. This might be undesirable, especially if the water from above is of a poor quality or potentially contaminated. Each situation should be evaluated individually, and the filter pack length should be designed specifically for that situation. In some cases, regulations may specify the minimum or maximum length of the filter pack above the screen.

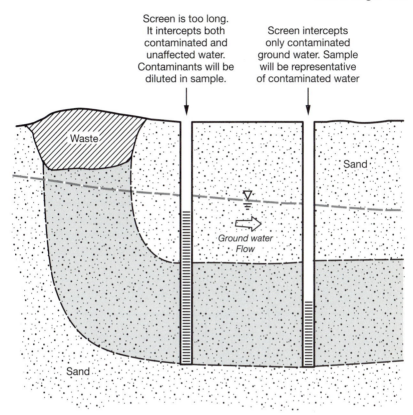

Fig. 8.5   A monitoring well screen that is too long will result in an inaccurate analytical result.

Natural Filter Packs.   A natural filter pack is constructed by installing a well screen and then simply allowing the formation to collapse about the screen. The well is then developed (see "Well Development" later in this chapter), which removes fine materials from the immediate vicinity of the screen. This leaves the coarse materials behind, making a natural filter.

A natural filter pack is appropriate only if the formation is coarse-grained, poorly sorted, and permeable. A poorly sorted (well-graded) formation will filter out fine-grained material better than a well-sorted (poorly graded) one, and thus makes a better filter pack.

To determine if a natural filter pack is appropriate for a particular well, a sample of the formation that is to be screened should be retrieved. The sample should be carefully examined, and if necessary, a grain size analysis of the sample should be performed.

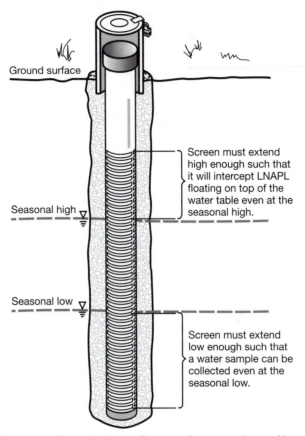

Fig. 8.6 Placement of a monitoring well screen when contaminant of interest is an LNAPL.

If the formation is composed of clay, silt, or fine sand, an artificial filter pack should be used. However, if at least 90% of the formation will not pass through a "10-slot" screen (which has an opening of 0.010 inch, or 0.25 mm), a natural filter pack may be considered (Aller and colleagues, 1989). If a natural filter pack is to be considered, the next step is to determine the coefficient of uniformity (C_u). (See Box 8.1 on the grain-size distribution test.) If C_u is greater than 3, a natural filter pack may be used. If C_u is less than 3, *or* if more than 10% of the material passes through the "10-slot" sieve, an artificial filter pack should be used.

Artificial Filter Packs. An artificial filter pack consists of granular material, usually sand, that is placed in the annulus of a well, about the screen. Artificial filter

BOX 8.1

Grain-Size Distribution Test

Performing a grain-size distribution test allows one to determine the sorting characteristics of a sample of sediment. For granular sediments, a sieve test is used. In this test, sediment is passed through a set of sieves with various sizes of openings. The amount of sediment that is caught by each sieve is weighed, and the results are plotted on a graph of grain diameter versus percent of total sediment that passed through each sieve.

A grain-size distribution test may be performed in a laboratory. However, if the purpose of the test is to determine where a screen should be placed, and if the borehole is open and drilling is in progress, a field test may be performed.

To perform a grain-size distribution test in the field, the following equipment is needed: sediment sample (about 500 grams), a water source, a balance and weighing containers, semilogarithmic graph paper having 3, 4, or 5 cycles, and sieves with the following size openings:

# 10	2 mm	0.080 in.	sand/gravel boundary
# 30	0.6 mm	0.024 in.	
# 40	0.46 mm	0.018 in.	coarse/medium sand boundary
# 60	0.25 mm	0.010 in.	10-slot screen

Protective gloves should be worn, with other necessary protective equipment used as indicated by the specific conditions. At some sites, a bucket or barrel will be needed to collect the water and sieved sediment for later proper disposal.

Procedure:

1. Dry and weigh a sample of about 500 grams.
2. Stack the sieves in order of opening size, with the largest opening at the top.
3. If sediments and water are to be contained, hold the sieves over the bucket or barrel.
4. Place the sample in the top sieve; disaggregate the clumps.
5. Shake the sieves and wash the sediment with water to move it through the sieves.
6. Dry and weigh the sediment retained on each sieve.
7. Determine the percent of total weight retained on each sieve. The lost sediment, sediment not caught on any sieve, is silt, clay, and fine sand. If the lost sediment amounts to more than 10% of the total weight, use an artificial filter pack.
8. Plot the cumulative percent retained versus grain size on semilog paper, with the percent retained on the arithmetic axis and grain size on the logarithmic axis. Connect the dots with a smooth curve. (See Fig. 8.7.)
9. Determine D_{60} and D_{10}. D_{60} is the diameter of grain at which 60% of the sample passed; D_{10} is at 10%. D_{10} is also referred to as the effective grain size.

(continued)

BOX 8.1 (*continued*)

Calculate the uniformity coefficient (C_u):

$$C_u = \frac{D_{60}}{D_{10}}$$

If $C_u > 3$, a natural filter pack may be used. If $C_u < 3$, the sediment is well-sorted, and the slot size of the screen is critical. To be on the safe side, use an artificial filter pack. ∎

packs are used in fine-grained or well-sorted aquifers, or whenever natural filter packs are inappropriate, as described in the last section.

Nielsen (1991) suggested that an artificial filter pack should be used in any of the following circumstances:

1. The formation is more than 10% fine sand, silt, or clay (more than 10% passes through the 10-slot screen)
2. The formation is heterogeneous
3. The well screen intersects layers of varying composition or grain sizes
4. If the coefficient of uniformity (C_u) is less than 3. (C_u is defined in Box 8.1.)

Sand used to construct an artifical filter pack must be carefully sized to the characteristics of the formation. To find the optimum grain size of the filter pack, Driscoll (1986) recommended multiplying D_{30} by a factor. The factors are as follows: for a fine-grained, uniform formation, the factor is 3, and for a coarse-grained, nonuniform formation, the factor is 6. Nielsen and Schalla (1991) cited several design techniques for choosing filter pack material. Other suggestions for designing filter pack grading and length are given in Driscoll (1986).

In specifying size of the filter pack sand, some workers refer to the "mesh" of the sand. For example, sand may be specified as "Colorado silica sand, 8–12 mesh." This means that it will pass through a number 8 sieve, but not through a number 12 sieve. Other common examples are "10–20 mesh," and "40–100 mesh."

Sand for artificial filter packs can be purchased in bulk bags. In addition, some well screens are constructed with the filter pack as an integral component, so that the sand need not be emplaced in a separate step. Silica sand is preferable to calcareous sand, as calcareous sand may dissolve over time in slightly acidic ground water conditions. Likewise, if a well should need to be acidified during refurbishment, silica sand will be more chemically stable than calcareous sand.

One final consideration should be checked when specifying the filter pack design. As described earlier, the borehole diameter must be large enough to install both the well screen and a filter pack that is 2 to 3 inches wide on all sides. Thus, if the well screen is 2 inches in diameter, the borehole must be 6 to 8 inches in diameter (Fetter, 1994).

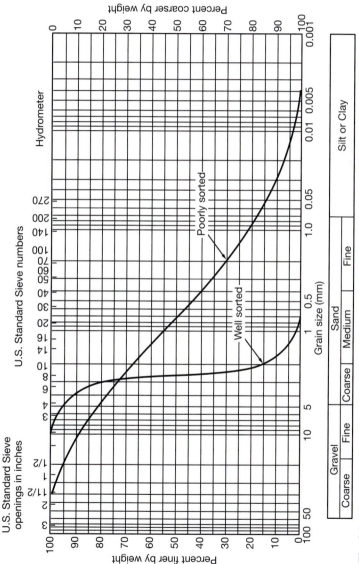

Fig. 8.7 Grain-size distribution curves.

251

Screen Slot Size and Type. Screens must have openings that allow water to move into the well, but block the movement of filter and aquifer materials into the well. The size and shape of the openings affect the ease with which water can flow into the well. The openings may be slots or perforations, and their size should be specified in relation to the grain size of the filter pack.

Screen openings are of three general types: sawn or slotted, wound (as in continuously wound), and bridge or louvered (Fig. 8.8). Sawn or slotted screens are made by sawing slots into a piece of casing. Machine-sawn screens are commercially available and are recommended for most applications. Homemade screens, slotted by hand with a hacksaw or cutting torch, are less expensive than commercial screens, but they do a poor job of optimizing water flow, preventing silt from entering the well, and maintaining screen strength. They therefore are not recommended.

Wound screens are constructed by winding threads of screen material around a skeleton of supports that run the length of the screen. This method of construction maximizes the space available for water to enter the well without diminishing screen strength.

Bridge screens and similar designs such as louver or shutter-type screens are constructed by punching holes in pipe metal. The metal is pushed outward to form an opening with a shutter-shaped louver or bridge-shaped piece above it. These screens should be used only in wells with artificial filter packs (Driscoll, 1986). They are more commonly used in supply wells than monitoring wells.

Slot Size for Natural Filter Packs. Slot size must be keyed to filter pack grain size and formation characteristics. Nielsen and Schalla (1991) recommended the following for monitoring wells installed with a natural pack:

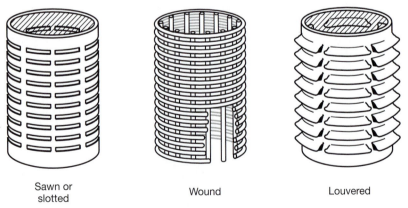

| Sawn or slotted | Wound | Louvered |

Fig. 8.8 Screen intakes: sawn or slotted, wound, and louvered.

Criteria		**Slot Size**
$C_u > 6$,	noncaving material above screened interval	D_{40}
$C_u > 6$,	readily caving material above screened interval	D_{50}
$C_u > 3$,	noncaving material above screened interval	D_{50}
$C_u > 3$,	readily caving material above screened interval	D_{60}

Layered formation materials, differing sizes • D_{50} of coarsest layer is less than four times D_{50} of the finest layer	Base slot size on the finest layer, use separate screens for each permeable zone

For extraction wells, Clark (1988) recommended using a slot size equal to the average D_{40} of six samples of filter pack material. If the aquifer is coarse or poorly sorted, D_{50} should be used instead. If the aquifer is very heterogeneous, the design should be based on the finer parts of the aquifer.

Slot Size for Artificial Filter Packs.   Choosing a slot size for a well with an artificial filter pack is a relatively simple matter, because the grain size of the filter pack is known. Nielsen and Schalla (1991) recommended the following: For extraction wells, a slot size that falls in the range of D_{15} to D_1 should be used. Fetter (1994) noted that many monitoring wells must be installed in silt or clay. For these situations, he recommended a slot size of 0.008 inch ("8-slot") and an appropriate filter pack.

Casing and Well Screen Material

Casing and well screen material should be chosen on the basis of strength and durability, chemical characteristics, cost, type of joints, and slot-type requirements. For monitoring wells, chemical characteristics are probably the most important consideration, as the chemical integrity of the sample is of greatest concern. For extraction wells, strength and durability may be the most important. Characteristics of common materials, polyvinyl chloride and steel, are outlined in this section.

Polyvinyl Chloride (PVC).   Polyvinyl chloride (PVC) is an ideal well construction material for many applications. It is relatively inexpensive, strong, lightweight, and resistant to corrosion. However, its characteristics limit its usefulness in some situations. Its strength limits its use to 1200–2000-foot depth (Nielsen and

Schalla, 1991), and if it comes into contact with some organic solvents, it may dissolve. Some studies have shown that PVC causes very low levels of vinyl chloride to appear in the water it contacts, and this might be undesirable in some monitoring programs (Barcelona and colleagues, 1983; Nielsen and Schalla, 1991). When PVC is used in a monitoring well, it should have threaded ends. Using glues or solvents to connect lengths to each other could result in the contamination of samples.

Steel. Four types of steel well screen and casing may be used: carbon ("black") steel, low-carbon steel, galvanized steel, and stainless steel. Steel well screen and casing is rigid, strong, and not temperature-sensitive. It is more expensive and heavier than PVC. Steel is inappropriate in some situations. For example, if a monitoring well is installed for the purpose of measuring metals concentrations in ground water, using steel may compromise the chemical integrity of the samples. In addition, because steel corrodes, if the ground water is highly saline, has high dissolved gas content, or is acidic, steel may not be appropriate (Nielsen and Schalla, 1991). Carbon steel may rust when exposed to air and water. Stainless steel is the most corrosion-resistant of the four types of steel; however, it also is the most expensive.

Annular Seal

The annular seal is composed of material that is placed in the annulus of the well, the space between the well casing or screen and the wall of the borehole. The annular seal keeps surface water from moving downward through the annulus. This prevents surface water from entering the well and prevents potential cross-contamination of aquifers. In nested piezometers, an annular seal closes off different high-permeability zones from each other.

Bentonite is a common material used for constructing annular seals. Bentonite is a commercially blended material containing montmorillonitic swelling clay derived from the weathering of volcanic ash. When wetted, it expands, and when dried, it shrinks. Used below the water table, bentonite makes an effective seal. Above the water table, it shrinks and cracks, and its use is not advisable.

Bentonite is commercially available in three forms: chips, pellets, and powder (Fig. 8.9). Chips set up rapidly and may be used for simple applications. Pellets are used when more working time is needed. Powder is used when the hole is more than about 30 feet or 10 meters deep. In this situation, the grout may be tremied into the hole. In the tremie process, a pipe or hose is inserted all the way to the bottom of the annulus, the bentonitic powder is mixed with water at the surface, and then the grout is pumped down the tremie pipe (Fig. 8.10). As the hole fills, the tremie pipe is gradually withdrawn. This process ensures complete filling of the hole. When chips or pellets are poured in, the possibility exists that they may bridge, forming air pockets and incompletely filling the hole.

Cement may be used as an annular seal in some situations. However, any cement used as a seal must be of the nonshrinking type. This reduces the likelihood that cracks or spaces may open up as the cement cures. Cement also has been associated with high pH in wells. As the cement ages, and as water leaches through

Fig. 8.9 Three forms of bentonite used in well construction: pellets, powder, and chips. (Courtesy of CETCO.)

it, water chemistry changes. Because of these limitations, a bentonite-cement mix may make a better annular seal than "neat" cement (cement with no additives).

A final consideration is the situation of an annular seal in a well with no screen. When the intake portion of a well is an open borehole, instead of a screen, how can the annular seal be placed? In these wells, the casing is cemented into place at the top of the intake portion of the well. This cement forms the base of the annulus. The annular seal is placed on top of the cement, filling the annulus above that point.

Surface Seals (Sanitary and Secure Seals)

The surface seal of a well is composed of two parts: the sanitary seal and the secure seal. A sanitary seal keeps surface water from entering the well; a secure seal prevents vandals from opening and possibly damaging the well.

The sanitary seal is placed above the annular seal. The sanitary seal should extend below the frost line; local building inspectors can provide information on this depth. If the seal is placed at too shallow a depth, freezing may heave the well out of the ground. The sanitary seal is made of cement or concrete placed in the annulus. Sides of this extended doughnut of cement or concrete should be vertical, not tapered, as a tapered shape will lead to frost heaving. The top of the seal should be sloped away from the well; this will encourage surface water and rain to drain away from the well.

A secure seal includes a well cap, protective housing, and a lock (Fig. 8.11). Either the well cap or the housing should be lockable. The well cap is placed directly on the well casing and keeps out rain and insects. It should be vented to prevent the possibly hazardous buildup of gases in the well.

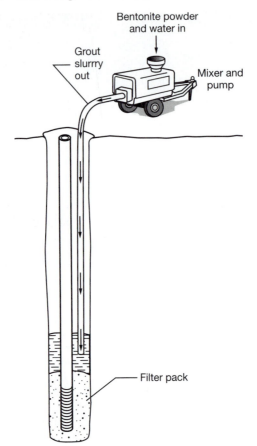

Fig. 8.10 Tremie grouting.

The protective casing may be flush-mounted (mounted even with the ground surface) or it may stick up a few feet from the ground. Wells in high traffic areas, such as monitoring wells at gas stations, should be flush-mounted. (In these installations, the cap should clearly say "DO NOT FILL" so that the well is not mistaken for the filler hole of an underground tank. In addition, a gasket or O-ring should be installed between the lid and the housing, to keep water out.)

Wells located in more remote areas, such as the middle of a grassy field, will be easier to find if they stick up a few feet, particularly if they are painted a bright color. Wells in areas subject to frequent flooding or standing water should be finished above ground. If an above-ground well is in danger of being hit by vehicles or heavy equipment, posts should be installed to protect the well. For example, this might be advisable for monitoring wells located near a roadway at a landfill.

Fig. 8.11   A secure seal includes a well cap, protective housing, and a lock. In this case, the housing lid has been unbolted and removed, and the well cap is in place and locked.

Pitless Adapters

In some areas, water supply wells are constructed with devices called pitless adapters. Pitless adapters connect the supply line of the pump with an underground distribution pipe that conveys water into a building (Fig. 8.12). The distribution pipe is buried to a depth below the frostline. To connect it to the well, a T-shaped fitting is installed in the casing. This fitting, the pitless adapter, allows the water to be pumped up and directed into the distribution line. Before pitless adapters became available, pits were dug around the wellhead, and the pump was installed in the pit, below the frostline, to keep it from freezing. Because this method increases the risk of admitting surface drainage and potentially contaminating the well, pitless adapters are preferred to pits.

Pump Depth

To determine the optimum depth at which the pump should be set for wells that will have a permanent pump installed, the water level and its likely fluctuations should be considered, as well as the expected drawdown. Clearly, the pump must be below the water level. However, it must be far enough below water so that in times of low water levels, or when pumping draws down the water level, there is still plenty of water above it.

The expected pumping demands must be considered. By using an estimate of the hydraulic conductivity (K) and the expected pumping rate, the drawdown in the well after several days of pumping should be calculated. For this calculation, using 3 days for confined aquifers, and 14 days for unconfined aquifers, is not unreasonable. The calculation should be made using the Theis equation, an

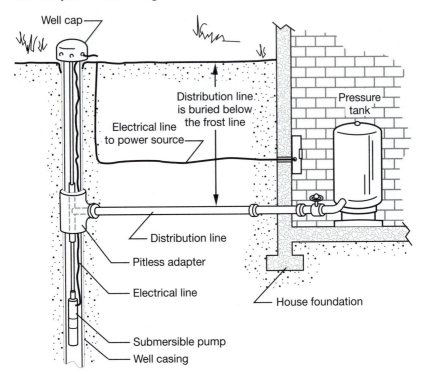

Fig. 8.12 A pitless adapter connects the well casing to an underground distribution line. Water is conveyed below the frost line directly into the building.

estimate of storativity, and estimating transmissivity as hydraulic conductivity times the length of the well intake. Once the expected drawdown has been calculated, it should be doubled to provide a safety factor. The pump should be placed below this level. If pumping test results are available, they should be incorporated into the design.

Well Design and Hydrogeologic Judgment

Many decisions and choices must be made in designing a well. Formulas and rules of thumb may be used as a guide. However, in each situation, the particular requirements and conditions of that situation should be considered. The final design should rely on the experience and best judgment of the hydrogeologist.

Well Installation

In general, well design is the responsibility of the hydrogeologist, and well installation is performed by the drillers, under the supervision of the hydrogeologist. Correct installation ultimately is the responsibility of the hydrogeologist. Because a good working relationship and clear communication with the drillers makes the task go more smoothly, the hydrogeologist should have an understanding of the steps involved in well installation.

Well installation begins while the borehole is being drilled. If the well is not to be screened, and instead, a portion of the borehole is to be left open, then the well casing is set during the drilling. In these situations, the hole is advanced to a certain depth, generally through the sediment and several feet into solid rock. Then, cement is poured at the bottom of the hole, and the casing is set into the wet cement. When it has cured sufficiently, drilling proceeds deeper into the rock by using a smaller bit size and drilling inside the casing, advancing the borehole to the required depth.

If the well is to be screened, the borehole is advanced to its total depth before well materials are installed. The installation begins by attaching a well bottom cap (or plug) to the well screen, attaching the screen to the well casing, and lowering the assembly down the hole. Note that if more sections of screen or casing are needed, as in deeper wells, then the first sections lowered downhole must be held with a clamp so that the next sections may be added. It may be tempting to try to hold them by hand, but several sections of screen and casing weigh a significant amount, and could quickly slip out of one's hands.

Sections of casing are added until the top of the casing is above the ground surface. A temporary cap is placed on the casing to keep sand and other objects from falling downhole during the remainder of the installation. The filter pack is installed next, unless a natural filter pack is specified. An artificial filter pack is installed by pouring the sand into the annulus, pouring an equal amount on all sides. The installer should be alert to the possibility of bridging. This occurs when a "bridge" of sand forms partway down the hole, leaving the space below empty. Bridging can be avoided by "wiggling" the casing from side to side (not up and down) or by tamping the sand with a small diameter pipe as the sand is being added.

When the top of the sand is the specified height above the top of the screen, the installer should stop pouring the sand. How can this point be determined? One of the following three methods may be used.

1. If the well is shallow, the top of the sand may be measured directly with a steel measuring tape.
2. In a deeper well, a small diameter pipe such as a tremie pipe may be placed downhole before starting to install the filter pack. A mark is made on the pipe at the level of the top of the casing. Then the small pipe is pulled. To test the height of the filter pack, the tremie pipe is reinstalled in the hole, and the difference between the marked point and the top of the casing is found.
3. The volume of the filter pack may be calculated mathematically. Because of the variability of natural materials in the field, this method should be used only if neither of the preceding methods works. The volume of the filter pack in the annulus is calculated, and then that volume of sand is added to the annulus. Assuming that the filter pack is to extend 3 to 5 feet above the top of the screen, the volume of the filter pack in the annulus may be found as follows:

Volume of filter pack in the annulus

$= \pi \times$ (borehole radius^2 – screen radius^2) $\times$ (length of the screen + 3 to 5 feet)

Next, the annular seal is installed. As described in the section on annular seals, this should consist of bentonite, cement, or a mixture of the two. Some regulatory agencies require that the first 2 feet of seal above the filter pack be of bentonite, and that a bentonite-cement mix may be used above that point. Using pure bentonite above the water table is not advisable because, as already discussed, it shrinks and cracks when dried.

The annular seal is installed by pouring the material down the annulus or by using a tremie pipe. The annulus is filled until the top of the material is just below the frost line.

Next, the surficial sanitary and secure seals are installed. A straight-walled hole is dug about the well, down to the level of the annular seal (below the frost line). By using a hacksaw or other cutting tool, the casing is cut off to the desired height, and the temporary cap is replaced on the well. The hole is partly filled with wet cement (or concrete), and then the protective housing is pressed into place. The remainder of the hole is filled outside of the housing with the cement. The cement is smoothed and shaped, sloping the surface so that rainwater will drain away from the well. The permanent well cap and lock are installed, and the cement is allowed to cure. The well construction diagram (see Chapter 6) should be completed while the details are still fresh in mind.

Well Development

Wells must be developed before they can be expected to provide consistent yields or representative chemical samples. Well development involves various postinstallation activities that maximize the performance of the well. It removes drilling fluids and water introduced during drilling, flushes out fine-grained material ("fines") from the aquifer, and stabilizes the filter pack. Monitoring wells that tap aquitards should not be extensively developed.

Several development methods may be used, depending on conditions. The simplest of these, overpumping, is also the least effective. Overpumping involves simply pumping water from the well at a high rate. This draws fines into the well, but overpumping alone will flush out only a portion of the fines. To increase effectiveness, the well also should be surged. Surging involves creating a wash and backwash action. One method of accomplishing this is to alternately turn the pump on and off. It is left on for a few minutes, then off for a few minutes. A pump used for surging should be of a type that will not be damaged by pumping silt and fine sand.

Another method of surging can be accomplished by a driller with a surge block. A surge block fits loosely within the well and is pumped up and down to create the desired surging action. This may also be accomplished by hand, using a bailer or slug.

Another development method is jetting, in which jets of water are directed horizontally at the sides of the well, from inside the well (Fig. 8.13). Finally, a well

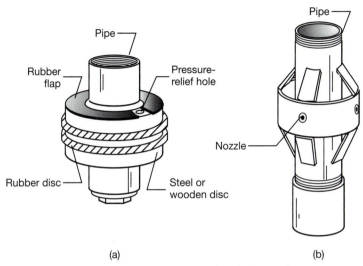

Fig. 8.13   Well development. (a) A surge block. (b) A jetting tool.

may be developed by acidification. In this method, acid is put in the well in order to dissolve calcite and to open cavities. This should not be done to monitoring wells, because it might change the chemistry of the formation water.

Water pumped out of a well during development should be disposed of according to pertinent regulations.

Well Abandonment

When wells are no longer needed, they should be properly abandoned. Old wells may pose a threat to public safety, they may serve as a conduit for contaminants from the surface to enter the subsurface, or they may promote cross-contamination of aquifers. Properly abandoning them minimizes these hazards.

Most states have abandonment regulations or guidelines. When no regulations exist, abandonment involves these general steps. When possible, the casing and screen should be pulled or drilled out. The well should be filled with bentonite grout or neat cement, or a mixture of the two. If the casing was not removed, it should be cut off below ground surface and the excavation refilled. More information on well abandonment is presented in Driscoll (1986). A sample well abandonment form appears in Chapter 6.

Geochemical Measurements

Chapter Overview

Hydrogeologists may be called on to investigate situations where the possibility of soil or water contamination exists, or where chemical quality of water is critical. In these situations, the hydrogeologist must use extreme care in taking samples.

Because decisions about pollution abatement and remediation are based on results of field tests and on lab analyses of field samples, the quality of field work may determine the success of the whole project. Sloppy field technique can result in misleading or meaningless results. Such mistakes may result in an expensive cleanup being undertaken when none is needed, or perhaps worse yet, no cleanup at all being done when such a cleanup is urgently needed. As a result, careful geochemical sampling and analysis are of immeasurable importance.

This chapter outlines steps for careful sampling of ground water for geochemical analysis. It also reviews the use of field instruments for analysis.

Sampling Ground Water

Why might one want to analyze the chemistry of ground water? In some cases, ground water in an area might be thought to be contaminated, and an analysis will help ascertain the extent or degree of contamination. Perhaps the ground water is uncontaminated, but needs a clean bill of health. Perhaps research is being done on the interactions between ground water and rock, and an analysis will help characterize those interactions. Perhaps a well owner is simply curious about what constituents may be found in the water.

In any of these situations, a sample of the water usually must be taken before an analysis can be performed. The sampling plan and sampling procedure should be meticulously developed and reviewed before any samples are taken. Because the sampling method may have surprisingly significant effects on the water chemistry, great care must be taken in collecting the sample. This section reviews components of a sampling plan and procedure.

Sampling Plan

Never take a sample without a plan. An unplanned sampling expedition is likely to result in too many samples of one type and not enough of another, incorrectly collected or preserved samples, inappropriate or forgotten sampling equipment, and ultimately, a second trip to the field to correct the mistakes of the first trip.

A plan for sampling ground water should have the following components: objectives of the sampling program; timing of the sampling trip; access; extraction methods and procedures; sample containers; field analyses, sample preservation, and storage; decontamination; transport and chain of custody; and lab analysis. These components are described here; considerations that will help answer the questions posed in the following sections are described later in this chapter.

Objectives of the Sampling Program

What is the reason for sampling? What kinds of analyses will be performed? What regulations, if any, apply to the site? What special objectives must be considered? The answers to these questions will dictate the protocol and timing of the sampling, the types of containers needed, the number and volume of samples obtained, and virtually every other component of the plan.

Timing of the Sampling Trip

When will samples be taken? Should sampling be conducted during a certain season? Seasonal conditions may change the geochemistry of the water. Because certain environmental regulations require it, many monitoring wells are sampled quarterly (four times per year). In most conditions, ground water moves very slowly, and so more frequent sampling is considered unlikely to reveal any new information. In some situations, however, more frequent sampling might be warranted. This would be the case if ground water were moving rapidly, for example, under a high gradient or through especially permeable materials (including cavernous limestones). More frequent sampling also might be appropriate for monitoring the effects of treatment or remediation activities on ground water.

Should sampling be conducted on a certain date or day of the week? Should the samples be taken at a particular time of day? *Holding time* is the maximum time the samples may be stored before analysis, without compromising analytical accuracy. For certain parameters, the holding time is short, perhaps a few days or even a few hours. Can the analytical lab perform the analyses before the holding time expires? A Friday afternoon may not be a good time to take samples if the holding times will expire while the lab is closed over the weekend. At some sites, pumping of nearby wells may affect the wells that are to be sampled. If this is the case, when might these effects be minimized? These questions must be considered before a sampling trip is undertaken.

Access

How can access to the wells be gained? Will special permissions of the landowners be required? Do the landowners want to be present during sampling? Might wildlife, angry watchdogs, poisonous plants, or livestock cause problems? Will keys be needed to gain entry to a gate, a building, or the wells? In some locations, local law enforcement officials should be notified that sampling will be taking place.

Extraction Method and Procedures

How will water be removed from the well for purging and for sampling? Will the method chosen affect the chemical characteristics of the water? Will the material, operation, or construction of the purging and sampling devices compromise the quality of results? For example, sampling with a top-emptying bailer or a vacuum pump might allow volatile components to escape. If the water will be analyzed for volatile organic chemicals, use of these sampling devices might result in an analysis that does not reflect the true composition of the ground water.

Will the extraction method work effectively in the particular wells to be sampled? For example, a 4-inch-diameter pump will not fit in a 2-inch-diameter well. A bailer with a 10-meter cord will not work if the water level is more than 10 meters below the ground surface. And if the water level is more than about 20–25 feet below the surface, a vacuum pump will not have enough lift to bring samples to the surface. If nonaqueous phase liquids (NAPLs) are found in the well, should they be sampled? How will purge water be dealt with? In what order will wells be sampled? What quality assurance and quality control (QA/QC) procedures will be followed? (QA/QC is discussed later in this chapter.) What personal protective equipment will be needed? Will the site be accessible to the sampling equipment or will more compact or more portable equipment be needed?

Sample Containers

What containers will be used? How many will be needed, and of what type should they be? Container type depends on which analyses are to be performed. If several different analyses are to be made, then several types of containers might be needed. In addition, several containers of a given type might be needed for each well. How large must the containers be? How many will be needed? Should preservatives be added to the containers before going to the field, or while in the field?

Field Analyses, Sample Preservation, and Storage

What analyses must be performed in the field? In what order? What instruments will be needed? What power sources will they require? Will flow-through cells (which allow monitoring of certain characteristics during pumping) be used? How will the samples be stored after the analyses are complete, and before transport?

Decontamination

How will field equipment be decontaminated? What decontamination solutions and supplies will be needed?

Transport and Chain of Custody

How and when will samples be transported to the analytical laboratory? What personnel and vehicles will be required? How will samples be stored during the transport? What chain-of-custody forms (forms documenting who holds the samples at any given time) will be used?

Lab Analysis

What parameters will be measured in the lab? Which analyses do the pertinent regulations require? Which laboratory will perform the analyses, and where is it located? How much will the analyses cost? When will they be performed? What is the risk that samples may be held too long before they are analyzed?

The sampling plan requires consideration of all of these questions. Suggested answers and solutions to potential problems are discussed in what follows.

Sampling Procedure

Once the sampling plan has been developed and finalized, the sampling itself may begin. Sampling activities should be conducted in a particular sequence at each well. Procedures should be specified in the sampling plan. Procedures also should be reviewed carefully and rehearsed, if necessary, before the sampling trip.

This section reviews steps in sampling procedure. For more information, consult Wilson (1995), which provided a detailed discussion of considerations for each step in the sampling procedure, or Nielsen (1991), which described the planning and installation of monitoring wells as well as the sampling process. Garrett (1988) published a convenient pocket-sized booklet summarizing the steps in sampling. For sites that are regulated, the technical guidance documents published by the regulatory agency should be consulted (e.g., see USEPA, 1986). Recent research on sampling techniques is published quarterly in the journal *Ground Water Monitoring and Remediation*, a publication of the National Ground Water Association.

Before Going Out to Sample

The following steps should be taken before the sampling trip begins:

Review the sampling plan.

Assemble maps and site plans.

Check the analytical procedures or with the laboratory for information about timing of analyses, containers, preservatives, and QA/QC procedures.

Gather equipment (see the following list).

Check on access to the site and wells.

Notify local law enforcement authorities, if necessary.

Equipment

The following items should be gathered and taken into the field for the sampling trip:

Sampling plan

Pumps or bailers for purging the well

Pumps or bailers for sampling the well

Calibrated bucket (or other means) for determining purge volume

Flow-through cell, if monitoring parameters during pumping

Sample containers and preservatives

Cooler and ice, plus thermometer for measuring temperatures in the cooler (a maximum/minimum thermometer is ideal)

Thermometer for measuring sample temperatures, pH and conductivity meters, plus calibration solutions

Eh and turbidity meters, if needed

Filtering apparatus and filters, if needed

Water-level measuring device

Personal protective equipment, if needed

Tarp on which to set out equipment

Container for storing purge or waste water, if needed

Field notebook and writing tools. Caution should be used when using markers that contain organic solvents, such as most felt-tip markers, because fumes from the markers may contaminate air or water samples

Keys to site and wells

Trip blanks (see the discussion of QA/QC later in this chapter)

Wrenches, screwdrivers, pliers, wire cutters, and duct tape

Decontamination equipment: buckets, lab-grade detergent, wash acid or solvent, sufficient deionized water to wash and rinse all equipment after sampling each well (as required)

Calculator

Water/NAPL interface probe, if needed

Chain-of-custody forms

Well logs and well completion diagrams

Sterile disposable gloves

Organic vapor analyzer (OVA), gas chromatograph (GC), or explosimeter, as needed.

Other field geochemical analytical instruments, as needed

Extra batteries for all battery-operated instruments, and other power sources (car battery, generator), as needed

Paper towels

After Arriving at the Wellhead

This section outlines the steps in purging and sampling. At the end of this section is a more extensive description of the processes of purging a well, sampling, filtering a sample, and preserving a sample.

Steps in Purging and Sampling.   Do not immediately open the well! Opening the well at this point may allow gases that have built up inside the well to escape. Once they have escaped, they cannot be measured. Do not open the well until air-monitoring equipment (if it is to be used) has been set up and is ready to measure the gases in this "headspace" of the well.

Check the site map and well records to ensure that the well is the one that is supposed to be sampled. An identification number might be found somewhere on the well; if one is found, it should be checked with the sampling plan to ensure that it is on the list of wells to be sampled.

Set up the first entry in the field notebook. See Chapter 2 for a list of information that should be included in the field notebook.

Inspect the exterior of the well, looking for damage or evidence of tampering.

If headspace analysis of the well is needed (analysis of the vapors trapped within the well casing, above the water), set up and calibrate the organic vapor analyzer (OVA) or explosimeter, or other appropriate gas-monitoring instrument. Take ambient air readings.

Remove the well cap.

If needed, use the organic vapor analyzer to take headspace readings. Take readings in the zone about the opening of the well. If a need for breathing protection is indicated, immediately put it on.

Measure the depth to water and the total depth of the well. Methods and instruments for making these measurements are described in Chapter 4. If the well contains a nonaqueous phase liquid (NAPL), an interface probe may be used to determine the depth to the top and bottom of the NAPL layer (Fig. 9.1). NAPLs may occur above (light NAPLs or LNAPLs) or below (dense NAPLs or DNAPLS) the column of water. As a result, the interface probe should be lowered all the way to the bottom of the well to check for DNAPLs.

If required by the sampling plan, sample any NAPL layers found in the well. To sample an LNAPL, carefully lower a bailer down the well until it touches the NAPL surface. Then, let it sink slowly for a distance equivalent to the NAPL thickness, plus about an inch. Carefully pull it to the surface. To sample a DNAPL,

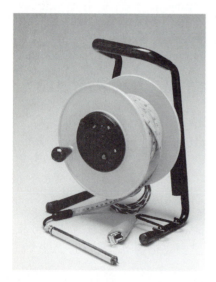

Fig. 9.1 Interface probes measure depth to the boundary between water and a nonaqueous phase liquid (NAPL). (Courtesy of Enviro Products, Inc.)

lower a double check valve bailer to the bottom of the well, then pull it carefully to the surface, or use a depth-specific sampler.

Purge the well. (See the section on "Purging" in what follows..)

Set up the sampling device. (See the section on "Sampling Devices" in what follows.)

Sample for the required constituents, filtering and preserving the samples as needed. Disposable sterile gloves should be worn. The appropriate size and type of sample container for each type of sample should be used (see Table 9.1). Samples should be taken in the following order.

Field Measurements.   Certain parameters may change rapidly on exposure to light, warmth, or air. If these measurements are to be made, they should be made in the field: temperature, pH, conductivity, dissolved gases (including dissolved oxygen), turbidity, or reduction-oxidation potential (Eh). These parameters may already have been measured during checks for purge adequacy. A description of tests of these parameters appears later in this chapter.

Volatile and Semivolatile Organic Constituents.   When sampling for volatiles and semivolatiles, the sample container should be filled completely, so that no air space remains at the top. A meniscus should appear at the top of the vial, so that the vial appears to be overfilled. The container should be capped and sealed securely. To check for air in the sample, turn the vial upside down and tap it to see if a bubble appears. If it does, discard the sample and try again, until no bubble appears. The sample should not be opened once it has been sealed, because opening it may allow the volatiles to escape.

PCBs and Pesticides.   A relatively large volume of sample is required for analysis of these components, and the container should be amber-colored to reduce the effect of light reaching the sample.

Metals.   Most samples taken for metals analysis should be filtered and acidified, although there is some disagreement among experts on this issue. Table 9.1 lists requirements for specific metals. Filtering is discussed in more detail in a later section.

Anions.   Samples taken for anions need not be filtered. However, certain anion analyses require that the samples be preserved. See Table 9.1 for specific requirements. Preservation of samples is discussed in more detail in what follows.

Radioactive Constituents.   Most samples taken for radioactive constituents should be filtered and acidified, with the exception of radon.

Bacteria.   Samples taken for bacterial analysis must be taken with sterilized equipment and be placed in sterilized containers. Disposable sterile gloves should be worn when taking any sample, but it is particularly important to wear them when sampling for bacteria.

TABLE 9.1

Sample Container, Volume Required, Sample Preservation, and Holding Time

Parameter Name	Container (P: polyethylene; G: glass)	Minimum Sample Volume Required to Perform a Single Analysis[a]	Preservation	Maximum Holding Time
Bacteria				
Coliform, fecal and total	P, G	100 ml	Cool to 4°C, and in presence of residual chlorine, 0.008% $Na_2S_2O_3$	6 hours
Fecal streptococci	P, G	100 ml	Cool to 4°C, and in presence of residual chlorine, 0.008% $Na_2S_2O_3$	6 hours
Aquatic toxicity	P, G		Cool to 4°C	6 hours
Inorganic tests				
Acidity	P, G	100 ml	Cool to 4°C	14 days
Alkalinity	P, G	100 ml	Cool to 4°C	14 days
Ammonia	P, G	400 ml	Cool to 4°C, H_2SO_4 to pH < 2	28 days
Biochemical oxygen demand (BOD)	P, G	1 liter	Cool to 4°C	48 hours
Biochemical oxygen demand, carbonaceous	P, G	1 liter	Cool to 4°C	48 hours
Boron	P, PTFE, or quartz	50 ml	HNO_3 to pH < 2	6 months
Bromide	P, G	100 ml	None required	28 days
Chemical oxygen demand (COD)	P, G	50 ml	Cool to 4°C, H_2SO_4 to pH < 2	28 days
Chloride	P, G	50 ml	None required	28 days

(continued)

TABLE 9.1 (continued)
Sample Container, Volume Required, Sample Preservation, and Holding Time

Parameter Name	Container (P: polyethylene; G: glass)	Minimum Sample Volume Required to Perform a Single Analysis[a]	Preservation	Maximum Holding Time
Inorganic tests				
Chlorine, total residual	P, G	200 ml	None required	Analyze immediately
Color	P, G	50 ml	Cool to 4°C	48 hours
Cyanide	P, G	500 ml	Cool to 4°C, NaOH to pH > 12, and in the presence of residual chlorine, 0.6 g/l ascorbic acid	14 days (24 hr if sulfide is present)
Fluoride	P	300 ml	None required	28 days
Hardness	P, G	100 ml	HNO_3 to pH < 2, H_2SO_4 to pH < 2	6 months
Hydrogen ion (pH)	P, G	25 ml	None required	Analyze immediately
Kjeldahl and organic nitrogen	P, G	500 ml	Cool to 4°C, H_2SO_4 to pH < 2	28 days
Metals				
Chromium VI	P, G	200 ml	Cool to 4°C	24 hours
Mercury	P, G	100 ml	HNO_3 to pH < 2	28 days
Metals, except boron, chromium VI, mercury	P, G	200 ml	HNO_3 to pH < 2	6 months
Nitrate	P, G	100 ml	Cool to 4°C	48 hours
Nitrate-nitrite	P, G	100 ml	Cool to 4°C, H_2SO_4 to pH < 2	28 days

(continued)

Parameter Name	Container (P: polyethylene; G: glass)	Minimum Sample Volume Required to Perform a Single Analysis[a]	Preservation	Maximum Holding Time
Metals				
Nitrite	P, G	50 ml	Cool to 4°C	48 hours
Oil and grease	G	1 liter	Cool to 4°C, HCl or H_2SO_4 to pH < 2	28 days
Organic carbon	P, G	25 ml	Cool to 4°C, HCl or H_2SO_4 to pH < 2	28 days
Orthophosphate	P, G	50 ml	Filter immediately, cool to 4°C	48 hours
Oxygen, dissolved				
Probe method	G bottle & top	300 ml	None required	Analyze immediately
Winkler method	G bottle & top	300 ml	Fix onsite and store in dark	8 hours
Phenols	G	500 ml	Cool to 4°C, H_2SO_4 to pH < 2	28 days
Phosphorus (elemental)	G	50 ml	Cool to 4°C	48 hours
Phosphorus (total)	P, G	50 ml	Cool to 4°C, H_2SO_4 to pH < 2	28 days
Residue (total)	P, G	100 ml	Cool to 4°C	7 days
Residue, filterable (total dissolved solids, TDS)	P, G	100 ml	Cool to 4°C	7 days
Residue nonfilterable (total suspended solids, TSS)	P, G	100 ml	Cool to 4°C	7 days

(continued)

TABLE 9.1 *(continued)*

Sample Container, Volume Required, Sample Preservation, and Holding Time

Parameter Name	Container (P: polyethylene; G: glass)	Minimum Sample Volume Required to Perform a Single Analysis[a]	Preservation	Maximum Holding Time
Metals				
Residue, settleable	P, G	1 liter	Cool to 4°C	48 hours
Residue, volatile	P, G	100 ml	Cool to 4°C	7 days
Silica	P, PTFE, or quartz	50 ml	Cool to 4°C	28 days
Specific conductance	P, G	100 ml	Cool to 4°C	28 days
Sulfate	P, G	50 ml	Cool to 4°C	28 days
Sulfide	P, G	500 ml	Cool to 4°C, add zinc acetate plus sodium hydroxide to pH > 9	7 days
Sulfite	P, G	50 ml	None required	Analyze immediately
Surfactants	P, G	250 ml	Cool to 4°C	48 hours
Temperature	P, G	1 liter	None required	Analyze immediately
Turbidity	P, G	100 ml	Cool to 4°C	48 hours
Organics				
Purgeable halocarbons	G, Teflon®-lined septum	5 ml	Cool to 4°C, and in presence of residual chlorine, 0.008% $Na_2S_2O_3$	14 days
Purgeable aromatic hydrocarbons	G, Teflon®-lined septum	5 ml	Cool to 4°C, and in presence of residual chlorine, 0.008% $Na_2S_2O_3$, HCl to pH < 2	14 days

(continued)

TABLE 9.1 *(continued)*
Sample Container, Volume Required, Sample Preservation, and Holding Time

Parameter Name	Container (P: polyethylene; G: glass)	Minimum Sample Volume Required to Perform a Single Analysis[a]	Preservation	Maximum Holding Time
Organics				
Acrolein and acrylonitrile	G, Teflon®-lined septum	5 ml	Cool to 4°C, and in presence of residual chlorine, 0.008% $Na_2S_2O_3$. If acrolein will be measured, adjust pH to 4–5 or analyze sample within 3 days)	14 days
Phenols	G, Teflon®-lined cap	1 liter	Cool to 4°C, and in presence of residual chlorine, 0.008% $Na_2S_2O_3$	7 days until extraction, 40 days after extraction
Benzidines	G, Teflon®-lined cap	1 liter	Cool to 4°C, and in presence of residual chlorine, 0.008% $Na_2S_2O_3$	7 days until extraction, 40 days after extraction
Phthalate esters	G, Teflon®-lined cap	1 liter	Cool to 4°C	7 days until extraction, 40 days after extraction
Nitrosamines	G, Teflon®-lined cap	1 liter	Cool to 4°C, and in presence of residual chlorine, 0.008% $Na_2S_2O_3$	7 days until extraction, 40 days after extraction

(continued)

TABLE 9.1 *(continued)*
Sample Container, Volume Required, Sample Preservation, and Holding Time

Parameter Name	Container (P: polyethylene; G: glass)	Minimum Sample Volume Required to Perform a Single Analysis[a]	Preservation	Maximum Holding Time
Organics				
PCBs	G, Teflon®-lined cap	1 liter	Cool to 4°C	7 days until extraction, 40 days after extraction
Nitroaromatics and isophorone	G, Teflon®-lined cap	1 liter	Cool to 4°C, and in presence of residual chlorine, 0.008% $Na_2S_2O_3$, store in dark	7 days until extraction, 40 days after extraction
Polynuclear aromatic hydrocarbons	G, Teflon®-lined cap	1 liter	Cool to 4°C, and in presence of residual chlorine, 0.008% $Na_2S_2O_3$, store in dark	7 days until extraction, 40 days after extraction
Haloethers	G, Teflon®-lined cap	1 liter	Cool to 4°C, and in presence of residual chlorine, 0.008% $Na_2S_2O_3$	7 days until extraction, 40 days after extraction
Chlorinated hydrocarbons	G, Teflon®-lined cap	1 liter	Cool to 4°C	7 days until extraction, 40 days after extraction
Pesticides	G, Teflon®-lined cap	1 liter	Cool to 4°C, pH 5–9	7 days until extraction, 40 days after extraction

(continued)

TABLE 9.1 (*continued*)
Sample Container, Volume Required, Sample Preservation, and Holding Time

Parameter Name	Container (P: polyethylene; G: glass)	Minimum Sample Volume Required to Perform a Single Analysis[a]	Preservation	Maximum Holding Time
Radiological parameters				
Alpha, beta, and radium	P, G	1 liter	HNO_3 to pH < 2	6 months

Sources: ASTM Standard D 4448. Code of Federal Regulations, Title 40, Part 136, Table II, July 1, 1996. USEPA (1979, 1983).

[a]To allow for repeated analyses, mishaps, and other analytical activities, the sample taken in the field should be several times larger than this minimum.

Decontaminate the sampling and analytical equipment.

When the cooler of samples is full, close it securely and latch it. The cooler should not be overfilled, and care should be taken while lifting it, as water samples are heavy.

Seal the cooler. In some situations, especially when required by regulations or when the analyses may be introduced as evidence in a legal proceeding, tamper-evident tape should be used to seal the cooler. When this tape is used, it becomes apparent if the samples have been disturbed.

Deliver the samples to the analytical laboratory directly. Confirm that the analytical lab has the proper facilities to keep them refrigerated until analysis.

Fill out and sign the chain-of-custody forms. When the samples leave the custody of one person and enter that of another, chain-of-custody forms should be signed by both parties. These forms state who had custody of a sample at what time, so that the custody steps may be retraced if need be. A sample chain-of-custody form is given in Chapter 7.

The steps in sampling procedure are reviewed in Box 9.1.

Purging

Why Purge?　The purpose of purging a well before sampling is to procure a sample that is representative of water in the formation. Before purging, a well holds water that originally came from the formation, but that has been in contact with air in the well and with well materials for some time. In addition, water that is situated above the screen, within the well casing, is stagnant, and has been out of contact with the formation for some time. Chemical changes may have occurred because of shifts in equilibria or biochemical changes, so this water is no longer representative of water in the formation. If it is purged out of the well, a representative sample may be obtained.

In purging a well, the nonrepresentative or "old" water is removed and "new" water is allowed to flow from the formation into the well. Unfortunately, it is not always easy to tell when the old water has been purged and only new water remains in the well. Formation transmissivity, hydraulic head, and the specifics of well construction affect the rate of recharge to the well. In some wells, recharge to the well is so slow that it may take days, or even weeks, for enough new water to flow into the well to sample. In these cases, the new water may be no more representative of the formation water than the old.

Low-Flow Purging.　Some recent studies (e.g., Barcelona and colleagues, 1994; and Puls and Paul, 1995; for a summary, see Stone, 1997) advocated very little purging of a well before sampling. These studies suggested that purging at high rates of flow introduces sediment into the well. Contaminants may have adhered to this sediment, and when the water is analyzed, the contaminants will appear to be present in higher concentrations than actually exist in the formation water. These studies showed that *low-flow purging*, or purging a well by setting the pump intake within the screened interval and pumping at a slow rate, typically 0.1 to 0.5 liters per minute, produces representative samples after very small amounts (less

BOX 9.1

Steps in Sampling Procedure

Do not open the well until everything is set up, including air-monitoring equipment.

Check site map and well records to ensure that you are at the correct well.

Set up the first entry in the field notebook.

Inspect and evaluate the condition of the well.

If headspace analysis of the well is needed, set up and calibrate the OVA, photoionization detector (PID), or explosimeter. Take ambient-air readings.

Remove the well cap. If taking headspace readings, take them immediately. Also take readings in the zone about the opening of the well. If a need for breathing protection is indicated, put it on immediately.

Measure the depth to water, the thickness of NAPL layers (if any), and the total depth of the well.

Take samples of the NAPL layers found in the well.

Purge the well, taking field measurements if so specified in the protocol.

Set up the sampling device (if it is different from the purging device).

Wearing disposable sterile gloves, take samples in the following order, filtering and/or preserving each sample as necessary:

Field measurements

Volatiles

Semivolatiles

PCBs and pesticides

Metals

Anions

Radioactive constituents

Bacteria

Place each sample, as it is taken, in a cooler at 4°C.

Decontaminate sampling equipment.

Seal the cooler (when full), fill out the chain-of-custody forms, and deliver the samples to the laboratory. ■

than one well volume) are pumped from the well. This has the added benefit of creating a much smaller volume of purged water that must be disposed of.

These studies used dedicated pumps, or pumps that are not removed from the well between sampling events. Using dedicated pumps minimizes the turbulence that takes place in the well during the introduction of a nondedicated pump or bailer. If a nondedicated device is to be used, it should be introduced into the well very slowly and carefully, with care to avoid mixing the stagnant water from above the screen with the water below. Chemical parameters of the purge water should be used to determine when the well has been adequately purged; this is further described in what follows.

If the sampling plan calls for purging the well at higher rates than just described, the following considerations apply.

Purge Rate and Purge Water Level.   Monitoring wells should be purged gently, to avoid stirring up sediment or introducing more sediment into the well. Low flow rates are desirable; however, this must be balanced with the need to produce a certain volume of water from the well before sampling.

When sampling for volatile organic chemicals or any constituents that might be affected by the concentration of oxygen or other dissolved gases in the water, the water level during purging should not drop below the top of the screen. If the water level drops below the top of the screen, air may be drawn into the well, changing the concentrations of these constituents. In addition, formation water may cascade into the well through the top of the screen, picking up more oxygen and further altering the water chemistry.

Purging a Slowly Recharging Well.   If, when one well volume is pumped out of the well, the water level drops to the level of the top of the well screen, and does not recharge to its original level within a half day or so, it may be difficult to obtain a representative sample from the well. For these slowly recharging wells, one approach, described by the ASTM in Standard Guide D4448-85a, is to empty the casing, let the well refill to its previous height of water, empty the casing again, let it refill again, and then sample. However, this may not be appropriate or practical, given the time it takes and the restrictions of the sampling program. Herzog and colleagues (1988) showed that when sampling for volatile organic chemicals (VOCs), it is sufficient to purge the well in the morning and then to return later in the day, or as late as the next day, to sample. Waiting for a longer time between purging and sampling for VOCs was likely to result in some of the VOCs being lost.

Purge Volume.   For wells that recharge more rapidly, three methods may be used to determine the appropriate purge volume. The first of these is to measure certain chemical parameters several times while purging, and to purge until the measurements stabilize. Temperature, pH, conductivity, dissolved oxygen, and turbidity are commonly used for this purpose; several of these parameters should be used in conjunction with each other. The assumption of this method is that formation water flowing into the well will yield consistent values of these parameters.

A second method for determining if a well has been adequately purged is to extract a specified number of well volumes before sampling. This method is conceptually simple. However, it is arbitrary and is not based on well-specific or formation-specific characteristics or conditions. Therefore, it does not apply to all situations equally well, nor is there agreement on what constitutes a well volume or the optimal number of well volumes that should be extracted.

A well volume may be loosely defined as the volume of water stored in the well before any is removed. However, as Wilson (1995) noted, the term well volume could be interpreted in several ways. The interpretation depends on which fraction or fractions of the water in a well are considered: the volume of the stagnant water in the casing (above the screen), the volume of the water in the screened interval, and/or the volume of the water within the filter pack. The volume of water in a given portion of the well may be determined by measuring the length of the water column in that portion of the well, before extracting any water. That length is then multiplied by the cross-sectional area of that portion of the well to give the volume.

To purge a certain number of well volumes, begin extracting water from the well, measuring the volume extracted by directing it into a precalibrated 5-gallon bucket or other suitable container. When the specified number of well volumes have been extracted, the well may be adequately purged, and sampling may begin. Contain the purged water if required to do so by regulations, by the sampling plan, or if the water is potentially contaminated.

How many well volumes constitute adequate purging? Wilson (1995) noted that various studies have suggested a range from less than 1 to more than 20. A common rule of thumb is 3 to 5 well volumes. The USEPA (1991) recommended that 3 to 10 casing and filter pack volumes should be purged, and ASTM Standard Guide D4448-85a suggested 5 to 10 well volumes, adding that the volume of the screened interval and/or filter pack could be included in the well volume calculation if it appeared that natural flow through these portions of a given well did not bring enough fresh formation water into the screened interval.

A third method for determining adequate purge volume and purge rate is to calculate how much water must be pumped, at what rate, in order to produce at least 95% formation water. This method may be used only if the hydraulic conductivity of the formation is known, perhaps as the result of a pumping or slug test. Barcelona and colleagues (1985) produced a set of curves that allow the user to estimate the percentage of formation water produced after a certain length of pumping for a given transmissivity.

Sampling Devices.　The sampling device used should be appropriate to the situation. The sampling device is likely to be a bailer or pump, chosen on the basis of construction materials, design, diameter, depth capabilities, and cost. These factors are examined in more detail here.

Bailers.　A bailer is essentially a long narrow cylindrical bucket on a string, with a hole in the bottom and a ball check valve to close it as the sample is retrieved (Fig. 9.2). Bailers may be made of PVC, stainless steel, or Teflon®. The bailer chosen should be made of a material that will not compromise the sample chemically. If the bailer and its cord cannot be decontaminated, disposable bailers should be used. Alternatively, bailers may be dedicated to the wells, meaning that one bailer is used in each well, and only in that well.

The bailer may be lowered rapidly into the well until it is just above the water level, and must be lowered slowly thereafter. If the sample is to be taken from a specific depth, a bailer with double check valves should be used: one check valve at the top and one at the bottom (Fig. 9.3). The bailer should be lowered to the desired depth and then retrieved.

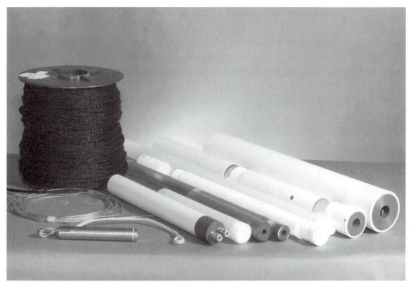

Fig. 9.2 Bailers may be made of PVC, stainless steel, or Teflon®. (Courtesy of Atlantic Screen & Manufacturing, Inc.)

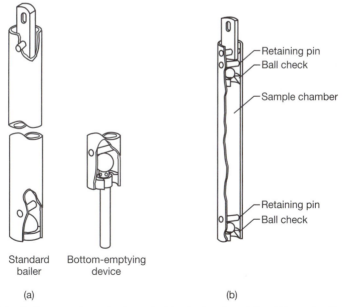

Fig. 9.3 Bailers may have single or double check valves (after Aller and colleagues, 1989).

When lowering or retrieving the bailer, care should be taken that the cord does not touch the ground, where it might pick up surface dust or contaminants and potentially introduce them into the well. Using a cord and reel assembly or the "windmill" method can help prevent this. To use the windmill method, the arms should be stretched out away from the sides and the thumbs should be hooked on the cord. Now the cord should be pulled up, hooking each thumb around it alternately in turn (Fig. 9.4). The wider the stretch, the more efficiently the bailer can be retrieved. Bringing it up slowly and smoothly will help avoid dislodging the check valve, spilling the sample, or volatilizing the gases within the sample.

Once the bailer is back on the surface, the water from the bailer should be transferred carefully to a sample bottle or filtering apparatus. This can be accomplished by simply pouring the water out of the top of the bailer. However, this may cause turbulence or off-gassing of volatiles, which may destroy the usefulness of the sample. A better procedure is to let the sample out of the bottom of the bailer by pressing a tube up into the hole at the bottom. This unseats the ball check valve, allowing water to run out the tube and into the sampling or filtering container. Some bottom-emptying devices have a stopcock that can be turned to regulate flow.

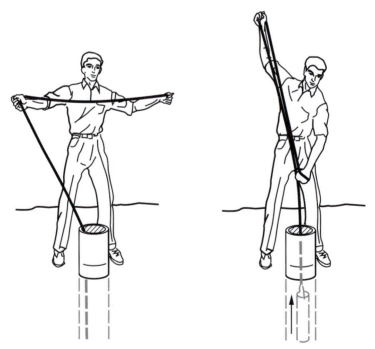

Fig. 9.4 Use the windmill method to retrieve a bailer quickly while keeping the cord off the ground.

If NAPLs are present in the well, a double check valve bailer should be used. It must be lowered slowly through the NAPL layer to avoid mixing the layers. When sampling for LNAPLs, the top of the bailer should not be lowered deeper than the bottom of the NAPL layer. The bailer should be retrieved slowly and carefully. If it is clear or translucent, the NAPL layer may be visible. If the layer is visible, the thickness of the NAPL layer should be measured before emptying the bailer.

Bailing is not always successful on the first try. It is not unusual for the bailer to return to the surface empty or only half full. This may happen because the bailer was not submerged fully in the water before retrieval, or it may happen because a grain of sediment became lodged in between the ball and seat of the valve. Depending on the size of the bailer and the depth to water, continuous bailing can quickly tire even the most energetic and committed of field personnel. This should be considered when planning the sampling trip.

Pumps. Pumps come in a wide variety of designs, sizes, and degrees of efficiency. Different kinds of pumps are available for purging, sampling, and for permanent installation in water supply wells. This section describes only sampling pumps.

Sampling pumps must extract water from a well in a way that does not alter the chemical characteristics of the water. For this reason, they usually are designed to work at low flow rates and with a minimum of turbulence. In a sampling pump, water should not come in contact with any materials that could chemically compromise the sample. Sampling pumps commonly run on electrical power rather than gasoline or other petroleum-based fuel. If they use air compressors, the compressor is usually oilless to eliminate the possibility of sample contamination. They may use supply tubing that is made of Teflon® or other polytetrafluoroethylene (PTFE) material. Sampling pumps are designed such that the sample only comes into contact with pump parts that can be cleaned and decontaminated. If funds are available and if sensitivity of the situation warrants it, a pump may be "dedicated" to a single well so that there is no chance of cross-contamination of wells. It is possible that dedicated pumps may be more cost-effective, if the well is to be sampled repeatedly.

Suction or vacuum pumps may be used only when the water level is less than about 25 feet below ground surface (or, more precisely, below the pump), because they cannot lift water more than that distance. These pumps sit at the surface and create a vacuum that draws water up into a tube inserted into the well. Vacuum pumps induce degassing of volatiles in the sample, and thus are inappropriate in situations where degassing will affect the constituents of interest. Vacuum pumps may be hand-operated, such as the familiar "pitcher" pumps with a handle that is cranked up and down, or they may be electrical or gas-driven. A "centrifugal jet pump" (Fig. 9.5) is a common type of suction or vacuum pump. These pumps generally are inappropriate for sampling, both because they induce degassing of volatiles in the sample and because they contain parts that may alter the chemical composition of the water. Peristaltic pumps (Fig. 9.6), in which the tubing and its contents are squeezed between rotating rollers that move the water along the

Fig. 9.5 A centrifugal jet pump. (Courtesy of Goulds Pumps, Inc.)

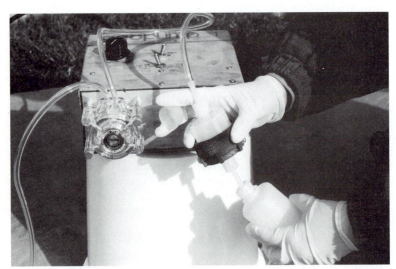

Fig. 9.6 A peristaltic pump.

tubing, are a somewhat better choice, because the sample never touches the pump mechanism. However, degassing still may occur.

A simple hand-operated vacuum pump (Fig. 9.7) may be used for sampling if the depth to water is not more than about 10 feet. To prevent the water from contacting the pump itself, a sample bottle with a stopper having two openings may be used. The sampling tube is connected to one opening and the pump to the other. When the pump is operated, water flows up the sampling tube and into the bottle, but does not pass into the pump. Degassing still is a problem with this method.

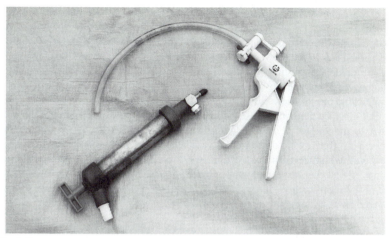

Fig. 9.7   Hand-operated vacuum pumps.

Downhole or submersible pumps (Fig. 9.8) positively displace water to force it to the surface. These designs include electric submersible pumps, gas displacement pumps, and bladder pumps. Each design has its own set of problems, mainly involving turbulence, contact of water with pump parts, or exposure of the water to air or other gases.

Fig. 9.8   A submersible centrifugal pump.

Electric submersible pumps use an electric motor to turn an impeller, a rotor, a threaded rod, or a set of gears (Fig. 9.9). The movement forces the water up a discharge line. If used to pump water with a high sediment load, these pumps might jam if sand is wedged between the gears (or impellers or rotor) and the pump housing. Some designs utilize throttles or controls that allow the user to regulate the pumping rate, but others do not. Pumping at a high rate may induce turbulence, which could alter sample chemistry.

Gas displacement pumps use a compressed gas (usually air or nitrogen) to push water to the surface. First, the pump is allowed to fill with water. Then, compressed gas from a tank or a compressor is allowed to fill the pump. This displaces the water, which exits through a discharge line. The main disadvantage of these pumps is that the gas comes into contact with the water, which can alter water chemistry significantly. Because of this, gas displacement pumps are not recommended as sampling pumps.

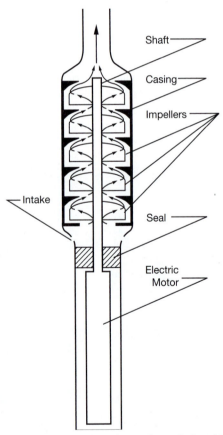

Fig. 9.9   This submersible pump operates by turning a shaft and impellers. Each impeller draws water in, then spins it upwards to the next impeller (after Price, 1996).

Bladder pumps operate on essentially the same principle as do gas displacement pumps. However, in a bladder pump, the sample never contacts the gas. This eliminates the major problem with gas displacement pumps. In a bladder pump, a flexible bladder alternately fills with water and then is squeezed by compressed air. When the bladder is squeezed, water is forced out a discharge line (Fig. 9.10). A controller must be used to regulate the length of the filling and discharge cycles for optimum performance.

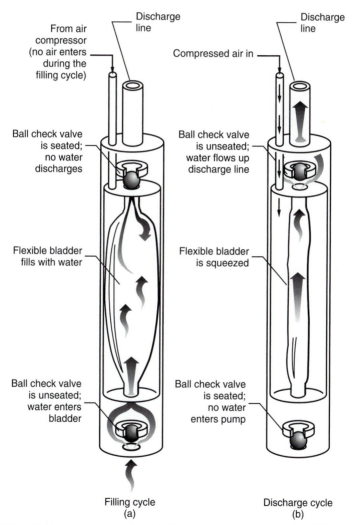

Fig. 9.10   Bladder pumps use a flexible bladder within a rigid housing. Water is (a) drawn into the bladder during the filling cycle, and then is (b) forced up a discharge line when compressed air squeezes the bladder.

Filtering

Why Filter? After a sample of ground water is extracted from a rock or sediment formation, it may change chemically. Some of these changes may occur immediately, and some may occur gradually. Because the goal of sampling and analyzing water from a formation is to determine chemical conditions within that formation, in certain conditions, described in what follows, it may be necessary to filter, acidify, or otherwise preserve the sample immediately after sampling.

Filtering removes suspended sediment, colloidal materials, and other particles from the sample. If the water is to be analyzed for dissolved constituents, then filtration ideally should not compromise the results. However, because any processing of the sample may result in an undesired chemical change, filtering should be performed only when necessary. Any regulatory requirements regarding when to sample and filter should be followed when they apply.

Filtering is particularly useful when the sample is to be acidified. Adding acid may dissolve the sediment or colloidal material in the sample; this may increase the apparent concentration of metals of other constituents in the water, yielding inaccurate results. This increased concentration sometimes is referred to as a "sampling artifact."

For a review of research on filtration of ground water samples, see Saar (1997).

When to Filter. A sample should be filtered and acidified if the sample will be analyzed for metals, dissolved phosphorus, phosphate, or radioactive constituents (except radon) (USEPA, 1986). A sample that is to be acidified should be filtered before it is acidified.

The question of which samples to filter or not filter is controversial, and not all experts agree on whether it ever should be done, much less when it should be done. Further research ultimately may resolve the issue.

How to Filter. To filter a sample, pass it through a 0.45-µm filter paper. This must be done in the field, immediately after retrieving the water from the well. In-line filters are available that can be attached to the discharge line of a pump (Fig. 9.11). One advantage of in-line filters is that they do not require that the water be exposed to air during filtering. Filter funnels or other filtration apparatus can be set up in the field, but are somewhat less convenient. Using one of these involves exposing the sample to air and perhaps losing volatiles. However, it may be the only choice for filtration of bailer samples, unless the bottom-emptying device on the bailer is fitted with a flexible tube.

After installing new filter paper, the first liter of water that passes through the paper should be discarded. This helps reduce contamination of the sample from the filter paper itself.

After collection of the samples, the filtering apparatus is cleaned, a fresh filter installed, and some distilled deionized water passed through it. The filtrate should be treated as if it were a sample, and the lab should analyze it for the same parameters. This quality check is performed when the same filtration unit is used for more than one sample, and when nondisposable portions of that unit come into

Fig. 9.11   In-line filters can be attached to the discharge line of a pump. (Courtesy of QED Environmental Systems, Inc..)

contact with the sample during the filtering process. The quality check is a test of how reliably the decontamination process removes remnants of the previous sample. This check need not be performed if the portions of the filtration unit that contact the sample are disposed of after each use.

Sample Preservation

Why Preserve Samples?   Samples must be preserved because chemical reactions will occur within the sample that may alter the concentrations of various parameters.

How to Preserve Samples.   All samples should be preserved by keeping them at 4°C immediately after collection. When certain parameters are to be measured, the samples may require additional preservation procedures.

A list of generally suggested preservation procedures is given in Table 9.1. However, each analytical laboratory may have its own specific suggestions for preservation, and environmental regulations may require different procedures. Check with the lab personnel before going out to sample. They can provide the appropriate sampling containers and, often, preservatives.

If a chemical preservative is to be used, the required amount is measured out and added to the sample container before the sample is added. For metals, a common preservative measure is to add 5 ml of concentrated nitric acid per liter of sample. Other acids or other preservatives may be appropriate; see Table 9.1. If an analytical laboratory has provided the sample containers, it may have already added preservatives to the containers. This should be checked carefully, and rinsing, spilling, or overfilling of these sample containers should be avoided. Extra bottles and preservatives should be carried in case of mistakes.

As soon as possible, each sample should be placed into a cooler with wet ice. The samples should be kept at 4ºC. A maximum/minimum thermometer placed in the cooler will tell to what temperature extremes the samples have been subjected.

Decontamination. Sampling equipment should be decontaminated between the sampling of successive wells, unless dedicated or disposable equipment is being used. Decontamination prevents samples from one well from being contaminated by material left over from the previously sampled well. In addition, decontamination prevents a clean well from being contaminated by putting dirty sampling equipment into it.

Decontamination involves washing the equipment with a laboratory detergent (e.g., Alconox), rinsing it with either a solvent like acetone (when organics are to be measured) or a weak acid like 10% HCl (when inorganics are to be measured), then rinsing twice with deionized water (Fig. 9.12). Any part of the pump, bailer, and bailer cord that touches the well or water and any probes inserted into the well must be decontaminated before they are used on another well.

Decontaminating a pump includes washing the outside of the pump and lines as well as pumping clean water through the pump to "decon" the inside. Bailers may be dunked into a decon vessel constructed by capping the ends of a length of PVC pipe and filling it with detergent or rinse solutions. Other equipment may be submerged in buckets of the solutions.

Special Considerations for Household Wells. Household wells are different from monitoring wells because they were constructed for an entirely different purpose. Household wells are designed to provide the maximum amount of water to the users, while monitoring wells are designed to facilitate collection of a

Fig. 9.12 Decontaminate any equipment which comes in contact with the water.

chemically representative sample of formation water. Sampling from household wells is possible, but the data may be less reliable than monitoring well data.

Several problems hamper the collection of reliable data from household wells. Well logs for household wells may be unavailable or indecipherable. Sometimes, screens in household wells extend across several aquifers or units, so water may be coming into the well from any or all of those formations (Fig. 9.13). Because the well is in use by the occupants, water levels cannot be considered static levels. To reduce this problem, try to sample household wells at a time when the pump has been idle for several hours. Pumps for household wells may be constructed of materials that will compromise the quality of the sample, and they may allow air to come in contact with the water, or degas the water by pumping with a vacuum. In addition, water heaters, pressure tanks, distribution lines, water softeners, and faucet aerators all cause chemical changes. Sampling should be conducted as far "upstream" of these devices as possible. Faucet aerators usually can be unscrewed. If water has been standing in pipes within the house, it may have warmed up; the faucet with the shortest possible distance from the wellhead itself should be used. Using an outdoor spigot, such as for attachment of a garden hose, may be convenient. These spigots are less likely than indoor spigots to carry softened water.

When sampling a household well, the water should be run for a long enough time to ensure that formation water is being produced, and several indicator parameters such as pH, temperature, and conductivity should be monitored to ensure stabilization before sampling. In addition to these parameters, dissolved oxygen and turbidity also may be used.

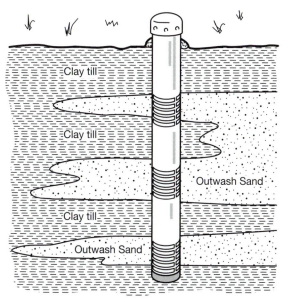

Fig. 9.13   Screens in household wells may extend through several aquifers or units.

Field Geochemical Measurements

Some measurements of chemical parameters must be taken in the field. This section describes those parameters and instruments, and also includes a discussion of procedures for quality assurance and quality control.

Measuring Chemical Parameters in the Field

Water chemistry can change rapidly once a sample is extracted from a formation and exposed to light, warmth, cold, air, or other environmental factors. Because of this, it is important to make measurements of some chemical parameters immediately after sampling. By the time the sample gets to the lab, it may be too late. In addition, field measurements of some geochemical parameters can be used to determine if a well has been adequately purged before sampling. This section describes instrumentation and procedures for making some field measurements.

Temperature

Temperature is a controlling variable: Once it begins to change, so do many other parameters. However, if the formation water temperature is known, it is easier to reconstruct what the chemical conditions might have been like in the formation. Sometimes temperature itself is the variable of interest. For example, this might be the case when tracking a plume of warm injected water or geothermal water, when differentiating two karst aquifer systems, or when delineating recharge and discharge areas.

Temperature is fairly simple to measure in the field. It requires a mercury thermometer or a temperature probe. Downhole probes yield measurements with low precision, but fairly high accuracy. After purging, water still in the well will be close to formation temperature. However, once it is withdrawn from the well, temperature may change rapidly. Downhole probes commonly give measurements to the nearest degree. If greater precision is required, use a mercury thermometer calibrated to 0.1°C. Armored thermometers are more durable for field use. Withdraw some water from the well, pour it into a thermos, and wait a minute to allow the thermos to equilibrate with the water temperature. Then discard this water, withdraw another sample from the well, pour it into the thermos, and cap the thermos with a rubber stopper with a hole in it. Insert the thermometer through the hole. Wait 2 minutes and read the thermometer (Garrett, 1988).

Conductivity and pH probes commonly have temperature probes built in. These are particularly useful if temperature is being used as an indicator of purging adequacy.

Electrical Conductivity

Electrical conductivity, sometimes reported as specific electrical conductance (or simply specific conductance), is a measure of the ability of a water to conduct electricity. In general, the higher the concentration of dissolved salts in the water, the easier it is for electricity to pass through the water. Electrical conductivity is used as a rapid, rough characterization of water chemistry. Like temperature, pH, and dissolved oxygen, electrical conductivity sometimes is used as an indicator of whether or not a well has been adequately purged before sampling.

Electrical conductivity usually is reported in micromhos (or microSiemens) per centimeter. Because they are temperature-dependent, measurements usually are corrected and reported as if they were made at 25°C. Most conductivity meters have the ability to automatically compensate for temperature, so the value is recalculated to the value at a temperature of 25°C before it is displayed on the meter. This should be indicated on the meter scale; some meters can display both the temperature-compensated and noncompensated values.

Electrical conductivity meters should be calibrated before each use. The probe is immersed in a solution of known conductivity, and the calibration on the instrument is adjusted until the display reads correctly (Fig. 9.14). The calibration solution should be of approximately the same conductivity as the water being tested. In order to select the calibration solution, an educated guess should be made as to what the sample conductivity will be. The information in Table 9.2 may help. If the measured conductivity of the sample places it outside the range of the calibration solution, recalibrate and remeasure the conductivity of that sample. Some instruments have two-point calibration; for these, two calibration solutions are used that bracket the expected conductivity of the water being tested.

Because electrical conductivity changes as the concentrations of dissolved gases in water change, it should be measured in the field immediately after sampling.

The simplest conductivity meter design is a measuring and display unit with an attached probe. After calibration, a sample is withdrawn, the probe is immersed in the sample, and the conductivity is read. Downhole probes have essentially the same design, except that the cable connecting the display unit and the probe is long enough to drop down a well so that downhole values may be measured.

A conductivity meter may be used with a flow-through cell. Water pumped from the well is directed through a discharge line, and a flow-through cell is fitted to the discharge line. The cell is essentially a closed beaker through which water

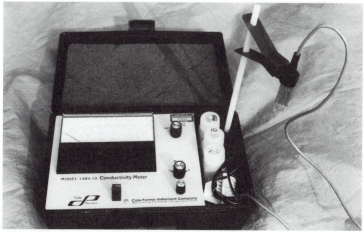

Fig. 9.14   A conductivity meter.

TABLE 9.2	
Typical Values of Electrical Conductivity	
Type of Water	Typical Values (µmhos/cm)
Rainwater	2–100
Ground water	50 to 50,000
Brines	up to 500,000
Ocean water	about 50,000
Landfill leachate	10,000

Sources: Cartwright et al., 1977; Hem, 1985.

flows as it leaves the well. The cap of the cell has a hole in it that allows insertion of a conductivity probe. Thus, conductivity can be measured continuously as the well is being pumped. This is particularly useful if conductivity is being used as an indicator of purging adequacy.

Fresh calibration solutions should be made up periodically. Prepare calibration solutions by using a stock solution of potassium chloride (KCl) as follows. Dilute 745.6 mg anhydrous KCl in distilled, deionized water and bring to a volume of 1000 ml at 25°C. This 0.01 N KCl stock solution has a conductivity of 1413 µmhos/cm. It may be diluted to produce reference standards of lesser conductivities (APHA, 1995). If the expected sample conductivity is greater than 1413 µmhos/cm, it will be necessary to make a reference standard with higher conductivity as well. To do so, prepare a stock solution of 0.5 N KCl by diluting 37,280 mg anhydrous KCl in distilled, deionized water and bring to a volume of 1000 ml. This solution has a conductivity of 58,640 µmhos/cm at 25°C, and may be diluted to produce reference standards with lesser conductivities.

pH
The negative logarithm of the hydrogen ion activity in a water is called the pH of the water. Like temperature and electrical conductivity, pH is an important chemical parameter that should be measured immediately after sampling. It gives a rough chemical characterization of the water. In addition, it can be used as an indicator of purging adequacy.

The pH of a water is reported in pH units. The pH scale generally is taken to range from 0 to 14. However, this full range is rarely encountered in the field; most natural waters fall somewhere in the range of 5 to 9. A pH of 7 is commonly assumed to be neutral, but this only applies to water at 25°C. Like electrical conductivity, pH is temperature-dependent. For example, at 60°C, the neutral point is 6.51; at 10°C, it is 7.26; and at 0°C, it is 7.47. As a result, every pH measurement should be accompanied by a temperature measurement. Values of pH that are lower than neutral indicate acidic conditions, and pH values that are higher than neutral indicate basic conditions.

Because the pH scale is logarithmic, a difference of 1 on the pH scale indicates a difference of a factor of 10 in the hydrogen ion activity. For this reason, pH is usually reported to at least one decimal point, and more where the precision of the pH meter warrants it.

All pH meters must be calibrated before every use. Calibration is performed using buffer solutions that have a pH in the same range as the water being tested. Most pH meters can accommodate a two- or even three-point calibration. This should be performed at least daily, bracketing the range of measurements expected. A one-point calibration should be performed before each measurement. To calibrate a pH meter, immerse the probe in a buffer solution. Gently stir the probe in the solution to allow adequate flow of solution about the probe. After 2 minutes, adjust the calibration control so that the display indicates the correct pH for that solution. After another 2 minutes, check to be sure that the display still reads correctly. If it does not, adjust the calibration again, and after another 2 minutes, check the display again. Under certain conditions, some pH probes take several minutes to equilibrate. If this occurs during calibration, ensure that during measurements, at least the same amount of time is allowed to pass before making a reading. When calibration is complete, rinse the probe with deionized water.

To make a reading, immerse the probe in the water sample (Fig. 9.15). Gently stir the probe in the water for about 2 minutes; then read the display. After another 2 minutes, read the display again. If the two values are within about 10% of each other, record the second value, remove the probe, and rinse it with deionized water. If the two readings are more than 10% different from each other, or if the calibration before the measurement took more than 4 minutes, continue stirring the probe in the sample for another 2 minutes, and check the display again. Continue this process until two sequential readings are within 10% of each other. Record the final reading, remove the probe, and rinse it with deionized water.

Some pH probes must be stored damp or even wet; check the manufacturer's instructions.

As with electrical conductivity meters, pH meters have various design features that may allow them to be used at the surface (including in a flow-through cell), or downhole.

Table 9.3 shows typical pH ranges for various types of waters.

Fig. 9.15 A pH meter.

TABLE 9.3

Typical pH Ranges for Water from Various Environments

Environment	Typical pH Range
Rain water	4–7
Freshwater lakes and streams	6.5–8.5
Ground water	6–8.5
Brines	Near neutral
Ocean water	7.8–8.4
Landfill leachate	Near neutral

Sources: Cartwright et al., 1977; Hem, 1985; Berner and Berner, 1996.

Eh

Eh is a measure of the reduction-oxidation (redox) potential of a sample of ground water. One method of determining Eh is to analyze the water to find concentrations of various redox pairs (e.g., Fe^{+2} and Fe^{+3}, Mn^{+2} and Mn^{+4}, and so on) and then to calculate the redox potential using standard formulas. Eh meters can give an estimate of redox conditions when analyses of those species are not available.

Eh meters rely on several key assumptions about equilibria in waters, and it is quite likely that those assumptions will not be valid in many field situations. As a result, field measurements of Eh are at best approximations, and should be interpreted only as rough indicators of redox conditions in water.

An Eh meter uses a platinum electrode and a silver/silver chloride reference electrode; when they are placed in the water sample, Eh is the potential difference between them. Implicit in the measurement is the assumption that the water achieves an equilibrium with respect to the electrodes. In addition, the method assumes equilibrium between all redox pairs in the water. This assumption is probably rarely valid for natural waters. However, it is more likely to be valid for ground waters than surface waters, because of the slower travel times in ground water and the increased contact times between water and minerals.

Eh meters consist of a probe and display/sensing unit (Fig. 9.16). Many pH meters are designed such that they will measure Eh, provided that an Eh electrode is used instead of a pH electrode. Before using an Eh meter for measurements, calibration of the meter should be checked periodically using a standard reference solution such as Light's or ZoBell's solution. Light's solution, which may be more stable than ZoBell's solution, has an Eh of +438 mV at 25°C. It is made by dissolving 48.22 grams of ferrous ammonium sulfate in 56.2 ml sulfuric acid in water and bringing the volume up to 1000 ml (APHA, 1995). Eh of the sample may be calculated as follows: Eh of the sample equals the measured Eh of the sample plus the theoretical Eh of the standard reference solution, minus the measured Eh of the standard reference solution (APHA, 1995).

To measure Eh, the probe should be immersed in the water sample and gently stirred. After several minutes, the display should be checked, then checked again after several more minutes. If the two readings are within 10 mV of each other (APHA, 1995), record the last measurement. If they are not, continue stirring and

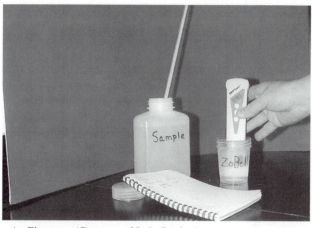

Fig. 9.16   An Eh meter. (Courtesy of Lois Ongley.)

checking the display every 2 minutes. When readings are complete, the electrode should be rinsed and stored according to the manufacturer's instructions. This may mean storing the electrode wet or damp.

Concentration of dissolved gases has a strong effect on Eh, and temperature has a strong effect on concentration of dissolved gases. For this reason, Eh of a water sample may begin changing the moment it is withdrawn from a well. Measurements should be made in situ or immediately after sampling. Using any type of suction or vacuum pump can cause degassing of the water, which in turn affects Eh. Likewise, pouring the water, such as from a bailer into a sample container, is likely to disturb the redox equilibria and change the Eh. Avoid pouring a sample that is to be used for an Eh measurement, and avoid allowing the sample any contact with air.

Table 9.4 gives typical Eh ranges of waters in various environments.

Dissolved Gases

Dissolved gases may affect important characteristics such as its corrosivity, taste, toxicity, potability, solubilities of other components in contact with the sample, and chemical equilibria of dissolved constituents. Dissolved gases may include oxygen, carbon dioxide, hydrogen sulfide, methane, radon, volatile organics, and others. Although tests for each of these gases vary, sampling requirements are the same: contact between the sample and air must be avoided. Therefore, when sampling for these constituents, avoid using a suction or vacuum pump to take the samples, as it may cause degassing. Likewise, the containers should be overfilled so that no air is trapped in the container. Disturbing or agitating the sample in any way should be avoided, and the samples should be kept at 4°C, to minimize changes in water chemistry. When sampling for dissolved oxygen measurements, the bottle must be amber or another dark color that minimizes the possibility of

TABLE 9.4

Typical Eh Values of Waters in Various Environments

Environment	Typical Eh Range (millivolts)
Rain water	+400 to +600
Freshwater lakes and atreams	+300 to +500
Ground water	−200 to +100
Oilfield brines	−300 to −600
Ocean water	+300 to +500
Acid mine drainage	+600 to +800
Water in wetlands/bogs	+100 to −100

Source: Garrels and Christ, 1965.

light reaching the sample. Light might encourage the growth of organisms that would use up the oxygen. Details of protocols and tests for specific gases may be found in a reference on standard methods, e.g., *Standard Methods for the Examination of Water and Wastewater*, published periodically by the American Public Health Association (1995).

Spectrometers and Colorimeters

Field spectrometers (or spectrophotometers) and colorimeters may give rapid field estimates of a host of parameters. They operate on the same principle: light of a certain wavelength is directed through a sample, it is absorbed by the constituent of interest, and a detector on the other side measures the amount of light that passed through the sample (Fig. 9.17). The higher the concentration of that constituent, the more light is absorbed, and the less light is measured by the detector. Various reagents may be added to the sample to enhance the sensitivity for particular parameters. Depending on the particular method, either deionized water or a "standard" of sample water to which no reagent has been added may be used. Another alternative is to use a set of standards to construct a calibration curve. Most methods for most parameters have potential interferences. This means that the presence of one chemical might inhibit the measurement of another. The manufacturer's instructions should be consulted to assess the potential for this.

Gas Chromatographs

Field-portable gas chromatographs (GCs) give a fairly rapid measurement of concentrations of various volatile organic constituents in ground water. Various designs have different types of detectors, display units, and sensitivities (Fig. 9.18). The simplest of these are "sniffer" devices, which sound an audible alarm when a certain level of organic vapors is detected. More complex devices display, in calibration gas equivalents, levels of organic vapors detected. Even more sophisticated devices have digital data analysis systems capable of integrating area under graphed data peaks and downloading data through a telecommunications port.

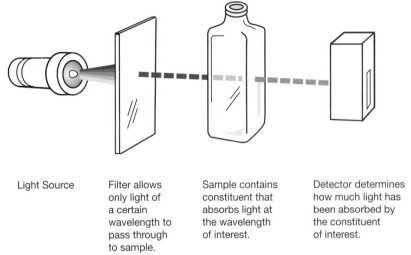

| Light Source | Filter allows only light of a certain wavelength to pass through to sample. | Sample contains constituent that absorbs light at the wavelength of interest. | Detector determines how much light has been absorbed by the constituent of interest. |

Fig. 9.17   Spectrophotometers and colorimeters operate on the principle that light is absorbed by constituents in the sample.

Common popular names for field gas chromatographs are OVA (organic vapor analyzer), PID (photoionization detector, which actually refers to a particular component of a certain type of GC, but is commonly used to refer to the whole GC), and HNu (a brand name, pronounced "H-New," which sometimes is used erroneously to refer to any field GC). The basic operation of these instruments is the same: A sample passes through a long column and organic constituents of the sample are retained for a period of time on materials within the column (Fig. 9.19).

Fig. 9.18   A field-portable gas chromatograph. (Courtesy of PE Photovac.)

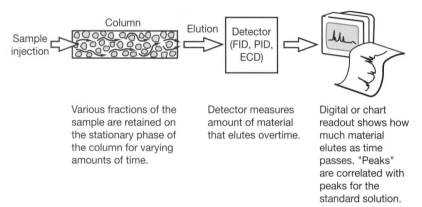

Various fractions of the sample are retained on the stationary phase of the column for varying amounts of time.

Detector measures amount of material that elutes overtime.

Digital or chart readout shows how much material elutes as time passes. "Peaks" are correlated with peaks for the standard solution.

Fig. 9.19   Gas chromatography.

Some of the constituents pass through and out the other end of the column (i.e., they "elute") fairly quickly, and others take longer. The time it takes for a particular vapor to elute is an indicator of what that vapor might be. Standards of particular chemicals are passed through the column under the same conditions, and their elution time is recorded. If elution time of a standard matches that of an unknown, the unknown may contain that chemical. The temperature of the column affects elution time, as do other variables. When the organic constituents elute, they are detected by one or more detectors: flame ionization detectors (FIDs) and photoionization detectors (PIDs) are common in both field and laboratory units. Electron capture detectors (ECDs) or Hall detectors rarely are used in the field, but may be used in the lab for particular types of analyses. The detectors register the amount of chemical that elutes from the column at a given time.

The concentration of organic vapor present in a sample is reflected in the size of the peak that results when a component elutes (Fig. 9.20). Standards of various concentrations are passed through the column, and their resultant peaks are compared with those of the sample. The higher the concentration, the bigger the peak.

At first glance, it might appear to be inappropriate to use a GC, which analyzes the composition of gas, to study the composition of water. However, GCs are particularly useful in two hydrogeologic applications. First, they commonly are used as tools to evaluate health and safety conditions. Because some organic vapors can be hazardous or toxic, GCs of various designs are used for air monitoring. For example, workers at the site of leaking underground storage tanks need to know if potentially harmful vapors remain in the tank, excavation, or in pockets of the backfill. They use hand-held GCs to monitor the air in these locations.

Second, GCs are used to determine concentrations of gases in headspace analysis. In this type of analysis, water containing volatile organic chemicals is placed in a closed container that also contains some air. The chemicals volatilize and the vapors collect in the air above the water. This "headspace" is analyzed using the

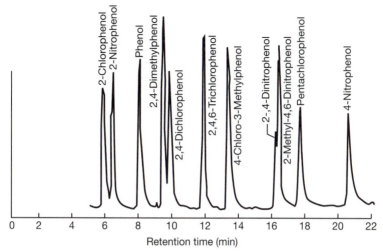

Fig. 9.20   A chromatogram (modified from 40 CFR Pt. 136).

GC. The amount of organic vapor in the headspace is proportional to the amount dissolved in the water, and so the amount in the water may be calculated.

Field GCs are convenient to use and may give excellent results. However, the successful operation of a GC requires some training and experience. Many variables can affect the quality of results, and an inexperienced operator may produce meaningless results. Finally, absolute identification of a compound is impossible using a GC. The operator can say with certainty only that a compound within the water elutes at the same time as a standard, not that it is the same chemical as that standard. Gas chromatography must be combined with mass spectrophotometry (MS) to identify a chemical absolutely. Although laboratory models of GC-MS systems are standard in analytical labs, field-portable GC-MS systems are only beginning to become commercially available. Still, field GCs fill an important need in air monitoring and some ground water chemistry applications.

Quality Assurance/Quality Control

Any time chemical measurements are made, whether in the field or in the laboratory, a certain amount of error is introduced into the measurements. It is important to try to minimize error, and it is equally important to try to make a quantitative evaluation of how much error exists in each measurement. In order to do this, quality assurance/quality control (QA/QC) protocol and procedures should be developed and followed carefully.

Selection of QA/QC procedures depends on the degree of precision and accuracy that is required in a given situation. The degree of precision and accuracy required depends on the intended use of the data. Each situation is unique. Described here are a few QC procedures that might be used in certain situations.

Replicate Samples

Replicate samples are two samples taken one immediately after the other, by the same sampling procedure, using the same sampling equipment. Replicate samples help to determine the precision of the analyses. At least one replicate sample per day, or 5% of the total number of samples each day, should be taken. To avoid any bias that might occur during the analysis, replicates should be labeled as if they are separate samples, not obviously labeled as replicates.

Under certain circumstances, it might be worthwhile to take replicates of all the samples, particularly if the consequences of having no sample at all are fairly severe. When the legal ramifications of the sampling and analysis are especially grave, when the timing of a sampling event is critical, or when the project is particularly sensitive, replicating all samples might be cost-effective. With replicates, if a container should leak or a sample be lost, the replicate may be used in its place. Once the necessary analyses have been completed, the replicates may be disposed of without analyzing them, as analysis may be the most expensive part of the sampling program.

Split Samples

Split samples are collected by taking one larger sample and dividing it into two or more smaller samples. These smaller samples may be analyzed as separate samples, or better still, at separate laboratories. This provides a check of analytical precision. At least one split sample per day should be taken. If many samples will be taken throughout the day, some 5% of the total number of samples during the day should be split. However, when degassing of samples may cause a problem, such as in samples taken for volatile chemicals or for dissolved gases, samples should not be split.

Blanks

Sampling blanks consist of distilled, deionized water that is analyzed just as if it were a sample. Ideally, blanks should not show measurable concentrations of any chemical constituent being tested. The purpose of a blank is to determine if some factor other than the *in-situ* water chemistry is operating to alter the analysis. Blanks should be packaged and labeled in such a way as to make them appear to be regular samples, so that no bias will be introduced at the lab. Several types of blanks may be used: trip blanks, field blanks, and rinsate blanks.

Trip Blanks.   Trip blanks consist of distilled, deionized water that is poured into a container at the laboratory before the sampling trip and sent along with the empty containers into the field. A trip blank container remains closed during the entire sampling trip. Its purpose is to help determine if the containers are contaminated or if some other process involved in the shipping and handling of samples might be altering water chemistry. Wilson (1995) noted that trip blanks usually are used only when sampling for volatiles.

Field Blanks.   Field blanks consist of distilled, deionized water that is treated as if it were a sample in the field. A field blank is created in the field by opening a new sample container and pouring the distilled, deionized water into it while in the field. This allows one to check to see if ambient conditions in the field might be affecting sample chemistry. Such conditions might include contaminants in the air, dust that gets into the sample containers, or other field conditions. At least one field blank should be taken on each sampling day. However, conditions may vary, and so each situation should be evaluated accordingly.

Rinsate Blanks.   Rinsate blanks are created by rinsing sampling equipment with distilled, deionized water, and then analyzing the rinsate as if it were a sample. A rinsate blank should be run through any equipment that is used for multiple samples, for example, bailers, pumps, and filters. The rinsate blank should be taken only after all equipment has been cleaned and decontaminated; it provides a check on the effectiveness of the decontamination procedures. At least one rinsate blank for each piece of equipment should be taken on each sampling day, or according to the requirements of the particular field situation.

Spiked Samples

Spiked samples are samples to which is added a known amount of some chemical, usually a chemical that is of importance at the site. Spiked samples are labeled and packaged to look like a regular sample, and they are sent to the lab without informing the lab that they are any different from any other sample. This provides a check on laboratory accuracy. The most useful type of spiked sample is a *matrix spiked sample*; these are made up by spiking samples of the ground water at the site. A matrix spiked sample is more similar chemically to the regular samples than is spiked deionized water. Spiked samples should make up at least 5% of all the samples taken.

More information on QA/QC is available in publications from the USEPA and the ASTM (see the References) as well as in Barcelona and colleagues (1985) and in Wilson (1995).

Hydrogeologic Mapping

Chapter Overview

What direction is ground water flowing? How deep will we have to drill to find an aquifer? How does water chemistry change along a flow path? Using tools and methods introduced earlier in this book, we can answer these questions and more. But how can we best display and convey to others the information gained using these tools and methods? Like other geologists, hydrogeologists use cross-sections and maps to communicate. This chapter describes the construction of these maps and diagrams, with particular emphasis on the display of hydrogeological and hydrogeochemical information.

Mapping Water Levels

Maps and cross-sections that show water level elevation are important to almost any hydrogeologic study. They may give information about depth to water, directions of flow, gradient, pressure conditions within an aquifer, and other hydrogeologic data. Whereas cross-sections and maps are used in many areas of geology, flow nets are unique to hydrogeology. Flow nets are cross-sections or maps to which hydraulic head contours and arrows have been added to indicate the direction of ground water flow. This section introduces the construction of hydrogeologic cross-sections, water-level maps, and flow nets.

Hydrostratigraphic Units

Constructing a water-level map or cross-section begins with an analysis of the geology of the area. How thick are the soils? How thick and of what composition are the sediments (if any) that overlie the bedrock? How variable are the soils and sediments, especially in their porosity and permeability? If the sediments are stratified, what is the thickness and composition of each layer? Of what is the bedrock composed? What is its structure and attitude? What is its degree of fracturing, and what is the orientation of those fractures? If the rock is stratified, what is the thickness and composition of each layer?

These questions should be answered before water levels can be plotted, because they help define the *hydrostratigraphic units* in the area. A hydrostratigraphic unit is a body of rock or sediment that has hydrogeologic characteristics that make it distinct from surrounding bodies of rock or sediment. Hydrostratigraphic units must be defined before water level mapping can begin, because only water levels from the same hydrostratigraphic unit can be used to construct a map of a specific potentiometric surface or water table.

In general, hydrostratigraphic units are of three types. Aquifers and aquitards (also called confining units) constitute two end points on a continuum, and leaky aquitards (also called leaky confining layers, or semiconfining layers) are units with some characteristics of each endpoint. Aquifers have hydraulic conductivities high enough that they can yield economically useful quantities of water (Fetter, 1994). Aquitards or confining units are bodies of rock with low hydraulic conductivity. They separate and confine aquifers, keeping the aquifers hydraulically isolated from each other. Leaky aquitards may form a boundary to an aquifer, but do not perfectly confine it. A small but significant amount of water flows through the semiconfining layer and into the aquifer.

One critical characteristic of hydrostratigraphic units is that their boundaries are not necessarily the same as formation boundaries. In fact, the boundaries of hydrostratigraphic units may actually *cross* formation boundaries, in some situations (Fig. 10.1). The primary characteristic that defines a hydrostratigraphic unit is hydraulic conductivity. The boundary of a hydrostratigraphic unit separates a unit with high hydraulic conductivity from a unit with low hydraulic conductivity.

In practical terms, this means that if two distinct stratigraphic formations, one overlying the other, both have high hydraulic conductivity, they may be mapped as one and the same hydrostratigraphic unit. Alternatively, if a stratigraphic unit of low hydraulic conductivity crosses a fault zone and becomes extensively fractured in that zone, resulting in high hydraulic conductivity, the fractured area may constitute its own hydrostratigraphic unit. The key in defining units is to look for hydraulic continuity and connection within units having relatively the same magnitude of hydraulic conductivity.

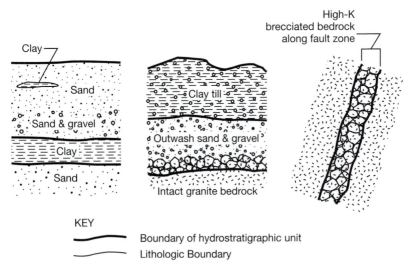

Fig. 10.1 Examples of situations in which the boundaries of hydrostratigraphic units do not coincide with the boundaries of formations.

Hydrogeologic Cross-Sections

Cross-sections are basic tools of geology, and every hydrogeologist should be able to construct and interpret them. Cross-sections show the view on a plane that is perpendicular to earth's surface, and show the view along a straight line. Fence diagrams are closely related to cross-sections, but they show the view along a sequence of points, from point to point, not necessarily in a straight line. In the following discussion, the term cross-section is used to refer to both, except where special mention of fence diagrams is needed.

A hydrogeologic cross-section differs from an ordinary geologic cross-section in that in addition to surface topography and position of formations, it shows hydrostratigraphic units and may indicate water levels as well. This section describes how to construct hydrogeologic cross-sections, beginning with the construction of a topographic profile, and then adding the geology and the hydrogeology.

Constructing a Topographic Profile

Construction of a cross-section begins with the construction of a surface topographic profile. To make this, a strip of paper is placed on a topographic map, its edge touching the two end points of the line of cross-section (Fig. 10.2). Ideally, the line of cross-section should pass through several "control" points (boring or well locations) that fall along the straight line. For fence diagrams, the points do not necessarily fall on a straight line, and when proceeding from point to point, the strip of paper must be rotated to intersect each successive point.

With the paper strip held firmly in place, a tick mark is made on the paper strip each time its edge touches a topographic contour line, a lake or stream, well or boring, major road, hill or mountaintop, or other important feature. It may be helpful to make brief notes on the strip that indicate exactly what each feature is. For example, when the strip touches a topographic contour line, the elevation of that line should be noted.

Next, the paper strip is placed horizontally across a clean sheet of gridded paper. A vertical scale is made on the grid, extending through the range of elevations encountered at the surface, and leaving room at the bottom to fill in information such as subsurface stratigraphy, well depths, and so on. Ideally, the

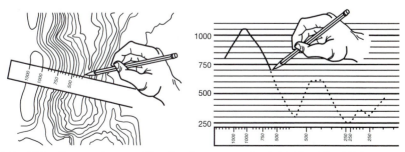

Fig. 10.2   Constructing a profile of surface topography (after Hamblin, 1995).

vertical scale should be the same as the horizontal scale. This is especially true if the cross-section is to be used to construct a flow net. In practice, however, vertical exaggeration is common. If vertical exaggeration is used, the exaggeration factor must be clearly marked on the cross-section.

While sliding the paper strip up and down, but not side to side, a dot is made at the appropriate elevation for each tick mark on the paper strip. Locations of the important geographic features indicated on the paper strip also should be marked on the grid. When all the points have been transferred to the gridded paper, a line is drawn connecting the points. This is the topographic profile. More information on constructing topographic profiles may be found in physical geology texts, particularly lab manuals.

Adding Geology to a Cross-Section

The topographic profile shows only the ground surface. To add geological information to the cross-section, the available information on subsurface geology first should be reviewed. This may include well or boring logs, surface outcrops, structural data, regional geologic data, or geophysical information, among other sources. This information is transferred onto the cross-section by considering the depth to each important feature (e.g., each lithologic or hydrostratigraphic contact). If well or boring log data are used, the location of the hole should be marked, as well as its total depth and the depths at which lithology changed as the boring advanced.

Driller's logs are a valuable source of information used in drawing geologic cross-sections. However, a hydrogeologist should be cautious about making overly literal readings of a driller's log. Drillers are professionals and are skilled at drilling. However, they are not hydrogeologists. Fine geological distinctions, such as the difference between limestone and dolomite, may not be worthy of note to a driller, and so may not be indicated on a driller's log. (In fact, both of these rocks may simply be called by a driller's term such as "lime rock.") As another example, it is not uncommon for a driller to give the name "slate" to what is actually a layer of shale. In unconsolidated formations, the log may simply show "clay" or "sand," without giving more detail on sorting, color, mineralogy, or other potentially geologically useful characteristics. The driller's logs should be interpreted in conjunction with regional geological information.

Once well log or boring log data are added to the cross-section, formation boundaries should be sketched in. This is done based on common sense, geological principles, and experience. The urge to draw extra formation boundaries at the slightest change in lithology should be resisted. For example, the fact that a sand unit becomes slightly more silty does not necessarily indicate a formation boundary. Care also should be taken to ensure that no structures are drawn, such as faults or folds, where no data warrant them. If appropriate, symbols or patterns may be added to indicate lithologies. Standard symbols should be used for the lithologies; these are given in Compton (1985). Drawing cross-sections is a skill that can be improved with practice, so a beginning geologist would be wise to practice whenever the opportunity arises.

Geologic information that falls near the line of the cross-section, but not precisely on the line, may be projected onto the line. Compton (1985) is a source of more information on projecting structures onto cross-sectional lines, as are some structural geology textbooks.

Adding Hydrogeology to a Cross-Section

Once a geologic cross-section is complete, hydrogeologic information may be added. Adding the hydrogeology to a cross-section involves including information on water levels, hydrostratigraphic units, and/or flow directions.

Water levels might be indicated only within the wells or borings, or they may be shown as a line extending across the whole area. When adding water levels, be consistent in using levels taken on the same date. In addition, water levels all should be taken at a similar point in the history of the well. Water levels taken during or just after drilling may indicate something about whether an aquifer is confined or unconfined. However, these levels will not be the same as those taken after a well is installed and developed, and those will not be the same as levels taken after a period of pumping. In fact, water levels may fluctuate rapidly enough that measurements taken a few days or even hours apart may not be comparable. Static water levels (nonpumping levels) will not be the same as levels taken while a well is being pumped.

When drawing a line to indicate a water level across an area on a cross-section, only those data points that come from wells tapping the same hydrostratigraphic unit, and the same portion of that unit (Figs. 10.3a and 10.3b), should be used. Likewise, only those levels that were measured on the same day, or, at worst, within a few days of each other, should be used. The use of data that indicate abrupt peaks or dips in water level should be considered carefully, because water levels usually are smooth surfaces that change gradually.

If vertical exaggeration has been used in constructing the cross section, this will complicate the use of the cross-section for drawing a flow net, as the relationships of flow lines, equipotential lines, and boundaries may no longer be as described earlier. Vertical exaggeration should be avoided in these cases, or should be taken into consideration as flow lines and equipotential lines are drawn.

Water levels are indicated on a cross-section by an upside-down triangle that points to the line of water level. The hydrostratigraphic unit to which the water level refers should be clearly indicated, for example, "Water Level, Upper Sand Unit," or "Potentiometric Surface, Silurian Dolomite."

Water-Level Maps

The concept of a water-level map is simple: A water-level map plots the elevation of a water table or potentiometric surface for a particular hydrostratigraphic unit. Constructing such a map can be simple and straightforward; after all, most students of physical geology can plot points and draw contour lines. But hydrogeologic mapping bears some traps that beginners or the unwary easily may be caught in. This section describes the process and highlights the traps.

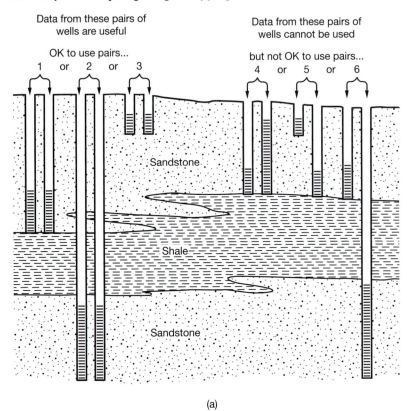

(a)

Fig. 10.3 (a) Water levels from each of the well pairs on the left may be used, but levels from the pairs on the right cannot, as they do not tap the same interval of the same hydrostratigraphic unit.

Constructing a Water-Level Map

Once hydrostratigraphic units are defined, water-level mapping may begin. Water levels in wells that tap a particular hydrostratigraphic unit are plotted on the map, and contour lines may be drawn to indicate the elevation of the water level. The following conditions must be met:

- Water levels must be defined in terms of elevation above a datum, *not* depth to water. Elevation of a measuring point on each well is generally surveyed using standard methods such as electronic total station, theodolite, transit, or alidade, and referencing all measurements to a benchmark or local datum. If water-level data are given in terms of depth to water, they must be converted to elevation before mapping. Elevation is found by subtracting the depth to water from the elevation of the measuring point at each well.

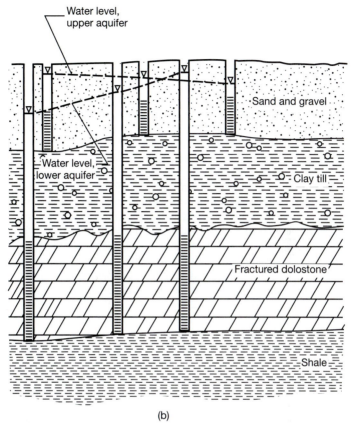

Water level,
upper aquifer

Sand and gravel

Water level,
lower aquifer

Clay till

Fractured dolostone

Shale

(b)

Fig. 10.3 (b)   Lines showing water levels are constructed separately for the shallow and deep aquifers.

- All water levels in the mapped area should have been measured at roughly the same time, ideally on the same day, but certainly within a few days of each other. This is because water levels are dynamic, and a water-level map represents water levels at a particular point in time.
- All water levels must have been taken from wells tapping (i.e., screened or open exclusively through) one and the same hydrostratigraphic unit. When a well intake taps several units, the water level may not be indicative of a single unit. Even if it is indicative of a single unit, the identity of the unit may not be easily deduced.
- For each well, the placement of the screened or open interval of the well must be consistent with respect to the unit boundaries. That is, if one well is screened through only the uppermost third of a unit, then all wells should be

screened through only the uppermost third of the unit. When this condition is not met, although the map may still provide some meaningful information, careful judgment must be used in interpreting it.

- If the purpose of the map is to show the elevation of the water table, then the wells used to provide data must have its intake situated such that this intake intersects the water table, but does not extend more than about 8-10 feet below it. This is because when very long intakes are used, the water levels in the well may reflect conditions deep within the aquifer instead of at the water table. If the hydraulic head deep within the aquifer is different from the hydraulic head at the water table, then the resulting water level will be misleading. This is particularly likely to occur when strong vertical gradients (up or down) exist.
- In the special case of a water table map, an additional source of data may be used: elevation of surface water bodies such as lakes, ponds, and streams. Because these water bodies are the surface expressions of the water table, they must be worked into the water table map along with data from water wells (Fig. 10.4). Care should be taken to first determine if the surface-water bodies are connected to the main water table, or if they are simply expressions of a perched water table (described in Chapter 4).
- Water table contours should not cross a surface body of water unless the water surface itself is inclined, as is true for streams. The water table contour line should cross the stream at the point where stream elevation equals water table elevation (Fig. 10.5). When crossing the stream, the contour lines should form shapes like the letter "V." The point of the V should point upstream if the stream is a gaining stream, and downstream if the stream is a losing stream.

In addition to these rules for water-level mapping, the following rules of contouring must be observed:

- At the very least, three points are needed to define a plane. However, it is desirable to use even more than three points. Using only three points to define ground water flow directions is risky, as described in this chapter.
- Contours may not extend outside the area enclosed by the data points.
- Contours may not cross or touch.
- Contour interval (the difference in value between two adjacent contours) should be consistent. When it is warranted, for example, in areas where more data are available, extra contour lines may be added. But they should be drawn with a different line pattern (e.g., dashes) and their presence should be noted in the key.
- Contours may end only by closing, or by extending off the contoured map area.
- Units of measurement must be indicated clearly, and should be consistent for all data reported on the map.

Following a few additional suggestions will help make the map easy to read:

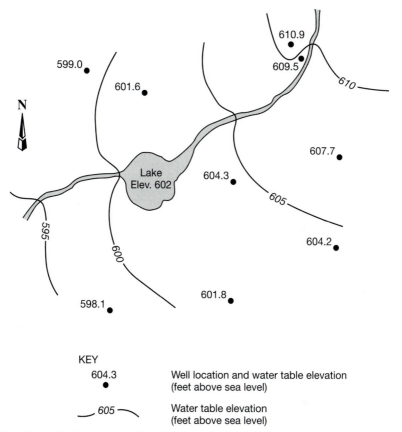

Fig. 10.4   Surface-water bodies, if they are hydraulically connected to the regional water table and are not perched features, should be incorporated into a water table map, as shown here. Note that the contours that cross the stream form V's that point upstream, because this is a gaining stream. For a losing stream, they would point downstream.

- If possible, use a contour interval that is a whole integer. Ideally, that integer should be numbers such as 1, 2, 5, or 10. If it is impossible to use a whole integer, then decimals should be used, ideally with an interval of 0.1, 0.2, or 0.5.
- Contour lines should be smooth and gently curving. Right-angle turns and other sharp bends are inappropriate in maps of water levels.

Once the map is constructed, check to be sure that the contours are smoothly drawn and that the interval is consistent throughout. Be sure that all points are contoured. Look for "bull's-eyes," multiple contours drawn about a single point,

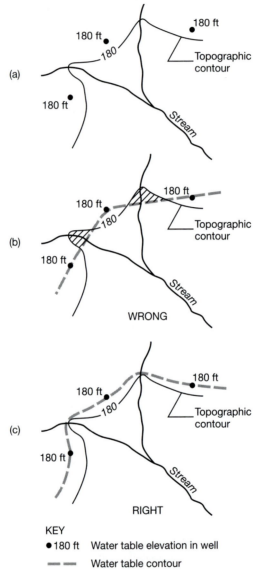

KEY

• 180 ft   Water table elevation in well

— — —   Water table contour

Fig. 10.5   (a) A topographic map showing the 180-foot elevation contour line. Three water wells in the area have water levels at 180 feet. Where should the 180-foot water table contour be drawn? (b) Wrong! This water table contour line does not take into account the fact that the elevation of the stream is the elevation of the water table. In the shaded areas, the water table is apparently higher than the land surface. (c) Right! The land surface is higher than the water table surface, except at the stream, where the elevations are equal.

showing that the value of that point is much higher or much lower than that of surrounding points. Bull's-eyes may indicate a true, unusually high or low point in a water surface, such as could be caused by a pumping or injection well. Alternatively, they may simply indicate a bad data point (Fig. 10.6). These points should be examined carefully to determine if the data are real or spurious. Bad data points might result from mistakes in reading water level in the field, or they might result from using a well or piezometer that is of different construction than that of the other wells or piezometers used to make the map.

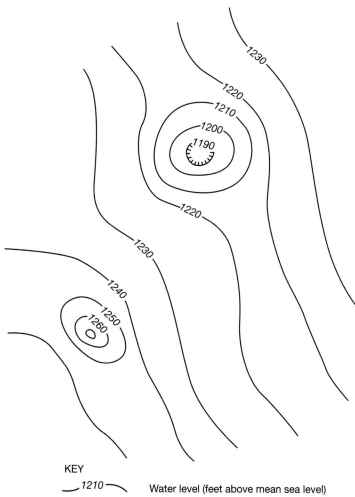

KEY

—*1210*—◝   Water level (feet above mean sea level)

Fig. 10.6   The two bull's-eyes on this water-level map indicate either truly high and low points or invalid data points.

Triangulation

Sometimes, a water-level map must be constructed from only three data points. This is not an ideal situation. At some sites, variations in water levels might be very small. When this is true, even a small error in a single water-level measurement might be enough to make a water level seem to be higher than the other two, when in fact it is lower, or vice versa. If only three points are used to define the water level, an error in only one of the three measurements might be enough to completely reverse the apparent direction of water flow.

A second reason to avoid making a water-level map from only three points is that water tables and potentiometric surfaces normally are not flat, planar features. If only three points are used, the water table or potentiometric surface will necessarily be constructed as a planar surface. If the three measurements are taken from a small enough area, however, a map made from three data points may give an approximation of local flow.

To make a water-level map from three points, a simple three-point problem must be solved. Various methods for doing this can be found in structural geology textbooks. One simple solution is presented here. Begin by drawing a light line from the point with the highest water level to the point with the lowest (Fig. 10.7). Using a ruler, measure the distance. Now select a water-level value that falls between these two points and that will have a contour drawn through it on the final map. For example, if contours will be drawn at 1-foot intervals, and the lowest and highest values are 21.4 and 23.7 feet, respectively, choose either 22 or 23 feet.

Where will these values fall on the line between the high and low points? To find the location of one of these points, use the following formula:

$$\frac{\text{Distance from low point to selected point (as measured with a ruler)}}{\text{Distance from low point to high point (as measured with a ruler)}}$$
$$= \frac{\text{difference between water levels at low point and selected point}}{\text{difference between water levels at low point and and high point}}$$

Keyed to Figure 10.7, this formula is as follows:

$$\frac{\text{Length of line } BC'}{\text{Length of line } AB} = \frac{\text{water elevation at } C' - \text{water elevation at } B}{\text{water elevation at A} - \text{water elevation at } B}$$

or,

$$\frac{\text{Length of line } BC'}{90 \text{ m}} = \frac{49.9 \text{ m} - 48.5 \text{ m}}{52.1 \text{ m} - 48.5 \text{ m}}$$

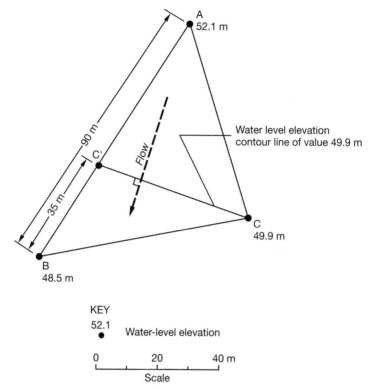

Fig. 10.7   Triangulation. Three water-level data points define a plane of constant slope. See the text for an explanation.

Fill in the known values and solve for the unknown. (In the example illustrated in Fig. 10.7, the unknown is found to be 35 meters.) Mark this distance on the map, and label it with its water-level value. Repeat this process on the other sides of the triangle. Connect points of equal value with straight lines.

Using Water-Level Maps to Determine Flow Direction

Ground water flows downgradient in the direction of greatest gradient. This is perpendicular to the contours on a water-level map. To determine direction of ground water flow, look for areas of high and low water levels on the map. Contour lines should pass between them. Water will flow from high water level to low water level, at right angles to the contours.

Ground water may flow in multiple directions on a single map. For example, in an area of a ground water divide, flow may radiate outward in all directions, away from a high point. In the area of a stream, flow on either side of the stream may be in opposite directions: toward the stream if it is a gaining stream, and away from it if it is a losing stream. Many more complex patterns may exist within a single site. For this reason, the more data points used, the better will be the quality of the final map.

Using Water-Level Maps to Determine Hydraulic Gradient

Hydraulic gradient is the change in hydraulic head between two points with respect to the distance the water travels between the two points. Use caution when calculating hydraulic gradient from a water-level map. Because maps are two-dimensional and flow occurs in three dimensions, a map may not give an accurate representation of flow. For example, what may appear on a map as flow in a straight line actually may be a flow line that curves in the third dimension (e.g., one that sinks and then climbs again) on a much longer flow path between two points than the map shows (Fig. 10.8).

Despite the fact that hydraulic gradient may be difficult to interpret quantitatively, a relative assessment of gradients may be made from a map. Areas where contour lines are spaced more closely together are areas of steeper gradient, and other factors being equal, areas of more rapid flow. Zones of lower conductivity will exhibit a similar variation in contour spacing. On the other hand, areas where contour lines are spaced farther apart are areas of lesser gradient, and possibly relatively slower flow or higher conductivity (Fig. 10.9). If a numerical value is to be estimated from a water-level map, it should be calculated as the difference in water levels from one point to another divided by the distance water travels between the same two points.

The reader should note that it is only meaningful to calculate a gradient between two points if water actually flows in that direction. Water will flow from

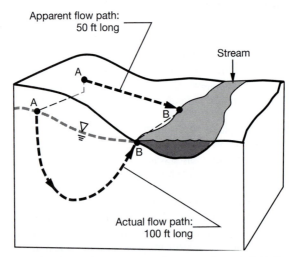

Fig. 10.8 In this situation, what appears on the map to be a straight flow line is actually a much longer, curving flow line when viewed in the third dimension.

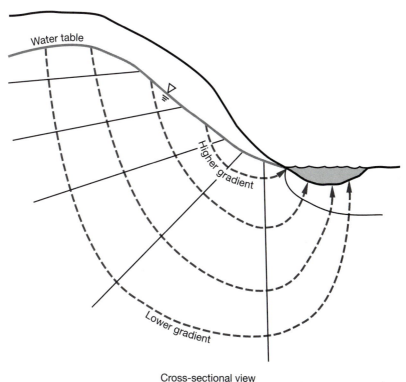

Cross-sectional view

Fig. 10.9 Close contour spacing indicates a higher gradient, whereas farther contour spacing indicates a lower gradient.

one point to another only if a line drawn between the points is perpendicular to the water-level contour lines. If the line drawn between the two points is not perpendicular to contours, calculating the apparent gradient between them will yield a number, but it will not be meaningful (Fig. 10.10). Finally, flow directions may be further complicated in karst aquifers, aquifers in which flow is mainly within fractures, and strongly anisotropic aquifers.

After a water-level map is drawn, it should be examined and inspected for errors. Suggestions for evaluating a water-level map are given in Box 10.1.

Flow Nets

Flow nets are water-level maps or hydrogeologic cross-sections on which are drawn lines with arrows indicating the direction of ground water flow. These lines are called flow lines. A second feature of flow nets is contour lines that connect points of equal hydraulic head. These lines are called equipotential lines.

Flow nets may be constructed in map view or in cross-section. Map-view flow nets are simply water level maps on which flow lines are drawn. As with a water-level

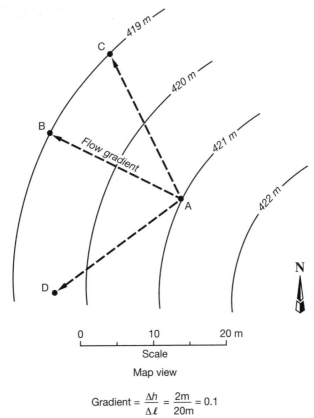

$$\text{Gradient} = \frac{\Delta h}{\Delta \ell} = \frac{2m}{20m} = 0.1$$

Fig. 10.10   Gradient is calculated along the flow path, in this case, line *AB*. Water does not flow in the direction along *AC* or *AD*, so the gradient should not be calculated along these lines, or any others except *AB*.

map, this type of flow net is capable of showing conditions only within one hydrostratigraphic unit. In addition, making this type of flow net assumes fairly isotropic conditions of hydraulic conductivity as one proceeds in the direction of flow. Advanced methods may be used in anisotropic and nonhomogeneous conditions.

Cross-sectional flow nets show contours of equal hydraulic head and, drawn at right angles to the contours, flow lines showing directions of water flow. Like a map-view flow net, this type of flow net assumes that hydraulic conductivity is relatively isotropic in the direction of flow. The primary advantage to a flow net drawn in cross-sectional view is that it can show vertical components of flow, including flow across hydrostratigraphic units.

This section provides rudimentary steps in flow net construction. For more information on constructing flow nets, see Freeze and Cherry (1979), Cedergren (1988), Fetter (1994), or Lee and Fetter (1994).

BOX 10.1

Evaluating a Water-Level Map

Frequently, a hydrogeologist must evaluate the quality of a water-level map. Anyone who draws a map will want to check it for mistakes. But evaluation of maps drawn by other people frequently occurs in hydrogeologic work, as well. For example, those who work in regulatory agencies must determine if plans drawn up by consultants are realistic and effective; consultants must evaluate maps drawn by previous workers at a site; and project managers must evaluate work done by their team members. Anyone evaluating a map should have experience in drawing maps.

In evaluating the quality and reliability of a water-level map, whether drawn by oneself or by others, the following key questions should be asked.

- Consider the water levels that are mapped. Are they the elevations of water above some datum (e.g., sea level, or a local datum)? They should be. A common mistake of beginners is to plot depth to water instead of water-level elevation.
- Are all the data points contoured from wells that tap (are screened or open through) the same hydrostratigraphic unit, and the same portion of that unit? They should be.
- Were all water-level measurements made on the same day, or at worst, within a few days of each other? They should have been.
- If the map is of the water table surface, are the data points taken from wells with intakes that intersect the water table and extend no more than 8–10 feet below the water table? They should be. Are surface-water bodies mapped as surface expressions of the water table? They should be, unless they are part of a perched water table.
- If the map is a water table map, do any contour lines cross constant-elevation lakes or ponds? They should not. Do any contour lines cross streams? They should, and they should cross at the precise point where elevation of the stream and elevation of the water table are equal. When crossing the stream, the contour lines should form shapes like the letter "V," pointing upstream for gaining streams and downstream for losing streams.
- Consider topography of the ground surface. Drop the tip of a pencil at random points anywhere on the map. Is elevation of the land surface higher than that of the water table? It should be, unless that point is covered by surface water. The reader should note that this test is not useful for maps of potentiometric surfaces of confined aquifers, because these surfaces may or may not be higher than the land surface.
- Are at least three data points used to draw contour lines? This should be the case. If only three points are used, contour lines should be straight and parallel. If more than three points are used, contour lines may curve.
- Do the contours form bull's-eyes? Are there places where contour lines make abrupt or sharp turns? If either of these features is present, some explanation should be obvious. For example, a pumping or injection well might cause a bull's-eye.

(continued)

BOX 10.1 *(continued)*

- Mentally draw lines from point to point along the outer edge of the data points, encircling the rest of the data points. Do any contour lines extend a significant distance outside these boundaries? They should not. Without more data indicating conditions in this region, little can be said about the water levels in these areas.
- Do any contour lines cross or touch? They must not.
- Is the contour interval consistent and easy to read? It should be.
- Are units of measurement clearly specified? They should be, for contour interval and water-level measurements as well as for map scale.
- Is the North direction clearly indicated? It should be.
- Is a scale given? It should be. ■

Constructing a Flow Net

When developing a flow net, a pencil should be used instead of a pen, because erasures are *de rigueur*.

Flow net construction begins with a determination of general direction of ground water flow. To do this in map view, a ground water contour map should be constructed, and areas of high and low ground water levels should be identified. Water will flow from relatively higher water-level elevation (higher hydraulic head) to relatively lower water-level elevation (lower hydraulic head). In cross-sectional view, high and low ground water elevations should be identified, particularly considering the locations of recharge areas, discharge areas, and constant-head boundaries such as lake beds. Again, water will flow from areas of higher hydraulic head to areas of lower hydraulic head.

Based on the relative differences in water elevations, a few lines showing direction of flow should be sketched. These lines are called flow lines. Flow lines should be drawn with arrows that show the direction of flow. Contour lines called equipotential lines should be sketched at right angles to the flow lines. These lines connect points of equal hydraulic head.

Once the basic shapes of the flow lines and equipotential lines are determined, the flow net should be refined. The intersections of flow lines and equipotential lines should make "squares," or roughly equidimensional shapes. Both types of lines should be smooth, since ground water does not make abrupt or sharp turns. Above all, the lines must meet each other at right angles. (An exception to this rule pertains when a flow net is constructed on a cross-section drawn with vertical exaggeration.) Use a T-square, right triangle, or other square corner to check that the flow lines and equipotential lines are perpendicular.

If values of hydraulic head are available for various points or surfaces in the flow net, it may be possible to make the flow net quantitative. This can be done by assigning a hydraulic head value to each equipotential line, using a regular contour interval. Known values of hydraulic head at various points should fit into the quantitative scheme; if they do not fit, the equipotential lines should be redrawn.

Flow lines and equipotential lines on a flow net have a clear and well-defined relationship to flow boundaries. These relationships are described in Table 10.1.

Table 10.1

Relationships of Boundaries to Lines in a Flow Net[a]

Type of Boundary	Relationship of Boundary to Flow Lines	Relationship of Boundary to Equipotential Lines
For cross-sectional or map-view flow nets		
Impermeable or no-flow boundary (e.g., cutoff wall, slurry trench, impermeable foundation of a structure, or other impermeable hydrostratigraphic unit)	Parallel	Perpendicular
Constant head (e.g., bed of a lake, bed of a constantly spilling reservoir of any type, a drain)	Perpendicular	Parallel
For cross-sectional flow nets only		
Water table	Oblique (nonperpendicular, nonparallel)	Oblique (nonperpendicular, nonparallel)

[a]This table applies only to flow nets with no vertical exaggeration.

This set of conditions applies only for isotropic, homogeneous media. In addition, if a cross-sectional flow net is drawn with vertical exaggeration, distortion of these relationships might result.

A flow net in cross-sectional view may show flow across hydrostratigraphic units. Where hydraulic conductivity changes as one proceeds along the flow path, contour spacing and direction will change. When water flows from one hydrostratigraphic unit to another, it is refracted, just as is light traveling through two types of glass. The lines in a cross-sectional flow net should be adjusted to show this refraction.

When a flow line passes through a boundary from a layer of high hydraulic conductivity to one of low hydraulic conductivity, it is refracted so that it travels in a direction that is closer to normal (perpendicular) to the boundary of that layer, thereby taking the shortest possible path through the layer. The opposite occurs when a flow line passes from a low-conductivity to a high-conductivity layer: It refracts so that it travels closer to parallel with the layer's boundary. See Fig. 10.11 for examples. More information, including methods to calculate the angle of refraction, is given in Freeze and Cherry (1979), Cedergren (1988), and Fetter (1994).

Mapping Hydrogeochemical Parameters

Hydrogeologic maps may be used to illustrate the spatial distribution of chemical constituents of the ground water. This may help illuminate trends across an area or zones of unusual water chemistry. Mapping hydrogeochemical parameters may be

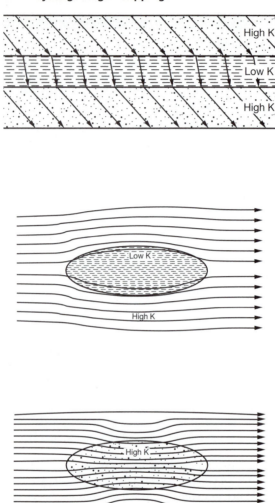

Fig. 10.11 Flow crossing a boundary between materials of two different hydraulic conductivities is refracted at the boundary.

accomplished by either of two methods: contouring values or adding hydrogeo-chemical diagrams to a map.

Contouring Hydrogeochemical Parameters

Contouring of hydrogeochemical parameters follows the same basic principles as contouring water levels or topography. Values of a particular parameter are plotted as point values on a map with sampling points identified. Then a contour interval is established, and contour lines are drawn connecting points of equal value. If the values contoured are concentrations, the contour lines are called isocons [Fig. 10.12(a)]. Contour interval and units of measurement must be indicated clearly on the map legend.

What values may be contoured? In general, they all should be measurements taken from the same hydrostratigraphic unit. They should be measurements with a reasonable precision: If a check of analytical accuracy reveals that analyses are questionable, those analyses should be excluded from the contouring effort. Ground water chemistry varies over time, but slowly; samples taken within a few weeks or months of each other are likely to be comparable. Of course, if some

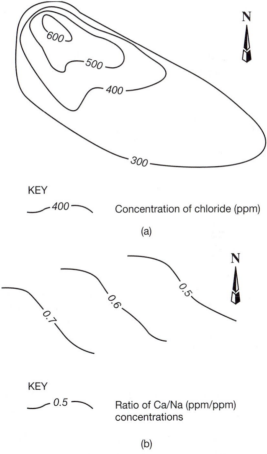

Fig. 10.12 (a) A contour map of chemical parameters shows isocons, lines connecting points of equal value. (b) In some cases, mapping ratios of various constituents is useful.

drastic change is known to have occurred that would affect the chemistry, then this should be considered in the mapping.

Other values besides concentrations may be contoured. It may be helpful, for example, to plot and contour *ratios* of various constituents, rather than their absolute concentrations [Fig. 10.12(b)]. This would be particularly true if two chemically different waters were mixing; in a case like this, the "fingerprint" of one type of water may be preserved in the ratios of constituents. For example, seawater has high concentrations of sodium and potassium. If rainwater mixes with seawater, the seawater will be diluted. If extensive mixing takes place, the seawater may become very dilute, and the high sodium and potassium concentrations will disappear. But even at this dilution, the ratio of sodium to potassium (and other ratios) in the mixture will be similar to that of seawater. Contouring this ratio may allow one to distinguish between waters affected by seawater intrusion and areas with no seawater intrusion.

In some situations, it may be appropriate to plot the *logarithms* of concentrations, rather than the concentrations themselves. This is most often true when the constituent is a contaminant, and when there is a very steep concentration gradient between the contaminant source and the uncontaminated water. In such a situation, concentrations may range over several orders of magnitude, and using a simple contour interval may not be effective. For example, imagine trying to contour BTEX (benzene, toluene, ethylbenzene, and xylene) concentrations that range from 0.1 to 10,000 part per billion (ppb). This is not an unusual range at sites where gasoline has contaminated the ground water. What contour interval should be used? Any regular interval will be either too fine or too coarse to reveal the structure in the data, whereas a logarithmic scale would show it clearly (Fig. 10.13).

Using Hydrogeochemical Diagrams in Mapping

Hydrogeochemical diagrams are graphical methods for illustrating and analyzing water chemistry data. Although many methods exist for plotting these diagrams, only a few are particularly useful for mapping. These are Stiff diagrams, kite diagrams, pie charts, and vector plots. Hem (1985) gave instructions for producing all of these diagrams. Any of them may be placed on a map next to sampling points to illustrate water chemistry at that point and to highlight changes across an area. Although of limited use for mapping, other types of diagrams, the trilinear and Piper diagrams, are useful for interpreting changes in water chemistry across an area. Construction of trilinear, Piper, and Stiff diagrams is described here.

Trilinear, Piper, and Stiff diagrams illustrate major ion chemistry, and all require that analyses be expressed in terms of equivalent weight units, for example, milliequivalents per liter (meq/l) or equivalents per million (epm). Because few chemical analyses performed for remedial investigations include all the major ions, it may not be possible to construct complete plots for most of these investigations. However, if analyses of even just the major cations are available, the construction of partial plots may be useful. The major cations are calcium, magnesium, sodium, and potassium, and the major anions are bicarbonate, sulfate,

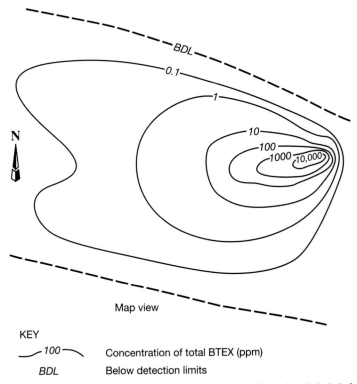

KEY

——100——   Concentration of total BTEX (ppm)

BDL   Below detection limits

Fig. 10.13   In some situations, to best show the structure of the data, it is helpful to draw contours at logarithmic intervals.

and chloride. If bicarbonate is expressed in terms of alkalinity (often expressed as "alkalinity as $CaCO_3$"), it must be converted into terms of HCO_3^{-1}. To do this, divide a measurement of "mg/l alkalinity as $CaCO_3$" by 0.8202; this will give the value of "mg/l HCO_3^{-1}" (Hem, 1985). For example, if an analysis reports that a water sample contains "148 mg/l alkalinity as $CaCO_3$", then the water contains 148/0.8202 or 180 mg/l HCO_3^{-1}.

To transform concentrations of ions into meq/l, multiply the concentration in milligrams per liter (mg/l) by the valence of the ion, and divide the result by the formula weight of the ion. If the original analyses are expressed in parts per million (ppm), then multiply the concentration in ppm by the valence and divide by the formula weight, and the result will be the concentration in epm. Valences of the major ions are as follows: Ca^{+2}, Mg^{+2}, Na^{+1}, K^{+1}, HCO_3^{-1}, SO_4^{-2}, and Cl^{-1}. Once the analyses are expressed in terms of meq/l or epm, they may be used to construct trilinear, Piper, or Stiff diagrams. An example of these conversions is given in Box 10.2.

BOX 10.2

Converting Ionic Concentrations to Units of Equivalent Weight

To convert from mg/l to meq/l:

$$\text{Concentration in mq}/1 = \frac{\text{concentration in mg}/1 \times \text{valence}}{\text{formula weight}}$$

To convert from ppm to epm:

$$\text{Concentration in epm} = \frac{\text{concentration in ppm} \times \text{valence}}{\text{formula weight}}$$

Example: An analysis of a water sample is given in the table. The concentrations in terms of mg / l are converted to units of equivalent weight by using the first preceding formula.

Ion	Concentration (mg/l)	Valence	Formula Weight	Concentration (meq/l)
Ca^{+2}	151	+2	40.08	7.53
Mg^{+2}	23	+2	24.31	1.9
Na^{+1}	165	+1	22.99	7.18
K^{+1}	19	+1	39.10	0.49
			Total Meq/l Cations:	17.1
HCO_3^{-1}	401	−1	61.02	6.57
SO_4^{-2}	58	−2	96.06	1.21
Cl^{-1}	345	−1	35.45	9.73
			Total Meq/l Anions:	17.51

■

Trilinear and Piper Diagrams

Trilinear and Piper diagrams generally are used to illustrate the major ion composition of water samples. Trilinear diagrams are triangular in shape, while Piper diagrams consist of two trilinear plots and a diamond plot. Either type can accommodate hundreds of sample points. They are particularly useful for detecting changes or trends in ground water chemistry across an area or through time (Fig. 10.14).

Trilinear Diagrams. Trilinear plots give relative concentrations of major ions as a percent of the total. To construct trilinear plots, for major cations and anions, do the following:

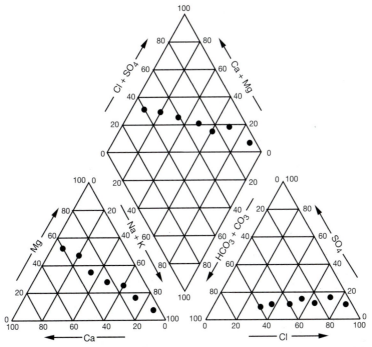

Fig. 10.14   Trilinear or Piper diagrams are particularly useful for detecting or illustrating changes or trends in ground water chemistry across an area or through time.

1. Express concentrations of the major ions in meq/l or epm.
2. (a) Find the total meq/l (or epm) of the cations.
   (b) Express the concentration of calcium as a percent of the total. Do the same for magnesium and for the sum of sodium and potassium; sodium and potassium concentrations are added together because they behave in a similar chemical fashion and because the concentration of potassium frequently is negligible compared to that of sodium.
3. (a) Find the total meq/l (or epm) of the anions.
   (b) Express each of the ion concentrations as a percent of the total.
4. Plot the results on two triangular graphs, with axes as shown for the triangular fields in Fig. 10.15. Use separate graphs for cations and anions.

   To plot a value on a triangular graph, consider each apex (or point) of the triangle to represent 100% of the total meq/l. The opposite side of the triangle represents 0% of the total meq/l. So an analysis that plots at the "magnesium" apex contains only magnesium (100% of the meq/l), and no calcium, sodium, or potassium. An analysis that plots on the baseline of the triangle is on the 0% magnesium line. In such a sample, there is no

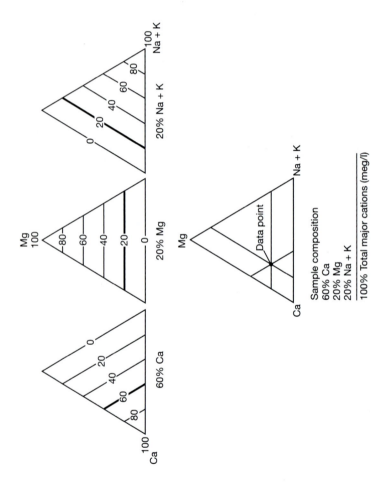

Fig. 10.15 Plot a point on a trilinear graph by considering each percentage separately. The top three triangles show how to find the line representing each percentage. The data point is plotted where the three lines intersect (bottom triangle).

magnesium, and calcium, sodium, and potassium together make up 100% of the total meq/l cations.

Begin plotting a value by determining on what line the magnesium value falls. Darken this line slightly. Then, turn the triangle on its side, and reorient. Determine on which line the calcium value falls, and darken this line slightly. Make a dot where the two darkened lines cross. This point represents the major cation chemistry.

To check the point, reorient the triangle by turning it again. Determine on what line the sodium + potassium value falls, and darken this line. It should pass directly through the dot. If it does not, recheck the magnesium and calcium values. All three lines should pass through the same point.

Follow the same procedure for the anion triangle.

Piper Diagrams. To make a Piper diagram, first construct the two trilinear plots for cations and anions. Position them as shown in the diagram, and construct the diamond field. Next, project two lines passing through the dot in each triangle, parallel to the outside edges of the diamond. Make a dot where these lines cross (Fig. 10.16). A complete Piper plot has three dots that represent one water analysis:

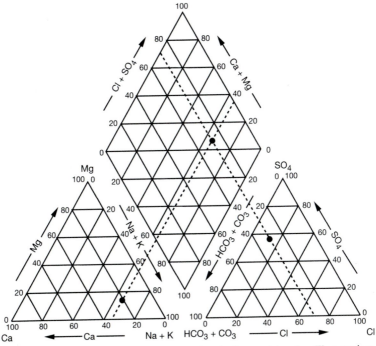

Fig. 10.16 On a Piper diagram, first plot points in the two triangles. Then project two lines into the diamond region, parallel with the outside borders of the diamond. Where these lines cross, plot a single point.

one dot in the diamond and one in each of the triangles. If complete major ion data are not available, either the cation or the anion triangle may be used independently of the rest of the diagram.

Stiff Diagrams

Stiff diagrams plot major ion chemistry in terms of meq/l (or epm) on two axes. One axis is for cations; the other is for anions. The axes increase outward from a common zero point (see Fig. 10.17).

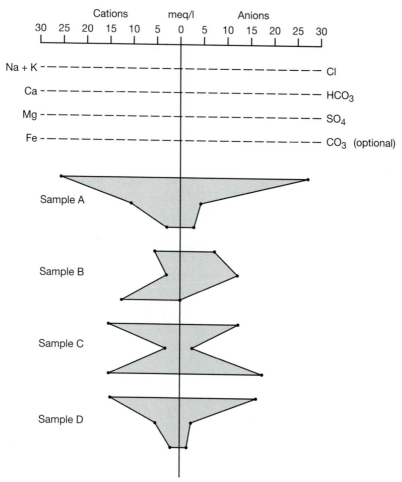

Fig. 10.17 Stiff diagrams for four water samples. Note that Samples A and D are chemically similar, despite the fact that Sample A has higher concentrations of the major ions.

To construct a Stiff diagram, set up a vertical zero line and define a horizontal scale appropriate to the values in the analysis. Then plot the following cation-anion pairs on horizontal scales, cations on the left and anions on the right: Ca^{+2} and HCO_3^{-1}, Mg^{+2} and SO_4^{-2}, and Na^{+1} and Cl^{-1}. Draw lines connecting the dots and closing the figure. For maximum visibility, which might be particularly important on a busy map, shade the figure.

Note that the shape of the figure illustrates the water chemistry. Although chemically similar waters may plot on figures with different scales, the similarity of shapes indicates similar composition.

Field Projects

Chapter Overview

What will an entry-level hydrogeologist do in the first year on the job? To a great extent, this depends on the type of employer. Some entry-level hydrogeologists find employment with federal, state, or local government regulatory agencies, such as the Environmental Protection Agency, or more research-oriented agencies such as the U.S. Geological Survey. Other new hydrogeologists work for private-sector consulting companies. These companies may handle hydrogeological work exclusively, or they may specialize in some other area (such as mining, construction, or engineering) and maintain a small staff of hydrogeologists who provide input as part of a team working on various ventures. Some companies take on a wide range of projects. Other companies attempt to fit a niche market and specialize in handling a select few types of projects.

Because hydrogeological problems are tremendously varied, it is impossible to predict exactly what assignments will be given to an entry-level hydrogeologist. In some companies or agencies, a new employee spends the first few weeks on the job reading and evaluating reports, accompanying more experienced personnel into the field, and learning about the workings of the organization. In other places, a new employee is immediately sent to the field and is expected to learn the job very rapidly on her or his own.

This chapter attempts to give an idea of the type of tasks and the scope of duties that might be assigned. Knowing something about the types of projects on which hydrogeologists commonly work will give an entry-level employee a framework for understanding what might be expected in the field. In addition, the more diverse a scientist's experience and knowledge, the more flexible that person will be, and in general, the more valuable she or he will be on the job. Toward those ends, this chapter outlines some common hydrogeologic projects and problems.

Information in this chapter also might be useful to a job seeker who is researching or interviewing with various employers. For example, at an interview, a candidate might ask the interviewer to describe some recent projects in which the company has been involved. Knowing something about typical hydrogeologic projects might help to put the answer in some perspective.

Although this chapter provides a few notes about these field situations, it does not describe the details. In addition, new hydrogeologic techniques are being devised and introduced on a regular basis, and it is important to stay current in the field. As a result, the reader is advised to consult other sources for more information. These other sources may include materials such as texts, professional journals, and trade magazines. An even better source of information may be conversations with more experienced hydrogeologists, such as one's colleagues or other contacts within the field. Joining hydrogeologic professional societies and participating in meetings, field trips, and other activities will help establish these connections. And finally, much can be learned by simply taking every opportunity to observe hydrogeologic projects underway (Fig. 11.1).

The projects described here are grouped under two major headings: "Site Investigations," which describes activities that precede any hydrogeologic undertaking, and "Using Hydrogeology to Solve Problems," which describes field projects in which hydrogeology might be applied.

Site Investigations

Whenever a hydrogeological project is undertaken, a site investigation generally is performed before other phases of the project take place. A site investigation, sometimes called a pre-construction evaluation, provides an assessment of the hydrogeological conditions at a site. It sets the stage for further investigations or operations at the site. A site investigation generally takes place before significant resources are committed to the project, so that any potentially unfavorable conditions are known in advance and any necessary adjustments may be made to the project plans. It is unlikely that an entry-level hydrogeologist would be placed in charge of such a project, but an inexperienced employee might be asked to perform several of the tasks involved in a site investigation.

Site investigations are performed as a part of virtually every hydrogeological project, whether the larger project is to evaluate ground water supplies in the region, characterize ground water chemistry, assess ground water contamination, devise a ground water remediation plan, or design a structure or excavation that requires geotechnical engineering. This section describes site investigations in general terms, noting what an entry-level hydrogeologist might be required to do. In addition, it focuses on two somewhat specialized types of site investigations: site assessments for real estate transactions, and wetland delineations.

Investigating Site Geology, Hydrogeology, and Hydrogeochemistry

Investigating the geology, hydrogeology, and hydrogeochemistry at a site requires a review of historical records and documents about the site. An important component of a site investigation is the collection and compilation of this data. Examining sources such as regional maps and cross-sections may give information on general topographic, geologic, or hydrogeologic conditions at the site. Site-specific maps and cross-sections may be available in some cases. The site investigation might also include gathering, evaluating, and interpreting boring logs, possibly using them to construct cross-sections. The sources of information described in Chapter 2 may be consulted during a site investigation. An entry-level hydrogeologist may be assigned the task of locating this information, reviewing it, and preparing reports that compile the data.

Once the general hydrogeologic setting of a site has been assessed, it often is necessary to obtain more site-specific information through a field investigation.

Fig. 11.1   What is happening in this picture? Note the Level B PPE, drilling method, and the various personnel. Simply observing field projects underway can be instructive to an entry-level hydrogeologist. (Courtesy of Stephen Lombardo.)

Field mapping may be performed. Drilling programs might be designed to evaluate soil conditions, obtain rock or sediment samples, install wells, take water samples, or gather other information about conditions at the site. Pumping tests, slug tests, or other tests of hydraulic conditions might be performed. When appropriate, geophysical exploration methods such as seismic refraction, seismic reflection, electrical resistivity, or ground-penetrating radar (GPR) may be used. The entry-level hydrogeologist might be assigned to any of these aspects of a field investigation. Ideally, the entry-level employee should be supervised by a more experienced hydrogeologist in the field until it is clear that the new person has the necessary field skills.

Site Assessments for Real Estate Transactions

Certain site investigations are performed specifically as part of a real estate transaction. Why might a site assessment be needed in these situations? Landowners may be held partly or wholly liable for environmental, health, safety, or other problems related to safety, or other problems related to materials or conditions on their land. Imagine buying a small parcel of land, closing the deal, and then learning that it needs a fifty thousand dollar environmental cleanup. Clearly, information on environmental conditions at the site is most useful before the transaction is completed.

Property buyers, sellers, and financial institutions all have interest in a site assessment. Those interested in buying property should learn about any possible environmental problems at the site. A property owner should assess the problems before selling, because they may have a fiduciary duty to disclose problems to a potential buyer. Before a bank or mortgage company agrees to loan money to the buyer for the purchase, the bank might insist that an investigation be conducted into potential environmental problems at the site so that the lending institution may minimize their financial risks. This is particularly true for commercial properties.

In all of these cases, the purpose of a site assessment is to avoid a situation in which the transaction is finalized before the problems are known. If the problems are not known, there may be serious consequences if it should be discovered later that the site will require large sums of money to clean up contamination.

Finding that a site has environmental problems does not necessarily mean that the sale will not be completed. In some situations, a site is known to be contaminated, but the value of the property outweighs the cost of the remediation. "Brownfields" transactions may take place when a site, particularly an industrial site, is known to be contaminated, but regulations permit further industrial uses of the land if the site is cleaned to an acceptable level.

Site assessments for real estate transactions generally are performed using a three-phase approach. Phase I of a site assessment includes a review of historical records and documents about the site and adjacent properties, a review of regulations that pertain to the site, and a field reconnaissance. In some projects, a preliminary field screening of soil, water, or air quality might take place, as well; however, this preliminary screening does not generally involve invasive procedures such as soil borings or well installation. It is not unusual for an entry-level employee to be assigned the task of tracking down and compiling information during a Phase I investigation. Box 11.1 describes the steps involved in conducting a Phase I site assessment.

Upon the completion of the Phase I investigation, a report is prepared that describes the activities that took place and delineates any areas of concern. Recommendations for further investigation might be made to characterize more definitively any potential problems at the site. Based on the results of the Phase I investigation, it might be apparent that some potential for environmental problems exists at the site. If the Phase I assessment shows the need for further investigations, a Phase II investigation may be performed, at the discretion of the client. Investigations at many, perhaps most, sites do not progress beyond Phase I.

Phase II of a site assessment involves field sampling and testing to further delineate and characterize areas of concern. Field and laboratory tests may be performed on soil, water, or air samples. In order to perform the work in this phase, some invasive procedures might be used. For example, soil borings may be performed, monitoring wells may be installed and water samples taken, and air testing equipment may be used. Phase III of a site assessment involves remediation of contamination at the site. More information on site assessment procedures may be found in ASTM Standard Practices E1527 and E1528 and in Colangelo (1991).

BOX 11.1

Conducting a Phase I Site Assessment for a Real Estate Transaction

APhase I review of historical records and documents generally includes gathering topographic, geologic, hydrogeologic, hydrologic, and land-use documents, aerial photographs, and other documents that provide information on activities that have taken place at the site. In addition to a review of these materials, a title search often is performed. This search gives information about previous owners of the property. Old site maps, zoning maps, plat maps, fire insurance maps, and other local sources of information should be consulted as well. Some sources of information are described in Chapter 2.

A Phase I review of pertinent statutes and regulations is performed by consulting the various government agencies that regulate the site, as well as checking the National Priorities List and CERCLIS (the CERCLA Information System; CERCLA is the Comprehensive Environmental Response, Compensation, and Liability Act of 1980). During a Phase I investigation, a list may be prepared of any sites near the property that might contain contamination that could potentially migrate to the site of interest.

A Phase I investigation also includes interviews with employees of companies on the site and neighbors who live or own property adjacent to the site. The purpose of these interviews is to learn more about the site history and the present and past activities at the site.

A site reconnaissance, or site walk-through, should be conducted during Phase I. This serves two purposes. First, it allows the hydrogeologist to make firsthand observations of conditions at the site, either confirming what the record and document review showed or possibly spotting something that did not appear during the record and document review. Second, the site walk-through allows the hydrogeologist to conduct a more intense scrutiny of areas that were highlighted as potential areas of concern during the record and document review. To allow for this, it is important that the site walk-through not take place until the record and document review is complete or at least well underway. (However, it sometimes is helpful to make an initial site visit before the record and document review, and then perform a more detailed walk-through after the review. This is particularly the case for large or comeplex sites.) A site reconnaissance is a challenge to a hydrogeologist's powers of observation, and should be conducted deliberately and with great care.

A site reconnaissance is conducted by walking about the entire site, observing and noting the environmental conditions. All features that might tell something about areas of potential environmental concern, or features that reveal the hydrogeologic conditions of the site, should be noted. At a minimum, the following features should be recorded.

Topographic features and conditions should be noted, including natural features as well as constructed ones, such as mounds, depressions, or drainage channels. Mounds and depressions should be observed carefully, as they may be relict waste piles or waste lagoons.

The *geology and soils* at the site may be mostly hidden from view or may be readily apparent, depending on the site. When observable, geologic features such as outcrops, structural features, and exposures of soil or sediments should be noted.

(continued)

BOX 11.1 *(continued)*

Particular attention should be paid to areas of unusual soil characteristics, discolored materials, oil-stained soils and oddly colored seepage faces, as these may be indicators of potential contamination. For example, they might indicate areas of buried drums or backfilled waste lagoons.

Surface water at the site may include features such as ponds, lakes, streams, wetlands, and surface impoundments such as retention basins. Surface drainage features also should be recorded, including ditches, pipes, sewers, grates, septic tanks, leach fields, outfall pipes, and other flow-directing devices. Color and odor of surface water should be noted. Culverts and low bridges built over dry channels or depressions may be indicators that considerable surface flow occurs during storms, even though there may be no water present at the time of the site visit.

Evidence of *ground water* conditions, such as seeps, springs, and areas of moist soil, should be noted. In general, no samples are taken during a Phase I unless the plan calls for sampling and testing. However, casual visual observation may reveal the moisture conditions. Wells, water tanks, and piping systems should be noted.

Notes on *vegetation* should include type, abundance, and condition of plants, and indications of stress, such as discoloration or stunted growth. Wooded areas should be investigated with care, as illegal dumpers sometimes prefer to dump in these secluded areas.

Transportation features, such as roads (paved or unpaved), bridges, parking lots, railroad tracks, or foot trails may be sources of contamination or may lead to it. Dead end roads are of particular interest, because they sometimes lead to illegal dump sites. Sometimes they are visible on air photos or topographic maps. Parking lots may be salted during icy conditions, and this may be a source of contaminants. Because waste dumping sometimes occurs along railroad tracks, and because trains carrying hazardous materials are a potential source of contamination, railroad tracks or spurs should be noted, whether they are currently in use or not. Likewise, old railroad beds should be noted, even if they currently hold no tracks.

Utilities and pipelines, whether buried, above ground, or overhead, should be noted. Electrical transformers are of particular interest because they may contain polychlorinated biphenyls (PCBs). The backfill around buried utilities may be a conduit for ground water contaminant flow. Locations of filling ports for underground storage tanks should be noted, as should their general condition.

Drums and tanks should be noted, whether they are above ground or buried. Storage and loading areas may be important, as well.

Buildings, other structures, and old foundations should be described. Activities that take place inside the buildings (e.g., industrial processes), as well as the construction materials of the building itself (e.g., asbestos), may be a source of concern. Describing and testing these features in detail is beyond the scope of this text.

General environmental conditions of *air, water, soil, and noise levels* should be noted. Evidence of odors, smoke, or dust should be recorded. ■

Wetland Delineation

In the last decade, interest in wetlands has increased dramatically. This is partly because of the development of knowledge about the role of wetlands in flood abatement and in providing habitat for wildlife, among their other benefits and functions. This increase in interest also is due to environmental legislation, particularly Section 404 of the Clean Water Act, or simply "Section 404." This legislation regulates activities that potentially will affect wetlands and establishes a permitting process.

Wetland delineation is described here as an example of a project on which hydrogeologists might work closely with other scientists as part of a multidisciplinary team. Most hydrogeologists are not well-versed enough in identification of wetland plants to perform the analyses of vegetation needed in a comprehensive delineation. However, hydrology is a critical aspect of delineations. As a result, hydrogeologists may be required to work with field botanists on such projects, and should have a general understanding of all aspects of the delineation.

Formal definitions of wetlands vary, and no single definition has been agreed on by all groups and agencies that work in wetlands science. However, the U.S. Army Corps of Engineers and the U.S. Environmental Protection Agency (USEPA) wrote a definition of a jurisdictional wetland, or a wetland that falls under the jurisdiction of Section 404. Jurisdictional wetlands are "those areas that are inundated or saturated by surface or ground water at a frequency and duration sufficient to support, and that under normal circumstances do support, a prevalence of vegetation typically adapted for life in saturated soil conditions. Wetlands generally include swamps, marshes, bogs, and similar areas" (40 CFR 230.3). Wetlands hydrologists might be hired by clients to delineate these jurisdictional wetlands, perhaps as part of a site assessment or as part of an effort to comply with Section 404 (Fig. 11.2).

The U.S. Fish and Wildlife Service, the USEPA, the Army Corps of Engineers, and the Soil Conservation Service cooperated to publish the *Federal Manual for Identifying and Delineating Jurisdictional Wetlands* (Federal Interagency

Fig. 11.2 In the foreground is a wetland, and in the background is an upland. Where is the boundary between them? A wetland delineation could locate it.

Committee for Wetland Delineation, 1989), or simply, the *Federal Manual*. As described in this manual, jurisdictional wetlands are delineated on the basis of three criteria. To be a wetland, an area must have all three of the following: (1) a prevalence of hydrophytic vegetation (vegetation that grows in water or in water-logged soils), (2) hydric soils, and (3) wetland hydrology. These criteria are defined in detail in the *Federal Manual*.

Procedures for delineating a wetland may be performed offsite or onsite. Offsite procedures, performed without visiting the site, yield a preliminary determination and may be used unless a precise wetland boundary must be drawn. Offsite procedures basically involve examining topographic maps, National Wetlands Inventory maps, aerial photographs, and soil surveys of the area. (See Chapter 2 for instructions on using a soil survey.) A national list of soils that often display hydric characteristics has been compiled by the U.S. Department of Agriculture (1991). This list should be consulted to determine if a soil that exists at the site appears on the list. Evidence of hydrophytic vegetation, hydric soils, or wetland hydrology are indicators of the possibility that a wetland exists at the site. For a more accurate determination, onsite procedures should be conducted.

Onsite procedures may be performed to any of three levels of complexity as defined by the *Federal Manual:* (1) routine, (2) intermediate level, or (3) comprehensive methods. Routine methods are used for small areas of relatively little complexity. Box 11.2 describes some of the indicators noted during routine onsite wetland delineation. Intermediate-level methods are used for large sites or sites that exhibit complex vegetative, soil, or hydrologic conditions. Comprehensive methods employ a fine sampling grid and are used when detailed documentation of the vegetation, soils, and hydrology is needed.

The *Federal Manual* is an excellent resource describing wetland processes and terminology. In addition, it contains an extensive bibliography of other publications on wetlands. Lyon (1993) presented a practical manual for using the criteria in wetland delineation. More information on wetland classification, and an excellent collection of photographs of various wetland types, appears in a Fish and Wildlife Service publication by Cowardin and colleagues (1979). Mitsch and Gosselink (1993) provided a comprehensive discussion of the geology, chemistry, and ecology of various wetland systems, as well as an introduction to wetland management.

Using Hydrogeology To Solve Problems

Hydrogeologic principles may be employed to solve a wide variety of problems. Entry-level hydrogeologists may be employed to work on projects in ground water quantity, such as finding new water sources, or explaining why wells have dried up. They may investigate problems in ground water quality, for example, finding and remediating ground water contamination. Geotechnical engineering and mining projects often deal with water as a major concern, and hydrogeologists can help bridge the gap between engineers and geologists. Ground water modeling may be involved in many types of hydrogeologic problems.

BOX 11.2

Delineating Wetlands in the Field

Asignificant amount of experience and expertise is necessary for accurate wetland delineations. The information in this box provides a general introduction to methods for wetland delineation, and is not a substitute for field experience. The material presented here is mainly drawn from the *Federal Manual,* which should be consulted for more detailed information.

For a routine onsite wetland determination, hydric soils and plant communities at the site are assessed, and potential indicators of wetland hydrology are investigated.

Soils

Hydric soils are soils that are saturated long enough during the growing season to develop anaerobic conditions. These conditions result in a number of chemical changes in the mineral and organic matter in a soil. For example, in anaerobic conditions, chemical reduction takes place, and the solubility of species such as iron is affected. Iron, in particular, becomes more soluble in reducing conditions. When and if the soil becomes reaerated, iron may reprecipitate as brownish-orange mottles (spots or streaks) within the pore spaces of the soil. Thus, mottled soil is one indicator of periodic soil saturation. These processes also give rise to the formation of gley or gleyed soil. Gley is indicated by a bluish- or greenish-grey color in mottles or concretions within the soil, and is caused by the periodic reduction of iron and manganese within the soil.

Other indicators of hydric soils include abundant undecomposed organic matter; odors of hydrogen sulfide (a rotten-egg odor), ammonia, or methane; and low chroma as defined by the Munsell Soil Color Charts (Munsell Color, 1994). The Munsell charts are a standardized set of color chips that are compared to soil samples so that an identification may be made of the hue, value, and chroma of the soil color. Chroma values of 1 or 2 in the Munsell charts are typical of hydric soils.

Vegetation

Vegetation at the site should be studied to determine if a prevalence of the vegetation is hydrophytic. Plants respond in various ways to the characteristics of wetland environments. The *National List of Plant Species That Occur in Wetlands* (Reed, 1988) categorizes plants on the basis of how well they are adapted to conditions in wetlands. In addition to the *National List,* some state or regional lists are available that categorize plants by the same scheme. The local offices of the Soil Conservation Service or the Fish and Wildlife Service may be helpful sources of regional wetland plant lists and other information.

The categorization of these lists is based on the following scheme. Some plants grow only when their roots are in water. These plants, which generally are found only in wetlands, are categorized as "obligate wetland" plants (abbreviated in the lists as OBL). Some plants cannot tolerate the waterlogged soil conditions found, however briefly, in wetlands. These plants generally are found only in uplands, not wetlands, and are categorized as "obligate upland" plants (UPL). Between these two end members is a continuum of plants that can adapt to either condition with varying degrees of success. These plants are called facultative plants. Facultative plants fall into three

(continued)

BOX 11.2 *(continued)*

categories: those found in wetland environments 67% to 99% of the time are "facultative wetland" plants (FACW); those found equally often in wetlands and uplands areas (34% to 66% of the time) are "facultative" plants (FAC); and those found in wetlands less than 33% of the time are "facultative upland" plants (FACU). If a prevalence of the plants at a site are on the OBL, FACW, and/or FAC lists, the site meets the vegetation criterion for a wetland.

Hydrology

The term "wetland hydrology" refers to permanent or periodic saturation of the soil. In a wetland, the soil is saturated long enough or to the degree that anaerobic conditions develop. These anaerobic conditions lead to the development of hydric soils and the success of wetland plants.

To assess whether wetland hydrology exists at a particular site, indicators of present or past soil saturation should be noted. If water is flowing or standing at or above the surface, a wetland condition exists, assuming that the observation is not made during unusual flooding conditions. Soggy, mucky soils underfoot are evidence of saturation. But suppose one visits the wetland during a dry period, when no water is visible, or when water is flooding only a portion of the total wetland. Near the upland boundary of a wetland, in particular, the evidence of inundation may not be obvious. What indicators might reveal the evidence that the area has been inundated in the past?

At various points along a transect or grid, holes should be dug or augered to a depth of 18 inches, and ample time allowed for water to flow into the hole. If it does, the water table is close to the surface, and a wetland condition probably exists. A portion of the soil taken from the hole may be squeezed; if free water can be squeezed out, the soil is saturated. (Heavy clay soils will not respond to these tests in the same way as sandy or organic soils.) Other wetland hydrologic indicators include evidence of rafted debris (e.g., water-deposited debris hanging from branches overhead); siltation or coatings on trees, grasses, and other plants; the development of shallow root systems, which is a plant's attempt to keep its roots above the saturated zone; high water marks on tree trunks, fenceposts, or other objects; and poorly decomposed plant material.

The information presented here should be used as a general guide to the concepts and elementary terminology of wetland delineation. In general, hydrogeologists are not familiar with field identification of plant species. However, hydrophytic vegetation is an important criterion in the delineation of wetlands. As a result, until and unless the required botanical expertise is gained, a hydrogeologist should not attempt to definitively identify a wetland boundary. ∎

Problems in Ground Water Quantity

In some regions, a common hydrogeological problem is exploring for, identifying, and characterizing ground water resources. People in these regions call hydrogeologists with problems such as these: "Our town needs a water supply; where and how deep should we drill, and how much can we pump?" "Why has my well gone dry?" The answers to these questions may emerge after a study of the ground water resources of the region.

Exploring for Ground Water Supplies

A ground water exploration project begins with the collection of regional and local geologic reports, meteorologic and hydrologic reports, well logs, well production records, and other hydrogeological data. This information is interpreted with the goal of identifying hydrostratigraphic units and evaluating potential aquifers.

If no cross-sections or water-level maps are available, it may be possible to use data such as well or boring logs to construct these diagrams. Once they are constructed, hydrostratigraphic units may be defined, and an assessment of aquifer properties may be made. In some situations, it may be useful to perform water-budget calculations. These calculations quantify the inflows (e.g., precipitation or aquifer recharge), outflows (e.g., evapotranspiration, streamflow, or pumpage), and changes in storage of water (e.g., a decline in ground water levels) in a hydro-logic system or in an individual hydrogeologic unit.

When needed, field mapping investigations may be performed. These investigations may involve examining outcrops and other exposures (such as exposures in mines and tunnels), surveying water levels in wells, exploring for springs or other hydrologic features, and surveying or gaging streams or lakes.

In many cases, more site-specific subsurface information is required. When it is cost-effective, and particularly for large or sensitive projects, test borings may be drilled and samples retrieved for testing of hydraulic properties (Fig. 11.3). Aquifer tests such as pumping or slug tests may be performed. Geophysical exploration methods may be used when appropriate.

Entry-level hydrogeologists might be involved in several phases of such a project. They might be asked to compile data from the reports that have already been published, making preliminary conclusions and identifying areas in which further study is needed. They might perform the field work, including field mapping, logging of drilling samples, or helping to conduct geophysical tests. They might be asked to construct cross-sections or water-level maps based on the compiled data. And they might be required to help perform field tests such as pumping or slug tests.

Investigating Declining Well Yield

In some situations, a well that has been performing adequately may experience a decline in well yield, a decrease in productivity, or slowed recovery rates. These may be due to problems with the well itself, such as siltation, biofouling, or clog-ging of the well intake. In these cases, the client should be referred to a well-servicing contractor.

In other cases, however, the cause of the problem is a regional decline in water levels due to overpumping or climatological factors (e.g., a drought), or inter-ference by increased pumping of nearby wells or excavation sumps. If the situation has been developing over time, and if it is pervasive throughout a region, it may be a part of a longer-term problem and may require the identification and development of a new water supply. If this is the case, the methods described under "Exploring for Ground Water Supplies" may be applicable. If the situation came on suddenly, or if it seems to be restricted to just a few wells in an area, the cause may be more local in nature. For example, newly constructed wells, wells where

Fig. 11.3 Test borings give site-specific information.

pumping rate has been increased, or an excavation, mine, or quarry in which the dewatering pumping rate has been increased all might have an effect on water levels in neighboring wells.

Problems in Ground Water Quality

For ground water to be useful, it must be present in abundant enough quantities, and it must be of an appropriate quality for the intended use. Depending on the cause of the problem, when a ground water supply is of poor quality, and no other sources are available and affordable, various treatment options may be employed, including such processes as aeration, water softening, ion exchange, and others. These processes may yield the desired result: water clean enough to use. However, in some cases, treatment is not enough or is not the best option. In these cases, it may be important to identify the cause of the poor water quality, and to remediate it.

Finding the Causes of Poor Quality

Before solutions may be developed for the problem of poor water quality, the cause or causes of the problems must be determined. Potential causes should be examined, and any contaminant sources should be identified, before remedial options can be considered.

Field Data Collection. Field data on ground water, soil, and soil vapor contamination can be collected from soil borings, direct push methods such as the Geoprobe® or Hydropunch® monitoring well installation and sampling, and measurement of field geochemical parameters. The entry-level hydrogeologist might be involved in any of these aspects of a field data collection program. Likewise, she or he might be assigned the tasks of helping to map or analyze the data to identify a contaminant source, discern trends, or delineate a contaminant plume.

Consider the Source. Potential causes of poor water quality may be natural or human-made. In some cases, we may be too quick to jump to the conclusion that a water-quality problem is caused by some kind of contamination, when the real root of the problem is natural. Natural causes of poor water quality may be related to interactions between ground water and mineral or organic material. For example, water may have naturally dissolved from the aquifer high concentrations of salt, iron or other metals, sulfur, or other materials that affect the taste and odor of the water. "Hard" water has high levels of calcium and magnesium, both of which are most likely to be natural constituents of the water. High concentrations of organic acids such as tannic acid may impart an undesirable color or taste to the water. Dissolution of these mineral and organic materials into the ground water may be affected by water temperature, reduction-oxidation conditions, pH, and solubilities, among other factors.

Although natural factors may cause poor water quality, contamination is the result of human activities. This may occur when some contaminant source is introduced into surface or ground water, or when water percolates through a contaminant while recharging the ground water. It also may occur when human activities cause some other undesirable change, such as increasing water temperature or changing the amount of a dissolved gas, such as oxygen, in a water source.

Human-made causes of poor water quality may emanate from point sources or nonpoint sources. Fetter (1993) reviewed the major causes of contamination in the U.S. today. Point sources are singular locations from which the contamination originates. They frequently are easily identified. An overturned rail car or tractor-trailer rig, a wastewater outfall from a manufacturing plant, a septic tank or cesspool, an injection well, a rupture in a chemical pipeline, and a leaking underground storage tank (LUST) all are examples of point sources of contamination. (For more information on LUSTs, see Box 11.3.)

Nonpoint sources, on the other hand, may be more difficult to identify, because they are often pervasive and insidious. Examples of nonpoint sources are agricultural fields onto which fertilizers, pesticides, and other chemicals have been applied; roads, highways, and large parking lots from which deicing salts, oils, and metals run off; urban areas from which stormwater runoff occurs, carrying a wide variety of urban contaminants; and many other sources.

Identifying Contaminant Sources. A quality problem may be discovered during a routine analysis of a routine sample, such as might be performed on water from a monitoring well, or during a periodic test of a drinking water well. For drinking water wells, the first sign of a water quality problem also might be a foul or unusual taste, odor, or color. Once a water quality problem has been discovered, a search for the source of the problem may ensue.

One of the primary steps in analyzing the situation should be to determine what constituent is responsible for the water quality problem. Sometimes the problem is characterized easily, such as an odor of gasoline or a taste of salt, and in these cases, the contaminant source may be fairly obvious. But in other situations, the problem constituent is not so easily identified, and the complaint about the water is that it simply "tastes bad." A chemical analysis of the affected ground water may reveal the problem constituent. However, the water must be analyzed for the particular constituent before it can be discovered that the constituent is present. For example, without performing an analysis for vinyl chloride, it is impossible to say for certain that vinyl chloride is present. A routine analysis for bacteriologic contaminants or major ions simply will not reveal the presence of the vinyl chloride. Once the identity of the problem constituent is known, the concentrations of that constituent may be checked against regulatory standards. If the concentrations exceed the standards, the water is contaminated, according to the regulations.

If the identity of the contaminant is known, the next step in the analysis may be to identify the potential contaminant sources. In a situation in which there is only one possible source of a particular contaminant, the identification may be fairly simple. In other cases, particularly at industrial sites, sites with dense population, or sites with many industrial or agricultural processes occurring in a small area, identifying the source may require some detective work.

Even in a situation in which the source appears to be obvious, it might be important to look a little deeper or do some further checking to see if there might not be more possibilities. For example, if water from a rural homeowner's well tastes salty, and an oil brine injection well is situated a hundred yards away, the injection well might appear to be the obvious source. But might there be other potential sources? Could the contamination come from roadway deicing salts used on nearby highways, or perhaps a poorly constructed septic tank in the homeowner's back yard? Multiple contaminant sources should be kept under consideration until they are convincingly ruled out.

BOX 11.3

Leaking Underground Storage Tanks

Underground storage tanks (USTs) hold gasoline, diesel fuel, jet fuel, heating oil, solvents, dry cleaning fluid, and any number of other chemicals (Fig. 11.4). USTs are a convenient way to store chemicals without taking up valuable space at the land surface.

For example, an ordinary gas station might have three underground tanks that store three grades of gasoline. It also might have a buried tank that holds diesel fuel, and in parts of the country where kerosene space heaters are popular, it might have a buried tank of kerosene. If the station has service bays and provides oil-changing service, it might have a buried tank that holds waste oil which is drained from customers' cars, then is held in the tank for later pickup by a waste oil company. A typical gas station, then, might have three to six USTs. If the tanks were not buried, the gas station would require much more land space. In addition, if the tanks were not buried, they would need thicker walls to prevent potential leaks and explosions caused by vehicles accidently crashing into them. USTs are safer than above-ground tanks in this regard.

Clearly, USTs provide some benefits. However, they pose risks, as well. Those risks include potential contamination of the ground water and soil from a leaking underground storage tank, or LUST. LUSTs are a serious concern in many parts of the United States. Many USTs were installed in the 1940s and 1950s. These tanks were, for the most part, made of steel. They had few, if any, of the leak protection devices, such as double walls and leak monitors, with which new tanks are equipped, nor did they have corrosion inhibitors. After half a century of contact with ground water and soil, the walls of many tanks have corroded, and leaks have formed.

When a chemical (often called "product" or "pure product") leaks from an underground tank, it may contaminate the sediment around the tank, and it may migrate through the sediment or into other nearby geologic materials. The chemical also may migrate into ground water. This is especially likely if the ground water table is high or if the material in which the tank is installed is permeable, or if the volume of chemical released is large.

Even if the sediment has a low hydraulic conductivity, the chemical may migrate. How? The chemical leaking from the tank first flows into the "backfill" around the tank. When the tank was installed, a hole was dug, the tank was placed in the hole, and the rest of the hole was filled, usually with the soil that was taken out as the hole was excavated. This "backfill" is likely to be looser, less dense, and more porous and permeable than the surrounding material, so it may provide somewhat of a conduit for flow even in low-permeability materials.

Further migration may take place. Many tanks have pipes that lead from the tank to some other outlet. For example, at a gas station, pipes lead from the underground gasoline storage tanks to the above-ground gas pumps at each pumping island. The soil around these "distribution lines" is a potential conduit for contamination, because it, too, is likely to be composed of backfill. If the distribution lines are situated close to other underground utilities, such as sewer, water, electrical, or natural gas lines, the contamination may migrate offsite via the backfill around these other utilities. At gas stations, there may be further contamination around the pumps, a result of spillage of gasoline onto the ground during fueling.

(continued)

BOX 11.3 *(continued)*

Site investigations for a potential LUST site should begin by determining the position of the tank. Old site plans should be examined, and site personnel should be consulted. In addition, some tanks may be registered with the office of the local fire marshall, which might maintain records of tank locations. A drilling rig or direct push device such as a Geoprobe® may be used to take samples of soil, water, or gas from locations near the tank. Some tanks may be buried within the foundation of a building, where access for a drilling rig is poor or impossible. In these situations, very small rigs or augers may be used, or even hand augers, where appropriate. Caution should be used during any subsurface drilling or excavation near a buried tank. Old site plans may be faulty, as may be the recollections of the site managers.

Samples taken during subsurface investigations may be scanned with a photoionization detector (PID), organic vapor analyzer (OVA), or other instrument that measures concentration of organic vapors (Fig. 11.5). If the tank held something other than organic chemicals (e.g., acids or bases), then a test appropriate to that material should be used.

In certain cases, a tank might be abandoned in place, by removing all remaining product, cleaning the tank, and filling it with some inert material. This might be the case, for example, if a tank is located within a building foundation, and removal of the tank would cause structural instability. If it is determined that a tank must be removed (a "tank pull," or, more colloquially, a "tank yank"), it must be done according to regulations and with the proper personnel present. For example, it might be required that a representative of the Fire Marshall's office be present during tank removal.

If contamination of soil or ground water has resulted from the leaking tank, a remedial plan might be put into action. This plan might involve steps such as further monitoring, soil excavation and disposal, "pump and treat" operations, bioremediation, or other options. Some remediation options are discussed elsewhere in this chapter.

LUST projects often are regulated by local, state, and/or federal statutes. Regulations specify how the site must be investigated, how the tank must be removed, how the contamination must be remediated, and what monitoring must take place. For example, regulations on monitoring might specify the number of wells that must be installed, their locations, their depths, the lengths of their screened intervals, the sampling frequency, and the constituents that must be monitored. Regulations also specify what forms and reports must be filed with which agency, and at what time intervals. It is important to be aware of which regulatory agency has jurisdiction over a site, so that the appropriate steps may be taken as the investigation and remediation proceed. ■

Contaminant sources may not be obvious if they are buried. For example, old leaking underground storage tanks may exist entirely unbeknownst to the current property owner. These objects sometimes may be located by examination of documents such as old site plans, historical aerial photographs, or fire insurance maps; by interviews with long-time residents of an area; or by using geophysical methods such as ground-penetrating radar (GPR).

Demonstrating that a particular source is responsible for causing a problem, or conversely, demonstrating that it is unlikely to be the cause, depends on the

Fig. 11.4   A leaking underground storage tank is removed.

hydrogeologist's investigation of two aspects of the ground water system: water flow directions and water chemistry. Flow directions may be interpreted from well-water level data and in some cases, surface-water level data, in combination with subsurface hydrogeologic data. Variations in flow direction may occur with variations in precipitation, stream discharge, pumping of nearby wells, and many other factors. These variations should be explored before definitively pinpointing a contaminant source.

Fig. 11.5   An organic vapor analyzer is used to measure vapors in soil samples taken from a LUST excavation.

Investigations of water chemistry might include sampling from several wells or surface water bodies located along the flow path, performing chemical analyses, and attempting to demonstrate a chemical similarity between the source and the contaminated well water. Chemically "fingerprinting" a contaminant source may be a challenging task for two main reasons. First, the concentration of the contaminant at the contaminated well may be anywhere from tens of times to billions of times lower than the concentration at the contaminant source. Second, chemical changes may take place as the contaminant moves along the flow path. These changes may include such processes as precipitation of salts or dissolution of aquifer material, hydrolysis, dissolution of a pure liquid into the water, acid-base reactions, reactions with soil gases, cation exchange on clay minerals, oxidation or reduction processes, biodegradation, volatilization, or radioactive decay. As a result, the chemical may not even be present in its original form once it reaches the affected well. A knowledge of chemical processes and chemical conditions within the aquifer will help the hydrogeologist analyze the situation.

Despite these difficulties, it sometimes is possible to "fingerprint" a contaminant and identify the fingerprints when they appear in water from an affected well. This fingerprinting usually is done by relying on ratios of various chemicals to each other. For example, when seawater intrudes into a freshwater aquifer, the resulting water is a mixture of both the freshwater and the salt water. As a result, the concentrations of sodium chloride are greatly reduced in the mixture. However, the *ratio* of sodium to chloride in the mixture is likely to approximate the *ratio* in the seawater, thereby providing something of a fingerprint that can be used in delineating the extent of the seawater intrusion.

As another example, consider gasoline, which is made of many different components. When gasoline is diluted in ground water, the actual concentrations of the components may decrease by a factor of tens of thousands or more. Nevertheless, if a gas chromatogram (see Chapter 9) of the diluted gasoline is compared to a gas chromatogram of the gasoline source, the two may appear very similar. The "fingerprint" may show peaks at the same points, because the composition is similar, even though the concentrations will be different. If some of the chemical changes previously listed have occurred, differences may appear. In the example of gasoline, volatilization, chemical degradation, or biological degradation might be likely to occur, depending on the age of the contamination event and other environmental factors. Chemical fingerprints must be interpreted with this in mind.

Mapping Contaminant Plumes.   An entry-level hydrogeologist might be asked to assist with the construction of a map of a contaminant plume by using data from a field and analytical investigation. A contaminant plume map is an attempt to portray three-dimensional information on a two-dimensional surface. As a result, some difficulties are likely to emerge during the process of constructing the map. Particular attention should be paid to well construction (particularly the depth and length of the well intake), sampling protocol, and analytical quality. More suggestions for producing a map of water quality data appear in Chapter 10.

Remediating a Water-Quality Problem

Once a contamination problem has been identified, remedial options may be designed. The entry-level hydrogeologist might be assigned the tasks of helping to implement a remedial plan or of monitoring the progress of the remediation activities once they are put into place.

Before a remediation project is initiated, steps should be taken to ensure that the contaminant source is removed, isolated, or contained, so that no more contaminant will be released. It might also be appropriate to install barriers to flow such as slurry walls or cutoff walls, which physically block movement of contaminants, or to create a hydrodynamic barrier by pumping and/or injecting water into wells strategically placed such that they isolate the contaminant from public water supplies. When it is impossible or is not cost-effective to remove the contaminant source altogether, as it might be in the case of a large waste pile, further introduction of the contaminants into an aquifer may be minimized by capping the waste with a barrier such as a clay cover. Other methods for source isolation and containment exist. For example, in some cases, contaminants may be "frozen" in place by in situ vitrification, which essentially involves passing a strong electrical current through the formation, partially melting it and turning it to a sort of glass.

Remediation options are many and varied, and new options are continually being devised and tested. Beginning hydrogeologists who expect to work in the area of remediation will do well to stay up to date on new remedial methods. This can be accomplished by reading research and trade journals, by attending professional conferences, by talking with other hydrogeologists, and by being aware of the status of environmental regulations. More information is available in the journal *Ground Water Monitoring and Remediation*, and sources such as Nyer (1992), Fetter (1993), National Research Council (1994), Nyer and colleagues (1996), and various documents of the U.S. Environmental Protection Agency.

Selection of a remedial method is based to a large extent on the cost and efficacy of the method. This in turn depends on factors such as the type of contaminant and its physical and chemical characteristics, the degree and extent of contamination, the hydrogeologic setting, and the toxicity and routes of exposure of the contaminant. In addition, the final design selection will be based on practical considerations such as the need for space for treatment devices, the need to maintain business operations during the remediation, and cost effectiveness. Finally, regulatory approval of the plan will be needed; in most cases, remediation designs must be approved by a regulatory agency such as the state or federal EPA. A few remediation designs are discussed briefly here.

Natural Attenuation. This option involves taking no action to remediate the contaminants, and simply allowing them to naturally attenuate over time. Monitoring wells should be installed and sampled regularly to evaluate the efficacy of the attenuation. This plan might be appropriate if there is little or no health or exposure risk, and if natural attenuation can be shown to be as effective as another remediation plan.

Excavation and Disposal. Contaminated soil may be remediated by excavating the contaminated portion and hauling it to an appropriate disposal site. If the depth and extent of the contamination is great, this may not be a cost-effective method. This method does not remediate contaminated ground water, but is limited to the vadose zone.

Pump and Treat. Pump and treat operations involve pumping contaminated water out of the ground, and treating it once it has been pumped out. Onsite treatment devices such as air strippers and filters are commonly employed for this purpose. In some cases, the treated water may be reinjected into the ground. Despite initial enthusiasm over this technology, the effectiveness of pump and treat operations has been questioned in recent years. The initial reduction in the concentration of contaminants may be great when this method is used. However, it often is difficult to remove the final fraction of the contaminant by pumping, especially when the residual saturation of the contaminant is such that the contaminant clings tightly to aquifer grains despite the pumping. In addition, the method may be ineffective in low-conductivity aquifers, or may be required to operate for extraordinarily long periods of time in order to be effective. As a result, pump and treat operations are sometimes used in combination or in sequence with other methods that are more effective at cleaning the residue.

Bioremediation. Bioremediation involves the cultivation of microorganisms that work in several ways to biodegrade the contaminants, reducing their concentration. In general, this method involves producing a combination of environmental factors that optimize the work of the bacteria. These factors include such things as nutrients, temperature, redox conditions, moisture conditions, and cometabolites (other compounds, in the presence of which the bacteria perform the desired tasks).

Soil Vapor Extraction and In Situ Air Sparging. Soil vapor extraction involves extracting volatile and some semivolatile compounds from the vadose zone by installing vapor extraction wells and placing them under a vacuum. The compounds are volatilized, move into the well, and are pulled to the surface, where they may be treated or disposed of. In situ air sparging involves injecting air into the ground to a depth that is below the contaminants. The air travels upward through the contaminants, volatilizing them, and then the air with the volatilized contaminants is extracted by a soil vapor extraction system. These systems are effective only at extracting chemicals that are fairly easily volatilized, for example, those present in gasoline spills.

Hydrogeology and Engineering Projects

The topics previously discussed relate to the abundance of supply and the chemical quality of ground water. But hydrogeological methods may apply in a number of other situations as well. This section outlines a few of these applications in - geotechnical engineering and mining.

Hydrogeology is important in many phases of a dam-building project. In such a project, pre- and postconstruction water budgets can be developed on the basis of meteorologic, climatologic, hydrologic, and hydrogeologic data. Collection of hydrogeologic data and identification of permeability and flow systems are important components of dam and reservoir engineering. Hydrogeology is considered in determining how the reservoir water levels will affect ground water levels and water levels in wells in the surrounding area, and in determining if the proposed reservoir will hold water or leak.

Stability of the dam and reservoir depends to a great extent on hydraulic conditions. Because pore water pressure affects the stability of earthen dams as well as concrete dams, hydraulic heads under and about any dam must be estimated as part of the design process, and flow nets must be developed for various scenarios of dam design and operation. At some dams, hydraulic head is monitored postconstruction as well. Hydraulic head values are a critical consideration during any engineered rapid drawdown of the reservoir, such as might be performed in order to make repairs on the structures. In addition to the stability of the dam itself, stability of the reservoir slopes is related to hydrogeology and must be considered during the planning and design phases. Hydrogeologists have input into all of these portions of the project. Entry-level hydrogeologists might participate particularly in the early phases of the project, gathering and compiling data in the office and in the field.

Foundation engineering is another area in which hydrogeologic considerations may be critical. Water affects the bearing capacity of a soil, and therefore is important in foundation design. For example, uplift pressures on a slab foundation may cause failure of the foundation, so these pressures must be anticipated in advance of construction. Pore water pressures should be considered in design of sheet pilings and retaining walls as well.

Construction operations of many kinds may involve ground water. In fact, very large construction companies may employ a few hydrogeologists on their permanent staffs. If a high water table exists at a construction site, then during the construction phase, dewatering may be necessary. Dewatering drains water from an area in order to make construction simpler and to help stabilize excavations. Dewatering often involves installing shallow well points in a linear series, attaching them all to the same header, and pumping out the ground water so that construction may proceed. Longer-term dewatering may be particularly important in zones of expansive soils, which swell when saturated and shrink when dried. In the construction of sanitary landfills, major concerns are the hydraulic conductivity of the liner material and the proximity of ground water. In addition, leachate drainage systems, and sometimes leachate treatment facilities, must be designed. Long-term monitoring programs must be constructed as well. All of these operations require the use of hydrogeologic principles.

The engineering of roadways must consider hydrogeologic conditions. A roadway must be adequately drained so that ground water seepage does not damage the roadway subbase and base courses. Installing drains below the roadway surface helps to control this problem. Draining surface water away from a roadway is important to prevent flooding during precipitation events and to prevent icing in cold weather. Design of this drainage should take ground water conditions into consideration.

In addition to roadway drainage, the stability of slopes along the roadway depends in part on ground water conditions. When a roadway is constructed by excavating through rock or sediment, the resulting slope must be engineered to be stable. Whether the slope is of rock or of sediment, ground water can contribute to slope instability, causing hazardous conditions. Installing drains can help prevent this problem, and hydrogeologic information can help facilitate the drainage design. Hydrogeology might also be used to determine what effect this drainage could have on nearby water supply wells.

Water quality also should be considered in the engineering of roadways. Water drained from roadways may contain deicing salts, metals from car and truck parts, oils and greases that drip from the vehicles, and components of rubber tires and brakes that wear away with use. This drainage should be dealt with in a way which does not compromise nearby water supplies.

Hydrogeologic conditions must be considered during the construction and operation of tunnels. Ground water may seep into a tunnel and must be drained appropriately. Tunnel wall stability might be significantly affected by pore water pressure from the seepage.

The engineering of deep mines and surface mines must consider hydrogeologic conditions. When present, water must be pumped out of working mines in order to allow work to proceed and to help prevent stability problems. Draining the mine efficiently might be a hydrogeological challenge in itself. However, in addition, draining the mine is likely to affect ground water levels in nearby water wells. This might disrupt or even destroy the usefulness of water-supply wells. Hydrogeologists might be involved in predicting these effects and remedying the situation by designing appropriate pumping schemes or locating new water supplies for the affected property owners.

The development of a mine might have an effect on ground water quality, as well. Mining brings rock or sediment materials to the surface, exposing fresh mineral faces to precipitation and surface runoff. Water draining from the mine may percolate through this fresh material, leaching it and picking up undesirable constituents. This might cause degradation of the quality of the water. For example, acid mine drainage is a water-quality problem that results from the mining of coal and some other materials. Water used in processing of mined material and spoil material may be significantly degraded in quality, and should be treated and disposed of properly.

Hydrogeologists might be employed for any of the projects described here, and more. An entry-level geologist is not likely to be put in charge of an engineering project. However, she or he might participate as a team member under the direction of a more experienced hydrogeologist. A new hydrogeologist might particularly be assigned tasks in the data-collection phase, including document review and field investigations, as well as the report-preparation phase of the project.

Ground Water Modeling

Ground water modeling is the use of physical, analog, analytical, or numerical models to predict the response of an aquifer or ground water system to various stresses. Ground water models help hydrogeologists predict what might happen to ground water levels or ground water chemistry in various scenarios. Today, the term "ground water modeling" is used almost exclusively to refer to numerical models, and the calculations usually are performed with the aid of a computer. Using a computer makes it easy for the investigator to alter variables and to make rapid assessments of a large number of potential scenarios.

Hydrogeologists investigating ground water quantity problems might employ models in various situations. For example, a model might be constructed to predict what would happen to water levels in an area if the pumping rate of a production well were to be increased. As another example, when a reservoir is to be constructed, the planners can use a model to predict how the changes in surface hydrology might affect ground water levels in the area. Quarries and subsurface mines that intersect water-bearing aquifers must be dewatered by pumping; ground water models can predict how this pumping will affect nearby wells, lakes, and wetlands.

Ground water models might be used to investigate problems related to ground water chemistry or contamination as well. For example, models can help to delineate the wellhead protection area around a public water-supply well. In a wellhead protection area, certain activities (e.g., waste disposal) might potentially cause contamination of the water supply, and communities might wish to delineate a zone within which these activities are not permitted. But how big should the protected area be? Modeling is one tool for finding an answer to this question. In other situations, ground water contamination might already be known to exist. In these cases, a model might be able to predict how long it will take for contaminated water to reach a water-supply well. Models can be applied to problems in ground water remediation as well. When contaminated water is to be pumped out of an aquifer as part of a remedial effort, a model might be used to predict how long pumping must continue in order to recover the contaminants.

Ground water models are mathematical representations of natural systems. Different modeling schemes are based on different assumptions and mathematical formulations. Depending on the characteristics of a given natural system, some modeling schemes may be better-suited to a situation than others. Even when well-suited to a particular field situation, any model is only as good as the data used to construct it. As a result, modeling is not a substitute for hydrogeological field work, but instead, relies on field work to provide input data. During the processes of model construction, calibration, and verification, field data on aquifer properties, recharge rates, water levels, and boundary conditions must be known or estimated with a reasonable degree of precision. Models that involve ground water chemistry also require the user to develop input data on the properties and behavior of the chemicals of interest in the subsurface. An entry-level employee might participate in the gathering of this essential input data.

Useful Formulas and Conversion Factors

Formulas for Geometric Figures
Area

$$\text{Area of a circle} = \pi \times (\text{radius})^2$$

$$\text{Area of a rectangle} = \text{base} \times \text{height}$$

$$\text{Area of a triangle} = 1/2(\text{base}) \times \text{height}$$

$$\text{Area of a trapezoid} = 1/2(\text{base} + \text{top}) \times \text{height}$$

$$\text{Area of a parallelogram} = \text{base} \times \text{height}$$

Volume

$$\text{Volume of a cylinder} = \pi \times (\text{radius})^2 \times \text{height}$$

$$\text{Volume of a sphere} = 4/3 \times \pi \times (\text{radius})^3$$

$$\text{Volume of a rectangular solid} = \text{length} \times \text{width} \times \text{height}$$

$$\text{Volume of a cone} = 1/3 \times \text{area of base} \times \text{height}$$

Conversion Factors

See the conversion tables for *Length*, *Area*, *Volume, Time, Flow* and *Length/Time Measurements* on the following pages.

Length

	mm	cm	m	km	in.	ft	mile
mm	1	0.1	0.001	1×10^{-6}	0.0394	3.28×10^{-3}	6.21×10^{-7}
cm	10	1	0.01	1×10^{-5}	0.394	0.0328	6.21×10^{-6}
m	1000	100	1	0.001	39.4	3.28	6.21×10^{-4}
km	1×10^{6}	1×10^{5}	1000	1	39400	3280	0.621
in.	25.4	2.54	0.0254	2.54×10^{-5}	1	0.0833	1.58×10^{-5}
ft	305	30.5	0.305	3.05×10^{-4}	12	1	1.89×10^{-4}
mi	1.61×10^{6}	1.61×10^{5}	1610	1.61	63400	5280	1

Area

	mm^2	cm^2	m^2	km^2	$in.^2$	ft^2	mi^2	acres
mm^2	1	0.01	1×10^{-6}	1×10^{-12}	1.55×10^{-3}	1.08×10^{-5}	3.86×10^{-13}	2.47×10^{-10}
cm^2	100	1	1×10^{-4}	1×10^{-10}	0.155	1.08×10^{-3}	3.86×10^{-11}	2.47×10^{-8}
m^2	1×10^{6}	1×10^{4}	1	1×10^{-6}	1550	10.8	3.86×10^{-7}	2.47×10^{-4}
km^2	1×10^{12}	1×10^{10}	1×10^{6}	1	1.55×10^{9}	1.08×10^{7}	3.86×10^{-3}	247
$in.^2$	645	6.45	6.45×10^{-4}	6.45×10^{-10}	1	6.94×10^{-3}	2.49×10^{-10}	1.59×10^{-7}
ft^2	9.29×10^{4}	929	9.29×10^{-2}	9.29×10^{-8}	144	1	3.59×10^{-8}	2.30×10^{-5}
mi^2	2.59×10^{12}	2.59×10^{10}	2.59×10^{6}	259	4.01×10^{9}	2.79×10^{7}	1	640
acres	4.05×10^{9}	4.05×10^{7}	4050	4.05×10^{-3}	6.27×10^{6}	43560	1.56×10^{-3}	1

Volume

	milliliters	liters	m³	in.³	ft³	yards³	gallons	acre-ft
milliliters	1	0.001	1×10^{-6}	0.0610	3.53×10^{-5}	1.31×10^{-6}	2.64×10^{-4}	8.11×10^{-10}
liters	1000	1	1×10^{-3}	61.0	0.0353	1.31×10^{-3}	0.264	8.11×10^{-7}
m³	1×10^{6}	1000	1	61,000	35.3	1.31	264	8.11×10^{-4}
in.³	16.4	0.0164	1.64×10^{-5}	1	5.79×10^{-4}	2.14×10^{-5}	4.33×10^{-3}	1.33×10^{-8}
ft³	2.83×10^{4}	28.3	0.0283	1.73×10^{3}	1	0.0370	7.48	2.30×10^{-5}
yards³	7.65×10^{5}	765	0.765	4.67×10^{4}	27	1	202	6.20×10^{-4}
gallons	3790	3.79	3.79×10^{-3}	231	0.134	4.95×10^{-3}	1	3.07×10^{-6}
acre-ft	1.23×10^{9}	1.23×10^{6}	1230	7.53×10^{7}	43,560	1610	3.26×10^{5}	1

Time

	seconds	minutes	hours	days	years
seconds	1	0.0167	2.78×10^{-4}	1.16×10^{-5}	3.17×10^{-8}
minutes	60	1	0.0167	6.94×10^{-4}	1.90×10^{-6}
hours	3600	60	1	0.0417	1.14×10^{-4}
days	86400	1440	24	1	2.74×10^{-3}
years	3.16×10^{7}	5.26×10^{5}	8770	365	1

Flow

	liters/sec	liters/day	m³/sec	m³/day	ft³/sec	ft³/day	acre-ft/sec	acre-ft/day	gal/sec	gal/min	gal/day
liters/sec	1	86,400	0.001	86.4	0.0353	3050	8.11×10^{-7}	0.0700	0.264	15.8	2.28×10^4
liters/day	1.16×10^{-5}	1	1.16×10^{-8}	0.001	4.09×10^{-7}	0.0353	9.38×10^{-12}	8.11×10^{-7}	3.06×10^{-6}	1.83×10^{-4}	0.264
m³/sec	1000	8.64×10^7	1	86,400	35.3	3.05×10^6	8.11×10^{-4}	70.0	264	15,800	2.28×10^7
m³/day	0.0116	1000	1.16×10^{-5}	1	4.09×10^{-4}	35.3	9.38×10^{-9}	8.11×10^{-4}	3.06×10^{-3}	0.183	264
ft³/sec	28.3	2.45×10^6	0.0283	2447	1	86,400	2.30×10^{-5}	1.98	7.48	449	6.46×10^5
ft³/day	3.28×10^{-4}	28.3	3.28×10^{-7}	0.0283	1.16×10^{-5}	1	2.66×10^{-10}	2.30×10^{-5}	8.66×10^{-5}	5.19×10^{-3}	7.48
acre-ft/sec	1.23×10^6	1.07×10^{11}	1230	1.07×10^8	43,560	3.76×10^9	1	86,400	3.26×10^5	1.95×10^7	2.82×10^{10}
acre-ft/day	14.3	1.23×10^6	0.0143	1230	0.504	43,560	1.16×10^{-5}	1	3.77	226	3.26×10^5
gal/sec	3.79	3.27×10^5	3.79×10^{-3}	327	0.134	11,550	3.07×10^{-6}	0.265	1	60	86,400
gal/min	0.0631	5450	6.31×10^{-5}	5.45	2.23×10^{-3}	193	5.12×10^{-8}	4.42×10^{-3}	0.0167	1	1440
gal/day	4.38×10^{-5}	3.79	4.38×10^{-8}	3.79×10^{-3}	1.55×10^{-6}	0.134	3.55×10^{-11}	3.07×10^{-6}	1.16×10^{-5}	6.94×10^{-4}	1

Length/Time Measurements*

	cm/sec	m/sec	ft/sec	km/hr	miles/hr	cm/day	m/day
cm/sec	1	0.01	0.0328	0.0360	0.0224	86400	864
m/sec	100	1	3.28	3.60	2.24	8.64×10^6	86400
ft/sec	30.5	0.305	1	1.10	0.682	2.63×10^6	26300
km/hr	27.8	0.278	0.911	1	0.621	2.40×10^6	24000
miles/hr	44.7	0.447	1.47	1.61	1	3.86×10^5	38600
cm/day	1.16×10^{-5}	1.16×10^{-7}	3.80×10^{-7}	4.17×10^{-7}	2.59×10^{-7}	1	0.01
m/day	0.00116	1.16×10^{-5}	3.80×10^{-5}	4.17×10^{-5}	2.59×10^{-5}	100	1
km/day	1.16	0.0116	0.0380	0.417	0.0259	1×10^5	1000
ft/day	3.53×10^{-4}	3.53×10^{-6}	1.16×10^{-5}	1.27×10^{-5}	7.89×10^{-6}	30.48	0.3048
miles/day	1.86	0.0186	0.0611	0.067	0.0417	1.61×10^5	1610
m/year	3.17×10^{-6}	3.17×10^{-8}	1.04×10^{-7}	1.14×10^{-7}	7.09×10^{-8}	0.274	0.0274
ft/year	9.66×10^{-7}	9.66×10^{-9}	3.17×10^{-8}	3.48×10^{-8}	2.16×10^{-8}	0.0834	8.34×10^{-4}
gal/day/ft²	4.72×10^{-5}	4.72×10^{-7}	1.55×10^{-6}	1.70×10^{-6}	1.06×10^{-6}	4.07	0.0407

*This table gives conversion factors for units of length per time [L]/[T]. In hydrogeology, several important quantities are expressed in these units, including flow velocity, specific discharge, and hydraulic conductivity. Although all of these quantities are expressed in the same units, the quantities themselves have very different meanings and are not interchangeable. Although in the strictest terms, the unit gal/day/ft² could be used to express a velocity, it is generally used only to express a value of hydraulic conductivity.

(continued)

Length/Time Measurements *(continued)*

	km/day	ft/day	miles/day	m/year	ft/year	gal/day/ft²
cm/sec	0.864	2830	0.537	3.16×10^5	1.04×10^6	21200
m/sec	86.4	2.83×10^5	53.7	3.16×10^7	1.04×10^8	2.12×10^6
ft/sec	26.3	86400	16.4	9.62×10^6	3.16×10^7	6.46×10^5
km/hr	24	78700	14.9	8.77×10^6	2.88×10^7	5.89×10^5
miles/hr	38.6	1.27×10^5	24	1.41×10^7	4.63×10^7	9.48×10^5
cm/day	1×10^{-5}	0.0328	6.21×10^{-6}	3.65	12	0.245
m/day	0.001	3.28	6.21×10^{-4}	365	1200	24.5
km/day	1	3280	0.621	3.65×10^5	1.20×10^6	24500
ft/day	3.05×10^{-4}	1	1.89×10^{-4}	111	365	7.48
miles/day	1.61	5280	1	5.88×10^5	1.93×10^6	39500
m/year	2.74×10^{-6}	8.98×10^{-3}	1.70×10^{-6}	1	3.28	0.0672
ft/year	8.34×10^{-7}	2.74×10^{-3}	5.19×10^{-7}	0.305	1	0.0205
gal/day/ft²	4.07×10^{-5}	0.134	2.53×10^{-5}	14.9	48.8	1

APPENDIX 2
Dimensional Analysis

Dimensional analysis refers to the inclusion of dimensions, or units of measurement, in a calculation. Dimensional analysis is important in hydrogeologic problem solving because it helps to ensure that the calculation is done correctly.

Most measurements made in hydrogeology are expressed in the dimensions of length [L] or time [T]. Various combinations of these are possible. For example, measurements of area are expressed in units of length squared, or $[L]^2$. Discharge measurements are expressed in units of length cubed per time, or $[L]^3/[T]$. Dimensional analysis essentially means writing the units of measurement at every step of a calculation, canceling them out where appropriate, including conversion factors where appropriate, and checking the units of the final answer to see that they are consistent with the required quantity.

The easiest way to understand dimensional analysis is to examine an example calculation. For this example, let us suppose that we wish to calculate the discharge through a pipe in gallons per minute when the radius of the pipe is known and it is known that water is flowing at a given velocity. For this example, let us assume that the radius of the pipe is 2.0 inches and the velocity of flow is 0.60 foot per second.

The first step in solving this problem using dimensional analysis is to set up the formula for discharge:

$$Q = V \times A$$

where Q = discharge, which is expressed in the dimensions $[L]^3/[T]$

V = velocity, which is expressed in the dimensions $[L]/[T]$

A = area, which is calculated as $A = \pi \times (radius)^2$ and is expressed in the dimensions $[L]^2$

The second step is to fill the known values into the formula:

$$Q = \frac{0.60 \text{ foot}}{\text{second}} \times \pi \times (2.0 \text{ inches})^2$$

Note that in setting up this equation, it is most helpful to write the quantities by using a horizontal line to indicate division, rather than a slash mark or diagonal line. A little practice with the method will soon make clear that this greatly facilitates the computations.

365

The third step is to perform the calculation and check to see that the dimensions are appropriate:

$$Q = \frac{7.5 \text{ feet} \times (\text{inches})^2}{\text{second}}$$

The dimensions are

$$\frac{[L] \times [L]^2}{[T]} \text{ or } \frac{[L]^3}{[T]}$$

These are the appropriate units for discharge. This tells us that the equation probably has been set up correctly. However, the work is not done yet.

The fourth step is to check to see that the units of the answer are easy to understand and appropriate to the specific situation. In this case, they are not. This is due to two reasons: first, the units are

$$\frac{\text{feet} \times (\text{inches})^2}{\text{second}}$$

This set of units is not easy to understand in practical terms because it is not internally consistent. But in addition, the problem asks that the discharge be expressed in units of gallons per minute. To achieve this goal, we will have to do some further calculating, using more dimensional analysis.

The next step is to return to the equation with the given values and to do further work on it to yield the desired result. In this case, we must convert the units of feet $\times$ (inches)^2 to gallons. In addition, we must convert seconds to minutes.

In dimensional analysis, conversions are done by using identities. An identity is a definition of one unit in terms of another. For example, perhaps it is necessary to convert a length measurement of 3 feet into units of inches. The identity that should be used is 1 foot = 12 inches. This identity would be set up in terms of a fraction in which the numerator (or top of the fraction) equals the denominator (or bottom of the fraction). When the numerator equals the denominator, the value of the fraction equals one. And when any quantity is multiplied by one, the result equals the original quantity. In the feet vs. inches example, using dimensional analysis, the conversion would appear this way:

$$\text{length} = 3 \text{ feet} \times \frac{12 \text{ inches}}{1 \text{ foot}}$$

$$= 36 \text{ inches}$$

It is important to note a few things in this simple example. First, note that the value of the fraction is one, so the value of the length measurement is not being changed by multiplying it by the fraction. Second, note that the answer is expressed in terms of inches only because the units of feet in the numerator and the denominator cancel each other out. This is key because it helps us determine how to write the fraction: We write it so that the unwanted units are canceled out.

Returning now to the original example, in which we attempted to calculate discharge in gallons per minute, we can apply the use of identities in the form of fractions which equal one. Using the identities 1 foot = 12 inches, 1 cubic foot = 7.48 gallons, and 1 minute = 60 seconds, the equation would appear this way:

$$Q = \frac{7.5 \text{ feet} \times (\text{inches})^2}{\text{second}} \times \frac{1 \text{ foot}}{12 \text{ inches}} \times \frac{1 \text{ foot}}{12 \text{ inches}}$$

$$\times \frac{7.48 \text{ gallons}}{1 \text{ foot}^3} \times \frac{60 \text{ seconds}}{1 \text{ minute}}$$

$$= \frac{23 \text{ gallons}}{\text{minute}}$$

Note that the identity 1 foot = 12 inches had to be used twice, because it was necessary to cancel out the units of inches squared (not simply inches).

The final step in dimensional analysis is to check it. Once again, we should check to be sure that the units are appropriate to the quantity being measured. In this case, they are units of $[L]^3/[T]$, which as we have already determined is appropriate for a discharge measurement. In addition, we should check that the units of our answer are the units required in the problem. In this case, the problem asked for units of gallons per minute, which is what we have determined. And finally, we should check the identity factors (conversion factors) to ensure that they have been written correctly. It is surprisingly easy to err by writing an incorrect factor such as "1 inch = 12 feet" or "7.48 ft^3 = 1 gallon" while performing a dimensional analysis.

Following these simple steps will greatly reduce the possibility of errors in computations. In addition, using dimensional analysis makes it simpler to explain to others exactly what has been done in a calculation.

Values of the Well Function $W(u)$ for Various Values of u

Values of the well function, $W(u)$, are calculated based on the following equation:

$$W(u) = -0.5772 - \ln u + u - \frac{u^2}{2.2!} + \frac{u^3}{3.3!} - \frac{u^4}{4.4!} + \ldots$$

See the table of calculated values of the well function on the following page.

u	W(u)	u	W(u)
1×10^{-7}	15.54	1×10^{-3}	6.33
2	14.85	2	5.64
3	14.44	3	5.23
4	14.15	4	4.95
5	13.93	5	4.73
6	13.75	6	4.54
7	13.59	7	4.39
8	13.36	8	4.26
9	13.34	9	4.14
1×10^{-6}	13.24	1×10^{-2}	4.04
2	12.55	2	3.35
3	12.14	3	2.96
4	11.85	4	2.68
5	11.63	5	2.47
6	11.45	6	2.30
7	11.29	7	2.15
8	11.16	8	2.03
9	11.04	9	1.92
1×10^{-5}	10.94	1×10^{-1}	1.82
2	10.24	2	1.223
3	9.84	3	0.906
4	9.55	4	0.702
5	9.33	5	0.560
6	9.14	6	0.454
7	8.99	7	0.374
8	8.86	8	0.311
9	8.74	9	0.260
1×10^{-4}	8.63	1×10^{0}	0.219
2	7.94	2	0.0489
3	7.53	3	0.0131
4	7.25	4	0.00380
5	7.02		
6	6.84		
7	6.69		
8	6.55		
9	6.44		

References

Aller, Linda, Truman W. Bennett, Glen Hackett, Rebecca J. Petty, Jay H. Lehr, Helen Sedoris, David M. Nielsen, and Jane E. Denne. (1989). *Handbook of Suggested Practices for the Design and Installation of Ground-Water Monitoring Wells*. Dublin, OH: National Ground Water Association.

American Public Health Association. (1995). *Standard Methods for the Examination of Water and Wastewater*, 19th ed. Washington, DC: American Public Health Association, American Water Works Association, and Water Environment Federation.

American Society for Testing and Materials. The ASTM publishes numerous standard practices for hydrogeological investigations, many of which are cited by number in this text. Standards are reviewed and revised regularly. A complete list of current standards may be obtained by contacting the ASTM at 100 Barr Harbor Drive, West Conshohocken, Pennsylvania 19428-2959 or by visiting the ASTM World Wide Web site at www.astm.org.

Amoozegar, Aziz. (1989). A compact constant-head permeameter for measuring saturated hydraulic conductivity of the vadose zone. *Soil Science Society of America Journal* 53:1356–1361.

Australian Drilling Industry Training Committee, Ltd. (1997). *Drilling: The Manual of Methods, Applications, and Management*. Boca Raton, FL: Lewis.

Ballestero, Thomas, Beverly Herzog, O. D. Evans, and Glenn Thompson. (1991). Monitoring and sampling the vadose zone. In *Practical Handbook of Ground-Water Monitoring*, ed. David M. Nielsen. Chelsea, MI: Lewis, pp. 97–141.

Barcelona, Michael J., James P. Gibb, John A. Helfrich, and Edward E. Garske. (1985). Practical Guide for Ground-Water Sampling, ISWS Contract Report 374, EP 1.23/2:600/2-85/104. Champaign, IL: Illinois State Water Survey.

Barcelona, Michael J., James P. Gibb, and Robin A. Miller. (1983). A Guide to the Selection of Materials for Monitoring Well Construction and Ground-Water Sampling, ISWS Contract Report 327, EP-600/52-84-024. Champaign, IL: Illinois State Water Survey.

Barcelona, Michael J., H. Allen Wehrmann, and Mark D. Varljen. (1994). Reproducible well-purging procedures and VOC stabilization criteria for ground-water sampling. *Ground Water* 32:12–22.

Berner, Elizabeth K., and Robert A Berner. (1996). *Global Environment: Water, Air, and Geochemical Cycles*. Upper Saddle River, NJ: Prentice Hall.

Brassington, Richard. (1988). *Field Hydrogeology*. New York: John Wiley

Buchanan, Thomas J., and William P. Somers. (1968). *Stage Measurement at Gaging Stations*. Techniques of Water-Resources Investigations of the U.S. Geological Survey, Book 3, Chapter A7. Washington, DC: U.S. Geological Survey.

———. (1969). *Discharge Measurements at Gaging Stations*. Techniques of Water-Resources Investigations of the U.S. Geological Survey, Book 3, Chapter A8. Washington, DC: U.S. Geological Survey.

Burger, H. Robert. (1992). *Exploration Geophysics of the Shallow Subsurface*. Englewood Cliffs, NJ: Prentice Hall

Campbell, Michael D., and Jay H. Lehr. (1973). *Water Well Technology*. New York: McGraw-Hill .

Cartwright, Keros, Robert A. Griffin, and Robert H. Gilkeson. (1977). Migration of landfill leachate through glacial tills. *Ground Water* 15:294–305.

Cedergren, Harry R. (1988). *Seepage, Drainage, and Flow Nets*, 3rd ed. New York: John Wiley.

Chow, Ven Te. (1959). *Open-Channel Hydraulics*. New York: McGraw-Hill.

Clark, Lewis. (1988). *The Field Guide to Water Wells and Boreholes*. New York: John Wiley.

Colangelo, Robert V. (1991). *Buyer Be(A)Ware: The Fundamentals of Environmental Property Assessments*. Dublin, OH: National Ground Water Association.

Compton, Robert R. (1985). *Geology in the Field*. New York: John Wiley.

Cowardin, Lewis M., Virginia Carter, Francis C. Golet, and Edward T. Laroe. (1979). *Classification of wetlands and deepwater habitats of the United States*, FWS/OBS-79/31. Washington DC: U.S. Fish and Wildlife Service.

Davis, Hank E., James L. Jehn, and Stephen Smith. (1991). Monitoring well drilling, soil sampling, rock coring, and borehole logging. In *Practical Handbook of Ground-Water Monitoring*, ed. David M. Nielsen. Chelsea, MI: Lewis, pp. 195–237.

Davis, Stanley N., Glenn M. Thompson, Harold W. Bentley, and Gary Stiles. (1980). Ground-water tracers—A short review. *Ground Water* 18:14-23.

Dawson, Karen J., and Jonathan D. Istok. (1991). *Aquifer Testing: Design and Analysis of Pumping and Slug Tests*. Chelsea, MI: Lewis.

Domenico, Patrick A., and Franklin W. Schwartz. (1998). *Physical and Chemical Hydrogeology*, 2nd ed. New York: John Wiley.

Driscoll, Fletcher G. (1986). *Groundwater and Wells*, 2nd ed. St. Paul, MN: Johnson Division.

Dunne, Thomas, and Luna B. Leopold. (1978). *Water in Environmental Planning*. San Francisco, CA: W.H. Freeman.

Everett, L. G., E. W. Hoylman, L. G. Wilson, and L. G. McMillion. (1984). Constraints and categories of vadose zone monitoring devices. *Ground Water Monitoring Review* 4(1):26–32.

Federal Interagency Committee for Wetland Delineation. (1989). *Federal Manual for Identifying and Delineating Jurisdictional Wetlands*. Cooperative technical publication. Washington, DC: U.S. Army Corps of Engineers, U.S. Environmental Protection Agency, U.S. Fish and Wildlife Service, and U.S. Department of Agriculture Soil Conservation Service.

Fetter, C. W. (1993). *Contaminant Hydrogeology*. New York: Macmillan.

———. (1994). *Applied Hydrogeology*, 3rd ed. New York: Macmillan.

Freeman, Tom. (1991). *Procedures in Field Geology*. Columbia, MO: FriendShip Publications.

Freeze, R. Allan, and John A. Cherry. (1979). *Groundwater*. Englewood Cliffs, NJ: Prentice Hall.

Garrels, Robert M., and Charles L. Christ. (1965). *Solutions, Minerals and Equilibria*. New York: Harper & Row.

Garrett, Peter. (1988). *How to Sample Ground Water and Soils*. Dublin, OH: National Water Well Association.

Hamblin, W. Kenneth. (1995). *Exercises in Physical Geology,* 9th ed. New York: Prentice Hall

Hatheway, Allen W. (1992). Don't forget the Sanborn maps. *AEG (Association of Engineering Geologists) News* 35(4):25–27.

Hem, John D. (1985). *Study and Interpretation of the Chemical Characteristics of Natural Water*, 3rd ed. Water-Supply Paper 2254. Washington, DC: U.S. Geological Survey.

Herschy, Reginald W. (1995). *Streamflow Measurement*, 2nd ed. London: Chapman & Hall.

Herzog, Beverly L., Sheng-Fu J. Chou, John R. Valkenburg, and Robert A. Griffin. (1988). Changes in volatile organic chemical concentrations after purging slowly recovering wells. *Ground Water Monitoring Review* 8(4): 93–99.

Holtz, Robert D., and William D. Kovacs. (1981). *An Introduction to Geotechnical Engineering*. Englewood Cliffs, NJ: Prentice Hall.

Hubbard, E. F., F. A. Kilpatrick, L. A. Martens, and J. F. Wilson, Jr. (1982). *Measurement of Time of Travel and Dispersion in Streams by Dye Tracing*. Techniques of Water-Resources Investigations of the U.S. Geological Survey, Book 3, Chapter A9. Washington, DC: U.S. Geological Survey.

Hvorslev, Mikael Juul. (1951). *Time Lag and Soil Permeability in Groundwater Observations*. U.S. Army Corps of Engineers Waterway Experimentation Station Bulletin 36, Vicksburg, MS: U.S. Army Corps of Engineers.

Johnson, A. I. (1967). *Specific Yield—Compilation of Specific Yields for Various Materials*, Water-Supply Paper 1662-D. Washington, DC: U.S. Geological Survey.

Johnson, Robert B., and Jerome V. Degraff. (1988). *Principles of Engineering Geology*. New York: John Wiley.

Kilpatrick, Frederick A. (1970). Dosage requirements for slug injection of rhodamine BA and WT dyes. In *Geological Survey Research*. Professional Paper 700-B. Washington, DC: U.S. Geological Survey, pp. B250-B253.

Kilpatrick, Frederick A., and Ernest D. Cobb. (1985). *Measurement of Discharge Using Tracers*. Techniques of Water-Resources Investigations of the U.S. Geological Survey, Book 3, Chapter A16. Washington, DC: U.S. Geological Survey.

Kilpatrick, Frederick A., and V. R. Schneider. (1983). *Use of flumes in measuring discharge*. Techniques of Water-Resources Investigations of the U.S. Geological Survey, Book 3, Chapter A14. Washington, DC: U.S. Geological Survey.

Krauskopf, Konrad B., and Dennis K. Bird. (1995). *Introduction to Geochemistry*, 3rd ed. New York: McGraw-Hill.

Kruseman, G.P., and N.A. deRidder. (1990). *Analysis and Evaluation of Pumping Test Data,* 2nd ed. Wageningen, The Netherlands: International Institute for Land Reclamation and Improvement.

Leap, D. I. (1984). A simple pneumatic device and technique for performing rising water level slug tests. *Ground Water Monitoring Review* 4(4):141–146.

Lee, David R., and John R. Cherry. (1978). A field exercise on groundwater flow using seepage meters and mini-piezometers. *Journal of Geological Education* 27:6–10.

Lee, Keenan, and C. W. Fetter. (1994). *Hydrogeology Laboratory Manual*. New York: Macmillan.

Lyon, John G. (1993). *Practical Handbook for Wetland Identification and Delineation*. Boca Raton, FL: Lewis.

Maidment, David R., ed. (1993). *Handbook of Hydrology*. New York: McGraw-Hill.

Manning, John C. (1997). *Applied Principles of Hydrology,* 3rd ed. Upper Saddle River, NJ: Prentice Hall.

Mitsch, William J., and James G. Gosselink. (1993). *Wetlands*, 2nd ed. New York: Van Nostrand Reinhold.

Munsell Color. (1994). *Munsell Soil Color Charts*. New Windsor, NY: Kollmorgen Instruments Corporation, Macbeth Division .

National Research Council. (1994). *Alternatives for Ground Water Cleanup*. Washington, DC: National Academy Press.

Nielsen, David M., ed. (1991). *Practical Handbook of Ground-Water Monitoring*. Chelsea, MI: Lewis.

Nielsen, David M. and Ronald Schalla. (1991). Design and installation of ground-water monitoring wells. In *Practical Handbook of Ground-Water Monitoring*, ed. David M. Nielsen. Chelsea, MI: Lewis, pp. 239–331.

Norton, D., and R. Knapp. (1977). Transport phenomena in hydrothermal systems: Nature of porosity. *American Journal of Science* 27: 913–936.

Nyer, Evan K. (1992). *Groundwater Treatment Technology*, 2nd ed. New York: Van Nostrand Reinhold.

Nyer, Evan K., D. Kidd, P. Palmer, T. Crossman, S. Fam, F. Johns II, G. Boettcher, and S. Suthersan. (1996). *In Situ Treatment Technology*. Boca Raton, FL: CRC Lewis.

Orient, Jeffrey P., Andrezj Nazar, and Richard C. Rice. (1987). Vacuum and pressure test methods for estimating hydraulic conductivity. *Ground Water Monitoring Review* 7(1): 49–50.

Price, Michael. (1996). *Introducing Groundwater*, 2nd ed. London: Chapman & Hall.

Puls, Robert W., and Cynthia J. Paul. (1995). Low-flow purging and sampling of ground water monitoring wells with dedicated systems. *Ground Water Monitoring and Remediation* 15(1):116–123.

Rahn, Perry H. (1996). *Engineering Geology: An Environmental Approach*, 2nd ed. Upper Saddle River, NJ: Prentice Hall.

Reed, Porter B., Jr. (1988). *National List of Plant Species That Occur in Wetlands: National Summary*. National Wetlands Inventory, U.S. Fish and Wildlife Service, Biological Report, I49.89/2:88 (24). Washington, DC: U.S. Government Printing Office.

Roscoe Moss Company. (1990). *Handbook of Ground Water Development*. New York: John Wiley.

Saar, Robert A. (1997). Filtration of ground water samples: A review of industry practice. *Ground Water* 17: 56–62.

Sara, Martin N. (1991). Ground-water monitoring system design. In *Practical Handbook of Ground-Water Monitoring*, ed. David M. Nielsen. Chelsea, MI: Lewis, pp. 17–68.

Shen, Hsieh Wen, and Pierre Y. Julien. (1993). Erosion and sediment transport. In *Handbook of Hydrology*, ed. David R. Maidment. New York: McGraw-Hill, pp. 12.1–12.61.

Smart, P. L., and I. M. S. Laidlaw. (1977). An evaluation of some fluorescent dyes for water tracing. *Water Resources Research.* 13(1):15–33.

Smith, Leslie, and Stephen W. Wheatcraft. (1993). Groundwater flow. In *Handbook of Hydrology*, ed. David R. Maidment. New York: McGraw-Hill, pp. 6.1–6.58.

Spangler, Merlin G., and Richard L. Handy. (1982). *Soil Engineering,* 4th ed. New York: Harper & Row.

Stone, William J. (1997). Low-flow ground water sampling—is it a cure-all? *Ground Water Monitoring and Remediation* 17(2):70–72.

U.S. Department of Agriculture. (1991). *Hydric soils of the United States*, 3rd ed. U.S. Department of Agriculture, Soil Conservation Service, Miscellaneous Publication 1491. Washington, DC: U.S. Government Printing Office.

U.S. Environmental Protection Agency. (1979). *Methods for Chemical Analysis of Water and Wastes*. Environmental Monitoring and Support Laboratory, Cincinnati, Research Reporting Series 4: Environmental Monitoring, EPA/1.23/5:600/4-79-020. Washington, DC: U.S. Government Printing Office.

———. (1983). *Technical Additions to Methods for Chemical Analysis of Water and Wastes*. Environmental Monitoring and Support Laboratory, Cincinnati, EPA/600/4-82-055. Washington, DC: U.S. Government Printing Office.

———. (1986). *RCRA Ground-Water Monitoring Technical Enforcement Guidance Document*. Washington, DC: U.S. Government Printing Office.

———. (1991). *Description and Sampling of Contaminated Soils: A Field Pocket Guide,* EPA/625/12-91/002. Washington, DC: U.S. Government Printing Office.

U.S. Geological Survey. (1977). *National Handbook of Recommended Methods for Water-Data Acquisition.* Office of Water Data Coordination, U.S. Geological Survey. Reston, VA: U.S. Department of the Interior.

————. (1980). *National Handbook of Recommended Methods for Water-Data Acquisition* (updates). Office of Water Data Coordination, U.S. Geological Survey. Reston, VA: U.S. Department of the Interior.

Viessman, Warren Jr., Gary L. Lewis, and John W. Knapp. (1989). *Introduction to Hydrology,* 3rd ed. New York: Harper & Row.

Walton, William C. (1987). *Groundwater Pumping Tests: Design and Analysis.* Chelsea, MI: Lewis.

Watson, Ian and Burnett, Alister D. (1995). *Hydrology, an Environmental Approach.* Boca Raton, FL: CRC Lewis.

West, Terry. (1995). *Geology Applied to Engineering.* Englewood Cliffs, NJ: Prentice Hall.

Wilson, James F. (1968, revised in 1986). *Fluorometric Procedures for Dye Tracing.* Techniques of Water-Resources Investigations of the U.S. Geological Survey, Book 3, Chapter A12. Washington, DC: U.S. Geological Survey.

Wilson, Neal. (1995). *Soil Water and Ground Water Sampling.* Boca Raton, FL: CRC Lewis.

Index